Rheology and Processing of Liquid Crystal Polymers

Polymer Liquid Crystals Series

Series editors: D. Acierno, Department of Chemical Engineering, University of Salerno, Italy
W.K. Brostow, Center for Materials Characterization, University of North Texas, USA
A.A. Collyer, formerly of the Division of Applied Physics, Sheffield Hallam University, UK

The series is devoted to an increasingly important class of polymer-based materials. As discussed in some detail in Chapter 1 of Volume 1, polymer liquid crystals (PLCs) have better mechanical performances, higher thermal stabilities and better physical properties than flexible polymers. They are more easily processable than reinforced plastics and can be used in small quantities to lower dramatically the viscosities of flexible polymer melts. PLCs can be oriented easily in shearing, electric and magnetic fields. They also have interesting optical properties, making applications in optical data storage possible, such as light valves and as erasable holograms. Current applications of PLCs include automobile parts. The areas of *potential* applications include the electrical, electronic, chemical, aircraft, petroleum and other industries.

One measure of the rapidly increasing interest in PLCs is the number of names that are used for them: *in-situ* composites, molecular composites, liquid crystalline polymers (LCPs), self-reinforcing plastics, etc. The very rapidly growing literature on the subject makes it more and more difficult for researchers, engineers, faculty members and students to keep up with the new developments. Newcomers to the field are typically overwhelmed by the complexity of these materials in comparison to traditional engineering plastics – as reflected in sometimes mutually conflicting conclusions in publications, phrased moreover in difficult terminology. The present book series solves these problems for people already in the field as well as for the novices. Experts in the field from all over the world have been called upon to clarify the situation in their respective areas. Thus, conflicting evidence is sorted out and general features are stressed – which becomes achievable after a uniform picture of the structures of these materials is provided.

Volume 1 gives an introduction to liquid crystallinity, describes characterization of LC phases including NMR studies, discusses lyotropic (produced in solution) as well as thermotropic (produced by manipulating the temperature) PLC phases. Volume 2 deals with rheology and processing. Volume 3 deals with mechanical and thermophysical properties of PLCs and PLC-containing blends, including inorganic PLCs, formation of PLC phases – also in non-covalently bonded systems – memory effects, phase diagrams, relaxation of orientations, creep and stress relaxation, thermoreversible gels, acoustic properties and computer simulations of PLCs. Volume 4 deals with electrical, magnetic and optical properties, including a discussion of displays and also of optical storage. Overall, the book series constitutes the *only truly comprehensive source* of knowledge on these exciting materials.

Titles in the series

1. **Liquid Crystal Polymers: From structure to applications**
   Edited by A.A. Collyer
2. **Rheology and Processing of Liquid Crystal Polymers**
   Edited by D. Acierno and A.A. Collyer
3. **Mechanical and Thermophysical Properties of Polymer Liquid Crystals**
   Edited by W.K. Brostow (forthcoming)
4. **Electrical, Magnetic and Optical Effects on Polymer Liquid Crystals**
   Edited by W.K. Brostow and A.A. Collyer (forthcoming)

Rheology and Processing of Liquid Crystal Polymers

Edited by

D. Acierno

Department of Chemical Engineering
University of Salerno
Italy

and

A.A. Collyer

formerly of the Division of Applied Physics
Sheffield Hallam University
UK

CHAPMAN & HALL

London · Weinheim · New York · Tokyo · Melbourne · Madras

Published by Chapman & Hall, 2–6 Boundary Row, London SE1 8HN, UK

Chapman & Hall, 2–6 Boundary Row, London SE1 8HN, UK

Chapman & Hall GmbH, Pappelallee 3, 69469 Weinheim, Germany

Chapman & Hall USA, 115 Fifth Avenue, New York, NY 10003, USA

Chapman & Hall Japan, ITP-Japan, Kyowa Building, 3F, 2-2-1 Hirakawacho, Chiyoda-ku, Tokyo 102, Japan

Chapman & Hall Australia, 102 Dodds Street, South Melbourne, Victoria 3205, Australia

Chapman & Hall India, R. Seshadri, 32 Second Main Road, CIT East, Madras 600 035, India

---

First edition 1996

Typeset in 10/12 pt Palatino by AFS Image Setters Ltd, Glasgow

Printed in Great Britain by TJ Press, Padstow, Cornwall

ISBN 0 412 59640 7

A catalogue record for this book is available from the British Library

Library of Congress Catalog Card Number: 95-83389

∞ Printed on permanent acid-free text paper, manufactured in accordance with ANSI/NISO Z39.48-1992 and ANSI/NISO Z39.48-1984 (Permanence of Paper).

Contents

Contributors

D. Acierno
Universita di Salerno, Dipartimento Ingegnaria Chimica E Alimentare, Via Ponte Don Melillo, 84084 Fisciano (SA), Italy.

F.N. Cogswell
10 Latimer Lane, Guisborough, Cleveland, TS14 8DD, UK.

A.A. Collyer
Flat 2, 9 Elmhyrst Rd, Weston Super Mare, BS23 2SJ, UK; formerly of Division of Applied Physics, Sheffield Hallam University, Pond Street, Sheffield, S1 1WB, UK.

B.Z. Dlugogorski
Department of Chemical Engineering, University of Newcastle, University Drive, Callaghan, Newcastle, NSW 2308, Australia.

M. Grmela
Department of Chemical Engineering, École Polytechnic de Montréal, 2900 Boul Eduard-Montpetit, Montréal, H3C 3A7, Canada.

J.B. Hull
Department of Mechanical Engineering, Nottingham Trent University, Burton Street, Nottingham NG1 4BU, UK.

W.-F. Hwang
Central R & D, 1702 Building, The Dow Chemical Company, Midland, MI 48674, USA.

A.R. Jones
Department of Mechanical Engineering, Nottingham Trent University, Burton Street, Nottingham NG1 4BU, UK.

V.G. Kulichikhin
Institute of Petrochemical Synthesis, The Russian Academy of Sciences, 29 Leninskii Park, 117912 Moscow, Russia.

G. Marrucci
Dipartimento di Ingegneria Chimica, Universita di Napoli, P. le Tecchio, 80125 Naples, Italy.

P. Moldenaers
Chemical Engineering Department, Katholieke Universiteit Leuven, W. de Croylaan 46, B-3001 Heverlee-Leuven, Belgium.

A.V. Semakov
Institute of Petrochemical Synthesis, The Russian Academy of Sciences, 29 Leninskii Park, 117912 Moscow, Russia.

V.F. Shumskii
Institute of Macromolecular Chemistry, Ukranian Academy of Sciences, Kharkovsloye Shosse 252119, Kiev, Ukraine.

K.F. Wissbrun
1 Euclid Avenue, Appartment 4E, Summit, NJ 07901, USA.

Preface

Liquid crystal polymers (LCPs) have many strange properties that may be utilized to advantage in the processing of products made from them and their blends with isotropic polymers. This volume (volume 2 in the series Polymer Liquid Crystals) deals with their strange flow behaviour and the models put forward to explain the phenomena that occur in such polymers and their blends. It has been known for some time that small additions of a thermotropic LCP to isotropic polymers not only gives an improvement in the strength and stiffness of the blend but improves the processability of the blend over that of the isotropic polymer. In the case of lyotropic LCPs, it is possible to create a molecular composite in which the reinforcement of an isotropic polymer is achieved at a molecular level by the addition of the LCP in a common solvent. If the phenomena can be fully understood both the reinforcement and an increase in the processability of isotropic polymers could be optimized. This book is intended to illustrate the current theories associated with the flow of LCPs and their blends in the hope that such an optimization will be achieved by future research.

Chapter 1 introduces the subject of LCPs and describes the terminology used; Chapter 2 then discusses the more complex phenomena associated with these materials. In Chapter 3, the way in which these phenomena may be modelled using hamiltonians is fully covered. Chapters 4, 5 and 6 deal with the practical results associated with the flow of LCPs, filled LCPs and LCP blends with isotropic polymers. These chapters refer mainly to thermotropic LCPs; the way in which these materials are processed is described in Chapter 7. The last two chapters examine exclusively lyotropic LCPs: Chapter 8 discusses the time-dependent effects occurring in this kind of LCP; Chapter 9 describes the phenomena that occur in lyotropic LCPs and the way in which they are processed, particularly with regard to the processing of molecular composites.

It is hoped that this book will be a suitable introduction to this fascinating and rapidly changing subject as well as providing a much deeper insight into the flow behaviour of these materials. Apart from Chapter 3, the mathematics used is of an A level or pre-university standard.

This work is of importance to all establishments in which rheological measurements are carried out on polymeric materials. Materials scientists, engineers or technologists in industry, research laboratories or academia should find this book invaluable in updating their information and understanding the processes involved in the flow of liquid crystal polymers and their blends.

D. Acierno and A.A. Collyer
1995

1

Introduction to liquid crystal polymers

A.A. Collyer

1.1 INTRODUCTION

A great interest in liquid crystal polymer (LCP) rheology and processing has been generated since the high modulus, high strength heat resistance of these materials became evident. Most engineering thermoplastics have tensile moduli and tensile strengths of about $3\,\mathrm{GN\,m^{-2}}$ and $140\,\mathrm{MN\,m^{-2}}$ respectively. From Table 1.1 [1], it can be seen that very high moduli may be obtained in drawn fibres of LCPs. In injection moulded LCPs, moduli in the range $10–40\,\mathrm{GN\,m^{-2}}$ are achievable in the orientation direction.

LCPs have, for a long time, been used in high-strength fibres and as fibre reinforcement for plastics matrices. Trade names such as Kevlar (Du Pont) and Twaron (Akzo) are fairly well known. It is comparatively recently that injection mouldable LCPs have appeared on the market, such as Vectra (Hoechst–Celanese) one of the better established tradenames.

Table 1.1 shows that the modulus and strength values of LCPs compare favourably with metals. For fibres, a specific modulus and specific strength (modulus or strength/relative density) are quoted in g per denier (g/d), where

$$\text{value in g}\,\mathrm{d}^{-1} = \frac{\text{value in GN}\,\mathrm{m}^{-2}}{0.087 \times \text{relative density}} \qquad (1.1)$$

Rheology and Processing of Liquid Crystal Polymers
Edited by D. Acierno and A.A. Collyer
Published in 1996 by Chapman & Hall
ISBN 0 412 59640 7

Table 1.1 Values of theoretical modulus and practical values for engineering materials [1]

Materials	*Theoretical modulus*		*Practical modulus*		*Tensile strength*
	($GN\,m^{-2}$)	($g\,d^{-1}$)	($GN\,m^{-2}$)	($g\,d^{-1}$)	(MNm^{-2})
Polyethylene	334	4000	1.67–4.18	20–50	27
Orientated polyethylene	334	4000	117	1400	3670
Polyethylene-terephthalate	146	1200	12.2	100	50
Polyamide 66	161	1600	5.0	50	80
Polyamide 66/30% glass fibre	–	–	8.0	66	160
Liquid crystal polymers	188	1500	125–175	1000–1400	140–240
Steel	–	–	217	320	460
Aluminium	–	–	71	300	80
Glass	–	–	77.6	350	3500
Carbon	–	–	240–400	1660–2765	2100–2800

Table 1.1 also gives values of theoretical modulus, which depends on the bond length and deformation constraints. Such a calculation assumes that the molecules are completely uncoiled, with no links. The theoretical modulus for polyethylene is double that achieved in practice for highly orientated fibres, and considerably greater than for samples with a random orientation. In the case of LCPs, the practical modulus is not much lower than that predicted by theory. The tensile modulus and tensile strength of polyphenyleneterephthalamide (PPT) (Kevlar) are about 127 and 2.6 $GN\,m^{-2}$ respectively.

To understand the differences between isotropic polymers (such as the usual engineering and speciality polymers) and LCPs, it is necessary to highlight the ways in which the two types of molecules behave and how this affects the bulk properties. In the isotropic case, the long molecules exist in a random coil configuration in the melt or in a solvent. In extrusion, injection-moulding and fibre-spinning processes the tensile and shear fields tend to align and uncoil the chains, but the chain continuity is low. On the removal of the stress, the molecules partially lose their orientation and tend to recoil.

Even the most recent equipment for drawing ultra-high modulus fibres from isotropic polymers is inefficient in aligning the molecules in the machine or draw direction but, as noted earlier, this is not true of LCPs, as the practical modulus is close to the predicted value. It is the property of liquid crystallinity or mesomorphism that provides the required alignment and chain continuity to approach theoretical values. Therefore, the solution is not in improving the efficiency of the drawing equipment but in using the inherent desire of molecules to align and remain in that alignment, which is the particular property of liquid crystallinity.

Frank [2] observed that the important requisites of a high-modulus plastics material are:

1. the individual molecules must be stiff;
2. their alignment must be nearly perfect; and
3. the ratio of the aromatic to aliphatic linkages must be high.

These conditions are met in LCPs such as aromatic polyesters, aromatic polyamides, polybenzobisthiazoles and related materials, and other architectures of lesser current importance.

LCPs consist of stiff lath-shaped molecules of width greater than thickness and length much greater than the other two dimensions. The mesogenic moieties that confer liquid crystallinity, the same types that give liquid crystallinity to small molecule liquid crystals, are incorporated into the backbone chain of the polymer for engineering LCPs. The LCP consists of a number of rigid units, some of which may be mesogens, separated by less rigid units or flexible units, X in Fig. 1.1(a). The X in the

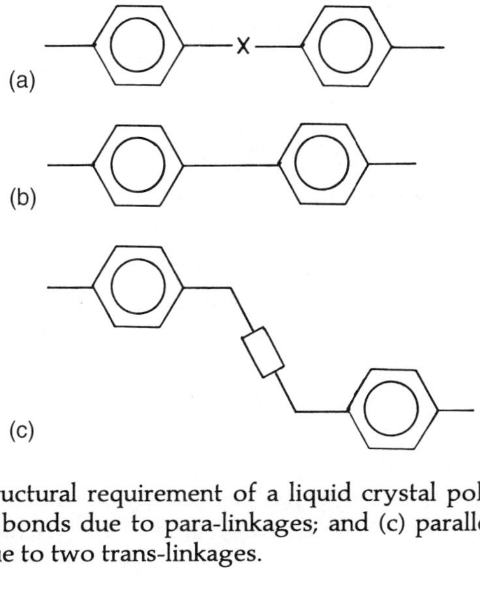

Figure 1.1 (a) structural requirement of a liquid crystal polymer; (b) parallel chain continuing bonds due to para-linkages; and (c) parallel, but oppositely directed bonds due to two trans-linkages.

structure has a profound effect on the behaviour of the LCP: it must promote either parallel chain continuity bonds due to para-linkages or parallel but oppositely directed bonds due to two trans-linkages. Moreover, the X in these structures gives rise to two different kinds of LCP. If X is stiff, the polymer may decompose before it melts and must be processed by dissolving in a solvent above a certain critical concentration. The solvent is later removed. The liquid crystallinity occurs due to the conformation of the main chains in the solvent and these solutions are called **lyotropic**.

When the LCP has at least a 30°C processing window between the melting and decomposition temperatures, or at least 100°C between the glass transition and decomposition temperatures (amorphous polymer), the material may be melt-processed, with the action of the mesogens in the main chain conferring alignment. These materials are called **thermotropics** as the mesomorphism occurs between two specific temperatures.

As the linkage X in lyotropic LCPs is stiffer than that in thermotropic LCPs, fibres spun from the former have higher moduli, strengths and continuous service temperatures.

Before looking in more depth at LCPs, it is instructive to build up a simplified picture of their flow mechanism. The molecules are long, stiff and lath-shaped and, as such, may be modelled by logs floating along a river. Groups of logs tend to align, but each group aligns differently from its neighbouring groups. This is conducive to setting up a log jam. If all the groups of logs are aligned, the logs will readily flow away.

In LCPs, the molecules form into domains of common molecular alignment. The average alignment direction in each of these regions is denoted by the director. There is generally no relationship between the directors of each domain. The order parameter S defines the degree of alignment of the molecules with the director.

$$S = 0.5(\overline{3\cos^2\Theta - 1}) \qquad (1.2)$$

where Θ is the angle between the orientation of a particular molecule and the director. For $S = 0$, there is a random distribution alignment; for $S = 1$, all molecules are parallel to the director. The value of S depends on the temperature, becoming zero at the isotropic melt temperature and about 0.8 at the temperature at which solidification occurs, T_g.

During processing all the directors are aligned and, under these conditions, the viscosity is very low, much lower than at a higher temperature when the isotropic state is reached, with its consequent increase in entanglement and molecular interaction. It is unusual that nature should provide two advantages: high stiffness and low viscosity.

The low viscosities of the thermotropic LCPs permit flow into thin-walled moulds and permit a high loading of glass fibres. This latter aspect is very useful in reducing the degree of anisotropy present in LCPs. At first, this may seem strange as the anisotropy gives rise to the desired high modulus, but unfortunately the anisotropy decreases the strength of weld lines. The addition of glass fibres gives a better multi-directional modulus and any loss in modulus due to a reduction in orientation is counteracted by the presence of a high weight percentage loading of glass fibres (approximately 50%). This is discussed fully in Chapter 5.

In lyotropic solutions, the viscosity increases with increasing polymer concentration, reaching a maximum value at the critical concentration above which the mesomorphism occurs. Further increases in concentration give a decrease in viscosity enabling processing and a higher level of liquid crystal polymer than in isotropic lyotropic dopes. This is discussed in depths in Chapters 8 and 9.

1.2 LIQUID CRYSTAL MESOPHASES

The introduction is intended to give a simple account of the salient properties of LCPs before analysing these properties in more depth. Some of the topics covered in this chapter will be developed more fully later in the book; others will be mentioned here only.

One of the important manifestations of liquid crystals is their melting or softening behaviour. When heating a crystalline solid that is not mesomorphic, it changes from the solid phase directly into an isotropic liquid phase at its melting point T_m. In liquid crystalline materials, several different mesophases may form before the isotropic melt phase occurs at the clearing temperature.

The mesophases are differentiated from each other by the different ways in which the molecules pack. This may give rise at the clearing temperature to a disordering of the molecules, either a positional disordering, an orientational disordering or a conformational disordering.

An explanation by Brostow [3] involves a description of the melting behaviour of the homologue series of n-paraffins. When methane melts, the regular spacing of the molecules is lost, giving **positional disordering**, with an attendant contribution to the entropy of fusion. When the next member in the series (ethane) melts, positional disordering occurs, but because two ethane molecules may be parallel to or perpendicular to each other, as well as a host of intermediate conditions, there can be a loss of orientation as well. This gives an additional degree of freedom: **orientational disorder**. N-decane, a longer paraffin molecule, has many more configurations due to rotations about single bonds. This permits another kind of disorder: **conformational disordering**. Each of these disorderings contribute to the entropy of fusion. Camphor, which contains 10 carbonations as does n-decane, has rigid and almost

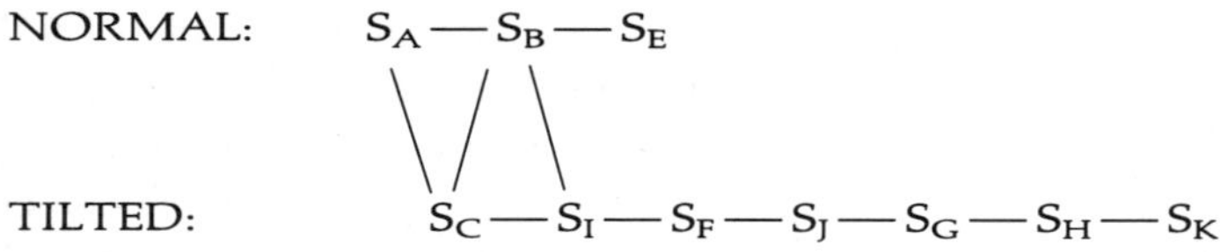

Figure 1.2 Ten common smectic polymorphs in chronological order of their identification.

spherical molecules, and no orientational or conformational effects can occur. As a result, the entropy of fusion of camphor is much less than that of n-decane. Thus, liquid crystallinity depends on molecular anisotropy and the manner in which the molecules pack. This is covered further in Chapter 3.

The long-range ordering required in the formation of Bravais lattices in crystalline solids is completely absent in gases, but a degree of long-range order occurs in liquid crystals. When a liquid crystal goes from a solid through various mesophases to the isotropic liquid phase, the degree of long-range order decreases and, in the case of orientational disorder, this is expressed by a decrease in the order parameter, S.

The mesophases occurring, in general, in liquid crystals may be divided into two categories: nematic (giving a low viscosity of interest to polymer processing) and smectic/cholesteric (giving a high viscosity mesophase).

The molecular organization in a **nematic** mesophase involves molecular alignment without a special regularity. This type of mesophase has been used in the spinning of lyotropic dopes to make fibres (such as Kevlar) and, more recently, has been used with thermotropic polymers to make melt-spun fibres and extruded and injection-moulded products.

In **smectic** mesophases, the molecules are aligned and stratified [4], which requires a greater degree of order. Figure 1.2 shows ten common polymorphic forms of a smectic mesophase in liquid crystals. These are shown in chronological order of their identification. The degree of order increases from left to right, and the polymorphs connected by lines show sequences observable on cooling, although a particular liquid crystal may skip many of the polymorphs.

The **cholesteric** mesophase comprises a helical nematic or smectic structure. The successive turns of the helix are maintained by steric hindrance or by intra-molecular bonding through, for instance, hydrogen bonding. As with smectic mesophases, a high viscosity results.

It is perhaps fortunate that the number of mesophases through which polymer molecules pass is considerably less than those encountered in liquid crystals of low molecular mass.

1.3 IDENTIFICATION OF MESOPHASES

With so many polymorphs in low molecular mass liquid crystals, identification is difficult. Several techniques in concert are used to achieve it, including:

1. mutual miscibility of known mesogens;
2. polarized light microscopy;
3. X-ray diffraction; and
4. differential scanning calorimetry (DSC).

1.3.1 Mutual miscibility of known mesogens

This method of identification was developed by Sackmann and co-workers [5, 6]. It is based on the assumption that if two mesogens are mixed and liquid crystal behaviour occurs over the entire compositional range, the same polymorph is common to both mesogens. The method, however, is not foolproof. It has been found that some mesogens that do not mix over the entire compositional range, do possess the same polymorph. Miscibility studies are not in themselves sufficiently discerning. This method may work with LCPs because the increased molecular mass leads to a reduction in miscibility and they possess fewer mesophase transitions. More detail is given in [7, 8].

1.3.2 Polarized liquid microscopy

When viewed between the crossed polars of a polarizing microscope fitted with a heating stage, small molecule liquid crystals cause patterns and formations to be seen, which alter with temperature and concentration. These patterns are termed **texture**; each polymorph shows different textures such that it is possible to chart the progress of a liquid crystal through various mesophases to an isotropic liquid [7, 9, 10]. Alderman and Mackley [11] have examined the effect of shear on the polymorphs in some LCPs.

The texture of nematic mesophases often consists of black lines and loops. These are referred to as **disclinations**, and they separate regions of differing molecular orientations. As the temperature is increased, the lines decrease in number and tend to shrink in length. The lines disappear at the clearing temperature when the isotropic liquid phase is reached. The sample then appears uniformly dark. Great care is needed in the interpretation of textures, particularly when identifying smectic polymorphs. Microscopy is useful in identifying smectic polymorphs of low order. Higher-order smectic polymorphs must be investigated using X-ray diffraction.

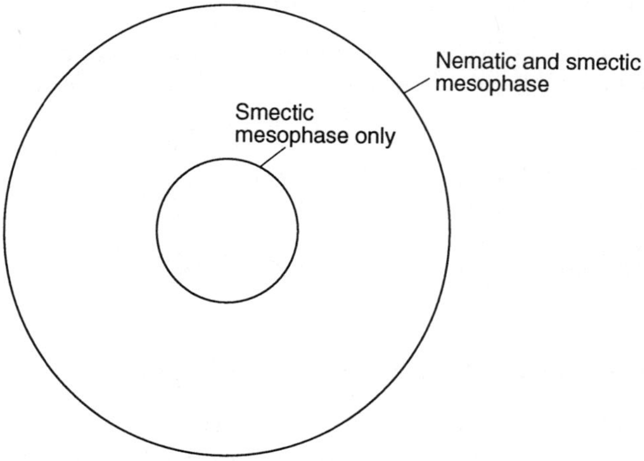

Figure 1.3 Generalized X-ray powder diffraction pattern from a smectic mesophase.

1.3.3 X-ray diffraction

The Debye–Scherrer technique on unorientated powder samples may be used to identify high-order smectic polymorphs. Figure 1.3 shows a typical diffraction pattern, consisting of an inner ring observed only in smectic mesophases, which gives information on smectic layer thickness, and an outer ring present in both nematic and smectic mesophases. This

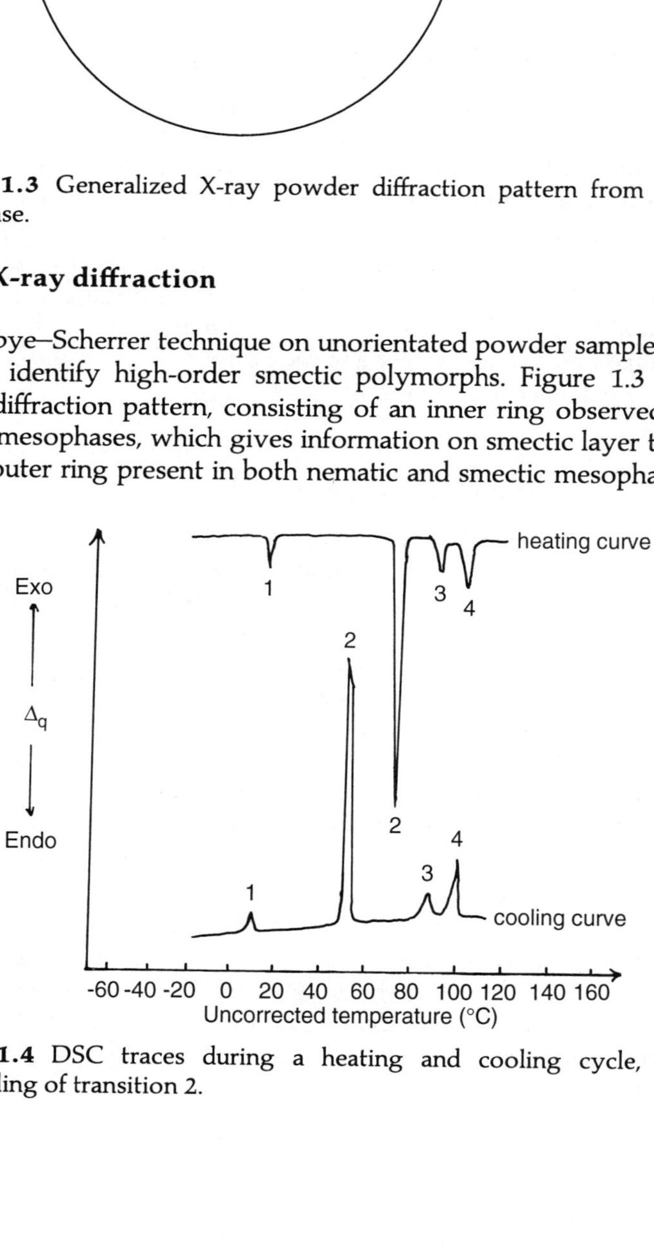

Figure 1.4 DSC traces during a heating and cooling cycle, showing supercooling of transition 2.

ring is generally diffuse and relates to the average distance between parallel rod-like molecules [4, 9].

The higher-order smectic polymorphs (S_E, S_B, S_H, S_G) give sharper rings at diffraction angles around 3° (represents 3 nm). S_F and S_I are of intermediate order between the above and N, S_A and S_C. X-ray patterns from LCPs with flexible spacers in the main chain are diffuse, making identification impossible. X-ray diffraction studies of orientated specimens provide much more information [12–14].

1.3.4 Differential scanning calorimetry (DSC)

The calculation of mesophase transition heats and entropies may be obtained from this technique as well as the transition temperatures. A typical DSC thermogram is shown in Fig. 1.4. All the mesophases are thermodynamically stable as they occur on both the heating and cooling cycles, and are termed **enantiotropic**. If a mesophase occurs only on cooling, it is said to be **metostable** and is termed **thermotropic** [4]. Transition 2 shows some supercooling. It is not usually possible to supercool the isotropic liquid.

A typical description of the transitions may be:

1. Crystalline to crystalline (K_1–K_2) at 283 K:

$$K_1 \xrightarrow{283} K_2$$

2. $$K_2 \xrightarrow{343} S_B$$

3. $$S_B \xrightarrow{360} S_A$$

4. $$S_A \xrightarrow{373} I$$

(I = isotropic)

DSC thermograms indicate whether the mesogenic moieties or liquid crystal units are incorporated into the backbone chain (as for engineering polymers) or in the sidechain (for polymers for electro-optical devices such as displays). In the case of main chain thermotropic LCPs with a single mesophase, a high temperature exotherm denoting a first-order transition isotropic to mesophase will occur with a lower temperature exotherm representing a first-order mesophase-crystal transition. At a lower temperature, T_g, the glass transition may appear as a base line shift.

With side chain LCPs, a glass transition is observed instead of a low-temperature crystalline transition, followed by a transition from the mesophase to the isotropic melt [15, 16].

It is not possible to identify polymorphs from DSC alone, but polymorphs with changes in enthalpy per repeat unit in the range 6.3–21.0 kJ mol^{-1} are usually smectic, with the higher orders having the higher changes in enthalpy. The nematic–isotropic transition enthalpy change is generally in the range 1.23–3.57 [7], but some values as high as 7.77 kJ mol^{-1} have been reported for nematic–isotropic transitions, making identification difficult [17, 18].

Using a combination of some of these techniques it is possible, with care and experience, to identify the polymorphs in a particular liquid crystal. Fortunately, the longer the polymer chain, the fewer the transitions, and in this work nematic mesophases, with their low viscosities, are of prime importance.

1.4 MOLECULAR ARCHITECTURE IN THERMOTROPIC MAIN CHAIN LCPS

Mesomorphism is induced into main chain LCPs by incorporating mesogenic moieties into the backbone chain such that they lie parallel to it. Such polymers are termed longitudinal LCPs and form only a small part of a gamut of different LCP architectures [19]. Frank [2] observed that, for LCP behaviour, the molecules must be stiff and lath-shaped. This is not entirely true as flexible polymers based on polysiloxanes, polyvinylether, polyphosphazene and polyethylene can show liquid crystal behaviour with the appropriate mesogens incorporated into the polymer. Semi-flexible polymers in which the stiff mesogens are separated in the backbone by flexible spacers may also show LCP behaviour. But, by far the most important molecules are the stiff ones because they give the highest moduli in the bulk material.

Figure 1.5 [2] shows the effect of molecular aromaticity on fibre modulus in aromatic aliphatic polyesters. The aromaticity is defined as the ratio of the number of SP_2 hybridized carbon atoms to the total number of carbon atoms in the repeat unit. For example, polybutyleneterephthalate fibres of low aromaticity have low modulus, whereas copolyesters containing diacids with aromaticity greater than terephthalic acid (TA) and ethylene glycol (2G) or 1,6 hexamethylene glycol (CG) produce fibres of tensile modulus intermediate between PBT and polyphenyleneterephthalamide (PPT).

In all cases, there is a compromise between high modulus and tractable melts. The inclusion of flexible units or swivel units, such as aliphatic units, leads to a reduction in modulus but promotes melt processing.

The above example was quoted from polyester structure, the currently favoured structure for thermotropic LCPs. Other architectures that give

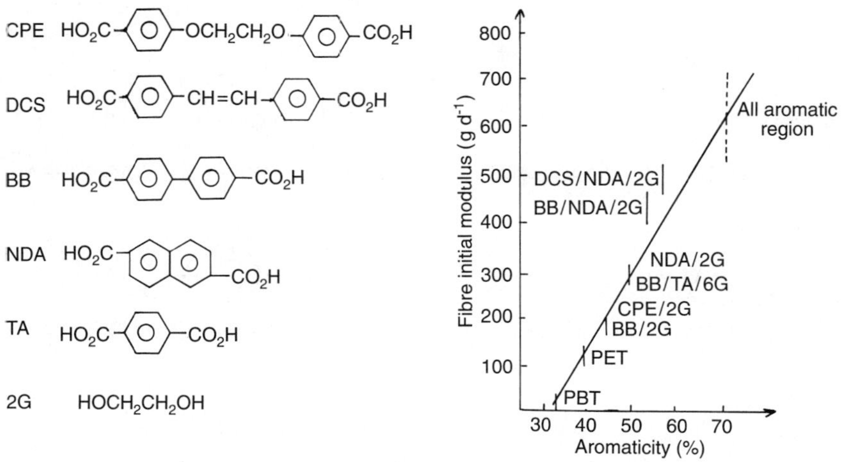

Figure 1.5 Effect of polyester aromaticity on fibre tensile modulus [1].

rise to thermotropic behaviour are polyesteramides, copolyazomethines and polyurethanes. Stiffer architectures, such as polyaramids, cellulosics, polyalkylisocyanates and polybenzazoles, give rise to lyotropic LCPs when dissolved in appropriate solvents.

The units used in experimental and commercial liquid crystal copolyesters are shown in Fig. 1.6. The melting temperatures of each of these units are above their decomposition temperature (approximately 450°C). HBA, TA, HQ, BP have collinear, para-linkages, whereas HNA has parallel but oppositely directed bonds. Other materials used have meta-linkages, such as resorcinol and meta-HBA. The reduction of T_m may be achieved by several different strategies, the idea being to introduce sufficient disorder in the system to frustrate crystallinity but not so much as to give an isotropic melt. The introduction of the right amount of disorder promotes mesomorphism, usually giving the desirable nematic mesophase. The methods by which this is achieved are now discussed, but the synthetic routes are described more fully in [1, 20]. Figure 1.7 shows various ways of reducing T_m.

1.4.1 Frustrated chain packing

Frustrated chain packing refers to mechanisms by which the arrangement of the molecules into a three-dimensional lattice is made difficult and yet maintains chain stiffness and linearity to give a lath-shaped structure for mesomorphism.

Random copolymerization of linear unsubstituted phenyl-based units reduces T_m, but the reduction is insufficient to give readily melt-

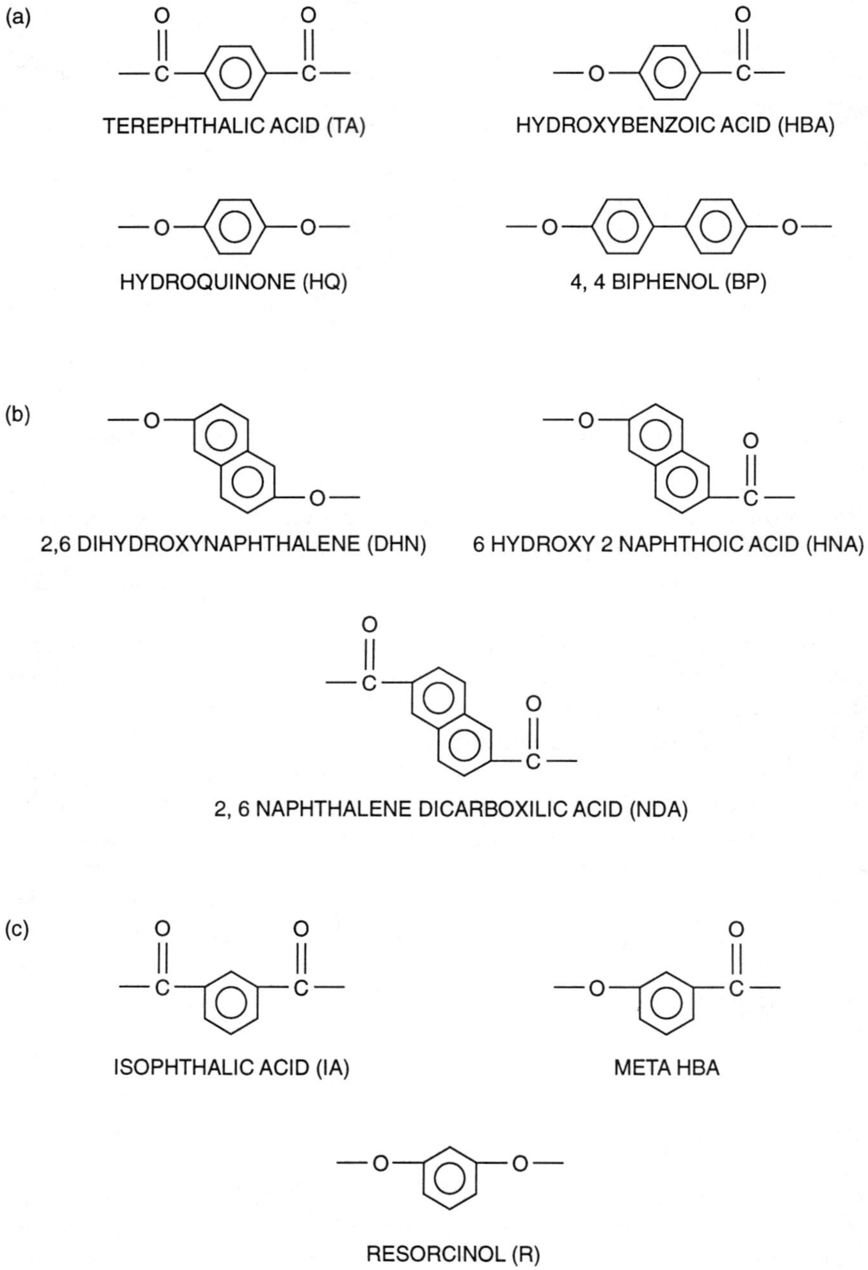

Figure 1.6 Architectures of units used in liquid crystal copolyesters: (a) swivel units; (b) crank shaft units; and (c) bent rigid units.

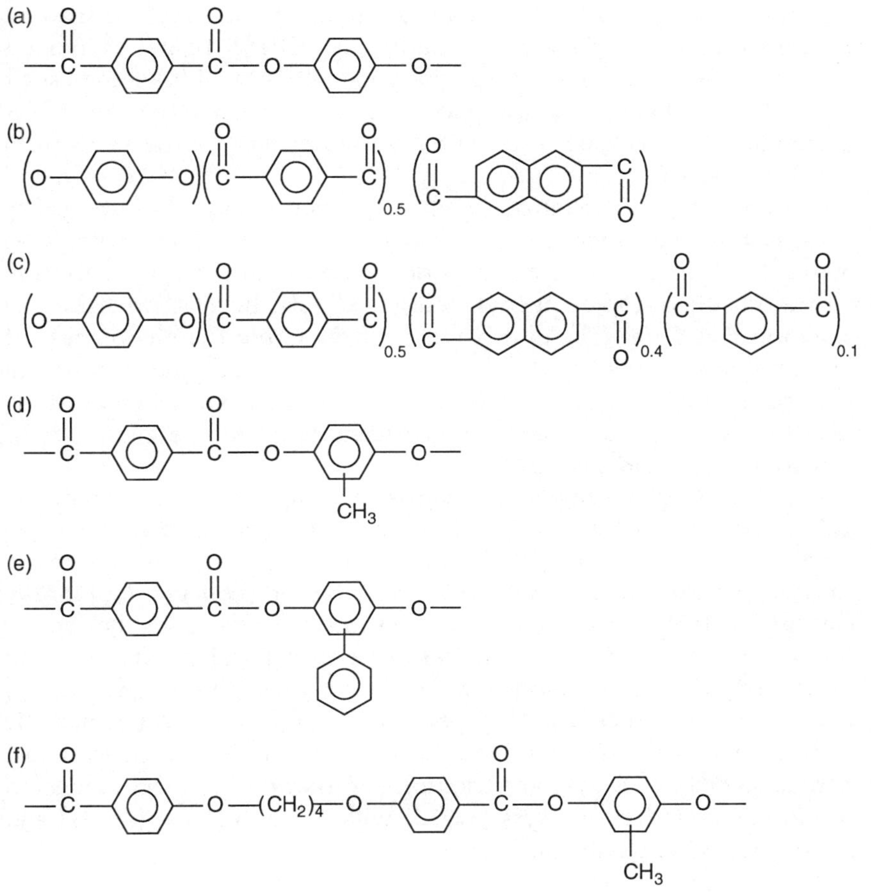

Figure 1.7 Strategies employed in commercial and experimental liquid crystal copolyesters to reduce the melt temperature [21, 22]: (a) copolymer ($T_m > 600°C$); (b) crank-shaft unit ($\sim 400°C$); (c) crank-shaft and bent unit ($\sim 350°C$); (d) small substituent ($> 400°C$); (e) large substituent ($> 340°C$); and (f) aliphatic unit ($\sim 210°C$).

processable material. For instance, from Fig. 1.7(a), the copolymerization of TA with HQ gives $T_m \simeq 600°C$, which is well above T_d.

A better way of using copolymerization is to incorporate a para-linked unit such as 4,4 biphenol (BP) as a chain extender [23]. This reduces T_m to approximately 400°C, as in the case of 'Xydar' (Amoco), which involves HBA/TA/BP. The melt temperature is still very close to T_d.

Naphthalene units have been the most successful in giving an acceptable amount of frustrated chain packing. Nematic mesophases are formed over a wide temperature range. These units are parallel-offset or crankshaft units with parallel but oppositely directed bonds. HBA is the one most often used in commercial thermotropic LCPs [1, 14, 24]. Figure 1.8 shows a comparison between copolyesters with and without crankshaft units, the T_m being much reduced by the incorporation of HNA ($T_m \simeq 250°C$). This type of disrupter is used in the 'Vectra' LCPs from Hoechst–Celanese.

Disorder may also be introduced by asymmetric ring substitution. The para-substituted phenylene rings have a plane of symmetry normal to the axis of the chain. If this symmetry is disrupted then, on copolymerization with another undisrupted phenyl mesogen, T_m will be greatly reduced, as shown in Fig. 1.7(e). The disorder arises from either the randomness of the copolymerization resulting from the head-to-head and head-to-tail isomerization of the asymmetrical unit [25, 26] or from the steric effect such as decreased coplanarity of adjacent units in the mesogen or increased chain separation [27–29].

This disruption of symmetry is effected by ring-substituted monomers such as chloro, methyl or phenyl substituted rings (see Fig. 1.7). The better effect is achieved with HQ rather than with TA. The size of the substituent rather than its polarity is important in reducing T_m [28, 29]. This implies that the steric effect of the substituent increases the separation of the mesogens and reduces molecular packing efficiency. In general, ring substitution alone is not sufficiently effective in reducing T_m, unless phenyl or other large groups are used [30], when a stable nematic phase is obtained [31]. CH_3 and Cl substituents can adversely affect thermal stability [30]. Asymmetric ring substitution has been pioneered by Goodman and co-workers [32, 33] with ICI and Payet [26, 31] and Schaefgen *et al.* [34] with Du Pont.

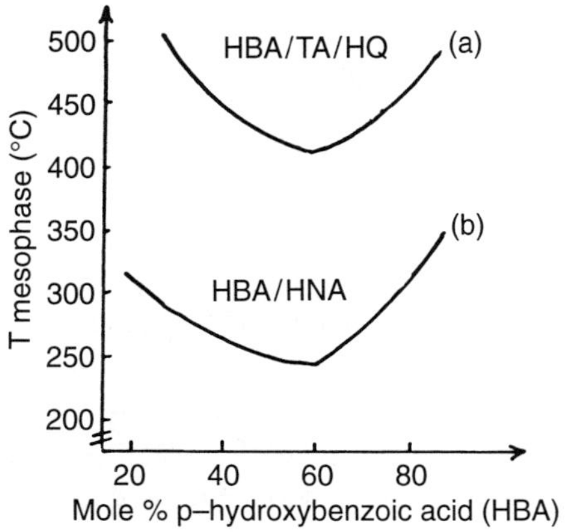

Figure 1.8 Variation of T_m with composition for two random copolyesters [1]: (a) without a crankshaft unit; and (b) with a crankshaft unit.

1.4.2 LCPs with flexible spacers

Spacers are either regularly distributed or randomly distributed.

Regularly distributed spacers

Nematic, smectic and cholesteric mesophases occur in many polyester systems involving regularly distributed flexible spacer units. The two main requirements are:

1. the rigid rod unit must exceed a certain critical length; and
2. the flexible spacer unit must be less than a certain critical length, depending on the structures of both units.

Flexible spacer units employed are generally methylene $-(CH_2)$ or $-(CH_2CH_2O)-$ types. If the spacer length is maintained constant T_l increases with rigid rod length up to T_d [35]. If the spacer length is increased, for a constant rigid rod length, the smectic mesophase becomes more stable than the nematic mesophase.

An odd–even effect occurs on increasing the length of the flexible spacer units, as shown in Fig. 1.9 [36]. Both T_m and T_l are higher for even

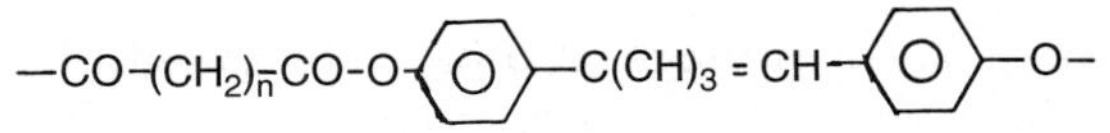

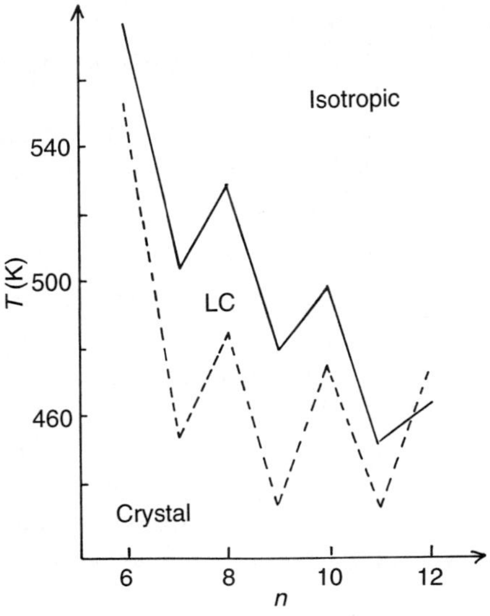

Figure 1.9 Variation of T_m and T_I with the number of methylene spacer units for homopolyesters of 4,4′ dihydroxy – methylstilbene and aliphatic acids [36].

numbers of methylene units, as are the entropy changes. The odd–even effect seems to be related to the trans-gauche conformation of the flexible spacer unit, which influences the mesophase stability [37]. The transition temperatures may be significantly reduced by replacing the methylene units with the more flexible silicone one [38].

Randomly distributed spacers

Mesomorphism also occurs when the rigid rod and spacer lengths are not constant. The effect is to reduce T_m below that of a similar but regular system, and the nematic mesophase exists over a larger temperature range [39]. The effect of the distribution of rigid rod lengths on T_l has not been established.

Eastman–Kodak produced the X7G LCP, which is a copolymer of ethylene terephthalate and HBA [40]. Over 30 mol.%HBA is necessary for mesomorphism and the maximum effect occurs at 60 mol.%HBA, the concentration in X7G, but the mechanical properties are inferior to the lyotropic polyamides or the rigid liquid crystal polyesters.

1.4.3 Non-linear units

Bent rigid units such as meta or other substituted phenyls, 1,6 or 2,5 linked naphthalenes or the incorporation of linked bonds disrupts the crystallinity to give mesomorphism. Popular units are meta – HBA, IA and resorcinol [1]. This method effectively lowers T_m but it is easy to go too far, and there is a penalty in the form of a reduction in chemical and hydrolytic stability [1].

Kinks may also be incorporated into the backbone chain by using nonlinear bonds such as anhydride [41] or carbonate [42], but the modulus of the bulk material is lower than with linear units.

1.5 LYOTROPIC MAIN CHAIN LCPS

The aromatic polyamide structure and the polybenzazoles (PBZ) structure form the bulk of commercial lyotropic LCPs, with the aromatic polyamide-hydrozides being of interest. In the main, they are spun into fibres but there has been much study of using PBZ molecules in molecular composites (discussed in Chapter 9 and [43, 44]). These molecules are dissolved in solvents to make lyotropic dopes. Two kinds of solvents may be used:

1. powerfully protonating acids such as 100% sulphuric acid, chloro-fluoro – or methanesulphonic acid, and anhydrous hydrogen fluoride;

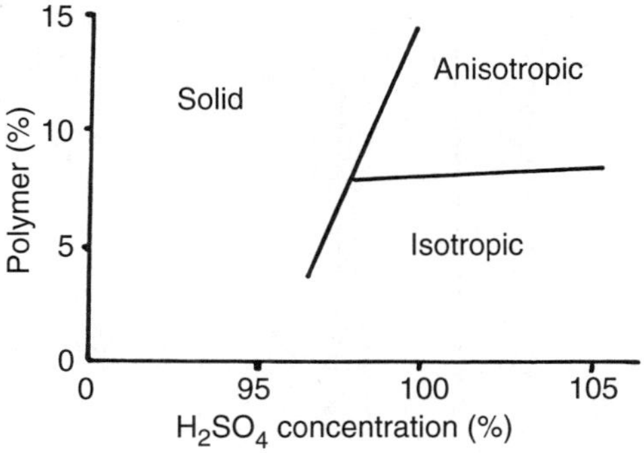

Figure 1.10 The phase diagram for PpPTA in sulphuric acid [45].

2. aprotic dipolar solvents such as dimethylacetamide containing about 2.5% of a salt such as lithium chloride or calcium chloride.

Some of the conditions required for obtaining a nematic mesophase rather than an isotropic solution include:

1. a polymer concentration being above a critical level;
2. a polymer relative molecular mass being above a critical value;
3. a temperature being below a critical level.

The critical concentrations and temperatures depend on the solvent. Most work has been reported on PpPTA in sulphuric acid and poly (p-benzamide) (PpPBA) in dimethylacetamide/lithium chloride. In the latter, the lithium chloride must exceed a certain concentration in DMA. Figure 1.10 shows the phase diagram for PpPTA in sulphuric acid [45].

This figure indicates that a direct transition from mesophase to isotropic liquid may exist, but this is not necessarily true, as it has been found that in some solutions the nematic mesophase and isotropic phase co-exist in equilibrium. In these circumstances, the nematic mesophase possesses locally a higher concentration of molecules of the larger molecular mass.

Figure 1.11 shows the variation of viscosity with polymer concentration for different RMMs, expressed as an intrinsic viscosity [46]. The viscosity of the solution drops rapidly above the critical concentration as the nematic mesophase forms. This is discussed more fully in [39, 44, 47, 48] and in Chapters 8 and 9.

1.5.1 Polyamides

Figure 1.12 shows the molecular repeat units of three of the best known lyotropic polyamides:

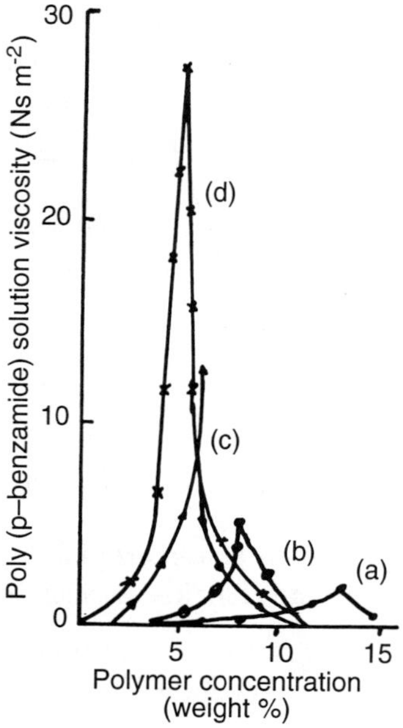

Figure 1.11 Variation of the viscosity of PpBA in dimethylacetamide/lithium chloride solutions with polymer concentration for various relative molecular masses expressed as intrinsic viscosities: (a) 0.47; (b) 1.12; (c) 1.68; and (d) 2.96 [46].

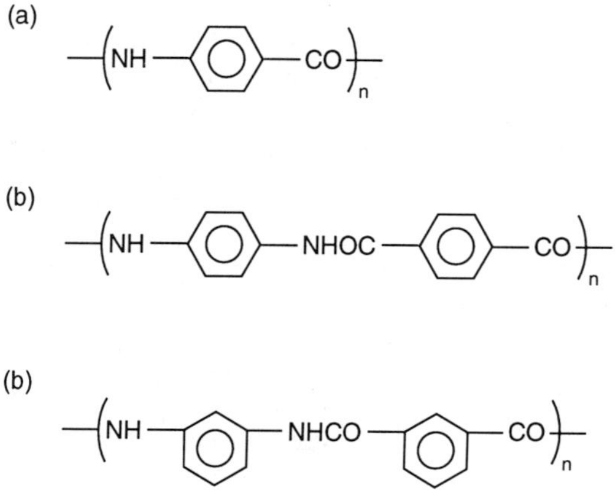

Figure 1.12 Chemical structures of some lyotropic polyaramids: (a) poly (p-benzamide); (b) poly (p-phenylene terephthalamide; and (c) poly (m-phenylene isophthalamide).

1. poly (p-benzamide) PpBA;
2. poly (p-phenylene terephthalamide) PpPTA; and
3. poly (m-phenylene isophthalamide) (PmIA).

Copolymerization of mesogenic units does not destroy mesophase formation, so that both random and alternating copolymers of PpBA and PpPTA will form mesophases, but copolymers with species that disrupt the lath shapes will cause a loss of mesomorphism at low concentrations of the disrupting species.

Substituents in the aromatic rings will permit mesomorphism provided they are not too large, but the critical parameters will be more stringent. Liquid crystallinity occurs in solutions involving poly (chloro-p-phenylene terephthalamide) and poly (p-phenylene chloroterephalamide).

PpBA was developed by Du Pont before PpPTA [45]. Fibres are made by either wet- or dry-spinning of lyotropic dopes of DMA or tetramethylurea containing lithium chloride [5]. PpBA was superseded by PpPTA but there is some evidence of PBA units with PpPTA in Kevlar 49 (Du Pont).

PmPIA is marketed by Du Pont under the trade name, Nomex. It is spun from hot DMA containing 3% calcium chloride. It can be wet- or dry-spun with the latter process giving the better mechanical properties [45].

The thermal stability is poorer than that of PpPTA as would be expected from the meta- rather than para-linkages (Fig. 1.12). Nomex materials are used in yarns and fabrics for protective clothing for firemen, policemen and workers with petrol or rocket-based fuel or molten metals.

The most widespread aramid fibre is PpPTA, sold under the trade names Kevlar (Du Pont) and Twaron (Akzo) [47]. This material is synthesized by a condensation polymerization involving p-phenylene diamine and terephthaloyl chloride. The resulting product is ground in water, filtered and washed to remove solvent and hydrogen chloride. The lyotropic dopes are often made from 20% PpPTA in 99.8% sulphuric acid. The dopes are solid at room temperature and so the spinning is carried out between 77°C and 90°C. The preferred process is dry-jet wet-spinning [45].

The grades available from Du Pont include Kevlar 29, Kevlar 49, Kevlar HT and Kevlar HM [48]. Another type, Kevlar T950, was developed for the rubber industry for bracing in radial tyres. PpPTA is often made into high-temperature laminates and honeycomb laminates for aerospace. The main applications are for protective fabrics (as for Nomex) and soft and hard armour (Kevlar HS). Further details are given in [44, 49].

(a)

(b)

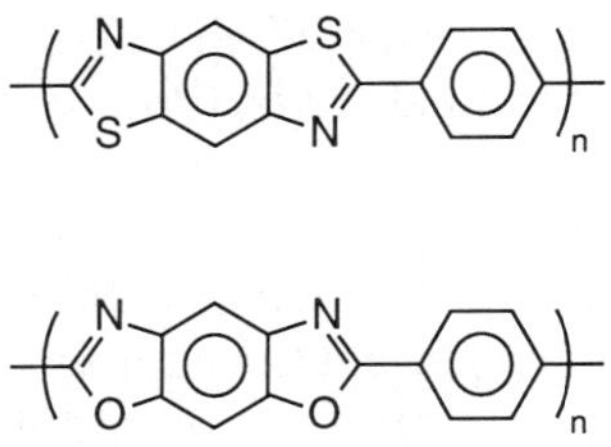

Figure 1.13 Structure of (a) trans-polybenzothiazole (PBT) and (b) cis-polybenzoxazole (PBO).

1.5.2 Polybenzazoles

Much less work has been carried out on these newer, stiffer materials, which have higher moduli and strengths greater than the lyotropic polyaramids. Much of the research has been carried out in the US Air Force. Ordered Polymers Programme which is described in [44, 50, 61] and in Chapter 9.

The PBZ materials are referred to as linkageless polyheterocylic polymers. As seen in Fig. 1.13, the phenyl rings in the backbone chain are joined by bonds rather than by atoms and hence are linkageless.

Trans-polybenzothiazole (PBT) and cis-polybenzoxazole (PBO) are synthesized in phosphoric acid adjusted with P_2O_5, which allows solutions of up to 20% PB to be made [50, 51], and similarly for cis-PBO [52].

These polymers are dry-jet wet-spun from either of two solutions:

1. a 5–6% polyphosphoric acid polymerization mixture; or
2. a 10% solution of the isolated polymer in methanesulphonic acid (MSA/chlorosulphonic acid) in the ratio 97.5 : 2.5.

The coagulant is water, and annealing is carried out at high temperature under nitrogen.

There are four groups into which PBZs may be divided, depending on the chain extension in solution, which affects the minimum or critical concentration for a stable nematic mesophase:

1. rigid rod homopolymers (Class 1);
2. mesogenic homopolymers (Class 2);
3. mesogenic copolymers (Class 3); and
4. non-mesogenic copolymers (Class 4).

Rigid rod homopolymers (Class 1)

The catenation angles in rigid-rod PBTs and PBOs are almost 180°, giving a lath shape. Polymers with an ideal rigid-rod structure may be described by an axis ratio x (ratio of the average contour length L to the width of the backbone d) [53]. For rigid-rod polymers X is the ratio of the persistence length l to d, and for these materials the persistence length l is equal to or greater than the contour length L [54], such that $X \geqslant x$. For such molecules, a critical concentration of around 5 wt% gives stable nematic mesophases. Such a low concentration allows the mixing of these materials in a solvent with random coil polymers to form a composite with rigid-rod reinforcement on a molecular scale. This unique property has promoted the study of these materials for such molecular composites [44, 50–52] and (Chapter 9).

These materials have poor compressive strength and applications are similar to those of PpPTA, namely films and fibres, heat-resistant clothes and bullet-proof vests; a future usage may be in high-temperature porous membranes for separation processes [43, 44].

Mesogenic homopolymers (Class 2)

The catenation angles for these rigid heterocyclic units lie between 150° and 165°, as shown for the trans- and cis-conformations of 2.5-PBO in Fig. 1.14, and the persistence length is thus reduced. It is assumed that the chain conformation for these materials is a planar zigzag (Fig. 1.14) in which the alternating bonds are co-parallel. This trans-conformation gives a much straighter molecule than the cis form, which is not mesogenic.

The two important members of the class are 2,5-PBO and 2,6-PBT; their repeat units are given in Fig. 1.13. Concentrations of almost three times that of Class 1 material are needed to form stable nematic meso-

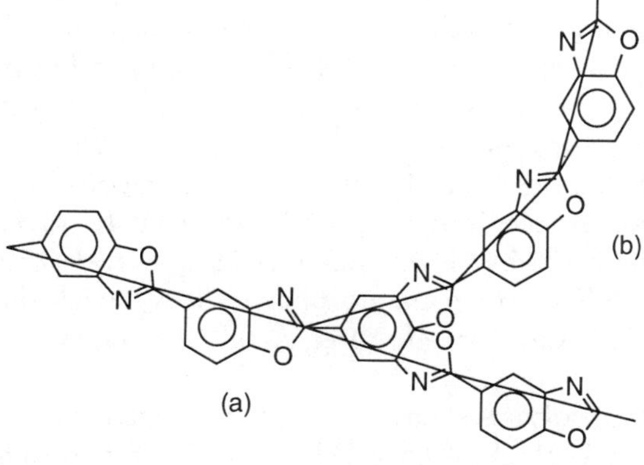

Figure 1.14 (a) Trans- and (b) cis-conformations of 2,5-PBO.

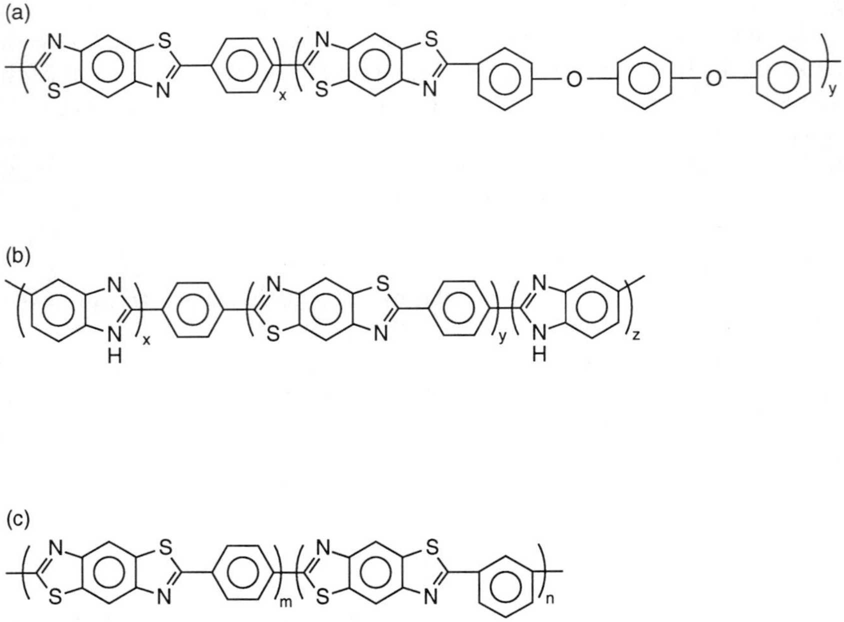

Figure 1.15 Types of mesogenic copolymer: (a) articulated rigid rod polymer; (b) ordered block copolymer (ABA) and (c) random block copolymer [50].

phases, with the straighter 2,6-PBT molecule requiring the lower concentrations [44, 50–55].

Mesogenic copolymers (Class 3)

As in thermotropic LCPs, these polymers consist of rigid rod segments in the backbone chain separated by more flexible units. The length of the rigid rod segment controls the degree of mesomorphism. The three main types are shown in Fig. 1.15 [50]. Articulated rigid rod polymers contain segments united by joints. The persistence lengths depend greatly on molecular architecture and nematic mesophases are formed.

In the articulated rigid rod structures, swivel units from flexible diacids have been incorporated into the backbone chain to give an extended chain conformation [56, 57]. As much as 25% swivel structure may be added to trans-PBT before mesomorphism is lost, which shows the capacity for meta-linkages for adopting a self-correcting extended chain conformation.

ABA block copolymers have been synthesized using A blocks of 2,5(6)-PBI with a B block of trans-PBT [58]. These copolymers have the advantage over homopolymers in the syntheses of rigid rod molecular composites in that the copolymers are more compatible than homopolymers in solution. Preliminary data show better tensile strengths from the copolymers than from the physical blends of the constituents of the copolymers [50].

Nematic mesophases occur in solutions involving random block copolymers when the higher concentrations are obtained using the P_2O_3 adjustment during reaction in polyphosphoric acid [51]. Little work has been carried out on these systems.

Non-mesogenic homopolymers (Class 4)

As the name suggests, these form isotropic solutions and are included for completeness. In each case, flexible spacer units are incorporated between rigid rod segments in the backbone chain. The rigid rod units may have catenation angles less than 150° or the rigid units are situated such that the catenating bonds cannot be coparallel. In all cases, the persistence length of the molecules in solution is insufficient to give mesophases.

1.6 FORMATION OF NEMATIC MESOPHASES

The phase separation behaviour of liquid crystals has been examined theoretically to obtain an estimate of the critical concentration at which a nematic mesophase forms. These theories apply equally well to thermotropic as well as lyotropic liquid crystals, and may be extended to LCPs.

The theories of Onsager [59] Isihara [60] and Flory [61, 62] require only that the molecules be asymmetrical; there is no need of attractive forces, in fact there will be intermolecular repulsion in that units of two molecules cannot exist in the same volume. In melts or in solutions, there is a limit to the number of rigid rod molecules that can be accommodated in a random arrangement. Above a critical concentration, the rigid rods will have to align, giving crystallinity or mesomorphism.

Crystallization may take place in two steps [63]:

1. co-operative chain alignment without a change in intermolecular interaction; and
2. an increase in intermolecular interactions due to the more efficient packing.

The second gives rise to phase separation, giving a mesophase or crystallinity. In lyotropic LCPs, solvent–polymer and polymer–polymer interactions are important, whereas chain regularity is important to both lyotropic and thermotropics.

In lyotropics, the nematic mesophase and isotropic phase may exist in equilibrium together and, as the concentration increases above critical, the region over which there is a mesophase increases. Onsager [59] applied cluster expansion methods to calculate the coefficients in the virial expansion. As a result of the complexity involved, he terminated the expansion at the binary cluster term. This limits the analysis but, nevertheless, he found that the asymmetric particles would separate into different phases in a solution, one isotropic and the other nematic with a slightly higher concentration of particles.

Onsager predicted phase separation when

$$C_{cr} = \frac{3.34}{BM} \tag{1.3}$$

where B is the second virial coefficient, M is the molar mass and C_{cr} is the critical concentration.

$$B = \frac{N_0 d L^2}{4M^2} \tag{1.4}$$

where d is the diameter and L the length of a molecule, and N_0 is Avogadro's number. For rigid rods $L \propto M$, B is independent of rod length and C_{cr} is inversely proportional to M or L. Isihara's [60] method was similar to Onsager's.

Oster [64], working with suspensions of the tobacco mosaic virus, applied Onsager's theory. Oster's value of 0.025 for the critical concentration meant that he had to increase greatly the molecular diameter in order to make the theory fit. He attributed this to the electrical double layer.

Flory extended the theory for higher concentrations [61] using a lattice theory developed by him [65, 66] and Huggins [67] for flexible chains and later extended by Flory [62] to semi-flexible chains. Flory assumed rigid molecules in a lattice and he computed the number of configurations and the entropy of mixing. On minimizing the free energy of mixing, he found that the system separated into two phases at a critical concentration C_{cr} that depended on the aspect ratio x $(= L/d)$ of the rigid rods, but was independent of the rod–solvent interaction. This theory could be extended up to zero solvent concentration and hence to thermotropic melts [39]. From this theory

$$C_{cr} = \frac{8}{x}\left(1 - \frac{2}{x}\right) \tag{1.5}$$

A minimum value of x of about 5.5 is necessary for mesomorphism. Flory's theory is limited by the approximations to the lattice formulation and both his and Onsager's theories do not show a temperature dependence. If the rods are not perfectly rigid, equation (1.5) gives the effective axis ratio X as discussed earlier and in Chapter 9.

Later Flory [68] extended his theory to semi-flexible particles, introducing the idea that the persistence length determines effective axial ratio for the particles. A temperature-dependent persistence length leads to a thermotropic behaviour in which temperature and concentration affect the phase transition.

Flory and Ronca [69] combined the original theory with an 'orientation-dependent energy', which was first suggested by Maier and Saupe [70]. The theory allowed the coexistence of a mesophase and an isotropic phase over a range of polymer concentrations as well as conditions in which no biphasic regions occur. As the concentration of the asymmetrical rods increases, the mesophase predominates. These excluded volume theories are discussed more fully in [52, 71–73].

The Maier–Saupe mean field theory [70] has been modified by Picken [74]. In this approach, the stability of the nematic mesophase is derived from an anisotropic potential. A molecule in a nematic region is assumed to sense the influence of the surrounding medium through this anisotropic potential. By equating the nematic mesophase, a first-order transition is obtained when

$$\frac{kT}{q} = 0.22 \tag{1.6}$$

where q is a constant describing the strength of the orientating potential, k is Boltzmann's constant and T is the temperature in Kelvin. The order parameter can be obtained from the theory.

The strength of the anisotropic potential described by q can be modified to account for concentration, and flexibility is incorporated into the theory using a contour projection length, which is the projection of the polymer chain in the direction of the first segment.

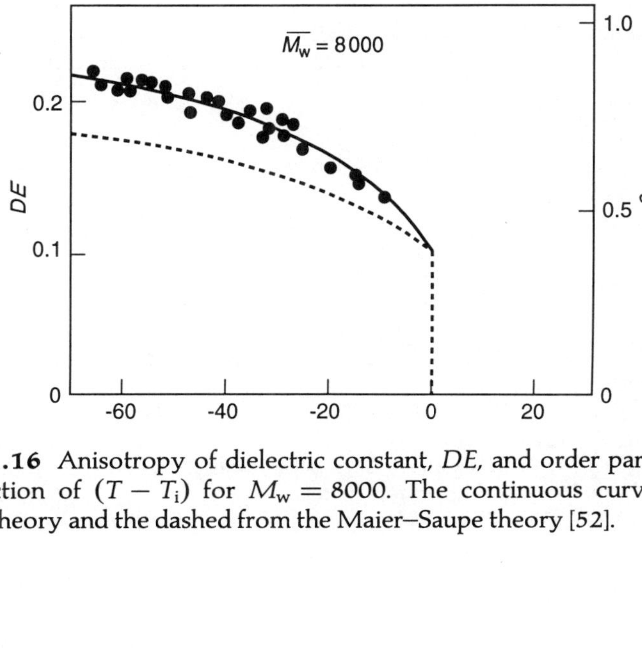

Figure 1.16 Anisotropy of dielectric constant, *DE*, and order parameter, *S*, as a function of $(T - T_i)$ for $M_w = 8000$. The continuous curve is from Picken's theory and the dashed from the Maier–Saupe theory [52].

Unlike the other theories, excluded volume is not taken into account. Picken's theory has good agreement with experimental values obtained from dielectric measurements. The anisotropy of dielectric constants, *DE*, is a good estimate of order parameter *S*. Figure 1.16 [52] shows the variation of anisotropy of dielectric constant and order parameter as a function of $(T - T_i)$. The continuous curve is from Picken and the dashed curve from Maier–Saupe. The experimental values agree well with those of Picken.

Similar approaches involving the combinations of a worm-like chain model and the Maier–Saupe mean field model have been carried out by Jähnig [75], Ten Bosch *et al.* [76] and Warner *et al.* [77]. More details of the modified Maier–Saupe theory is given by [52].

1.7 CONCLUSION

It is hoped that this chapter has introduced the subject of liquid crystallinity or mesomorphism sufficiently for rheologists to appreciate the unusual properties and that the above description will act as a firm basis for the following chapters.

ACKNOWLEDGEMENT

The author would like to thank Mr M. Furniss of the department for the provision of the diagrams.

REFERENCES

1. Calundann, G.W. and Jaffe, M. (1982) *Proc. 26th Conference on Chemistry Research, Houston TX, Nov 1982*, The Robert A. Welch Foundation, Ch. VII.
2. Frank, F.C. (1970) *Proc. Roy. Soc.*, **A319**, 127.
3. Brostow, W. (1992) in *Liquid Crystal Polymers: From Structures to Applications*, A.A. Collyer (ed.), Elsevier Applied Science Publishers, London and New York, Ch 1.
4. Griffin, A.C., Vaidya, S.R. and Steele, M.L. (1985) in *Polymeric Liquid Crystals*, A. Blumstein (ed.), Plenum Press, New York and London, 1–20.
5. Arnold, H. and Sackmann, H. (1960) *Z. Phys. Chem. (Leipzig)*, **213**, 145.
6. Sackmann, H. and Demus, D. (1973) *Mol. Cryst. Liq. Cryst.*, **21**, 239.
7. Krigbaum, W.R. (1985) *J. App. Polym. Sci.: App. Polym. Symp.*, **41**, 105–18.
8. Demus, D. and Richter, L. (1978) *Textures of Liquid Crystals*, Verlag Chemie, Weinham.
9. Noël, C. (1985) in *Polymeric Liquid Crystals*, A. Blumstein (ed.), Plenum Press, New York and London, 21–64.
10. Noël, C. (1992) in *Liquid Crystal Polymers: From Structures to Applica-*

tions, A.A. Collyer (ed.), Elsevier Applied Science Publishers, London and New York, Ch 2.
11. Alderman, N.J. and Mackley, M.R. (1985) *Faraday Disc., Chem. Soc.*, **79**, 149–60.
12. Chivers, R.A., Blackwell, J., Gutierrez, G.A. *et al.* (1985) in *Polymeric Liquid Crystals*, A. Blumstein (ed.), Plenum Press, New York and London, 153–66.
13. Blackwell, J., Gutierrez, G.A. and Chivers, R.A. (1985) *Ibid.*, 167–82.
14. Blackwell, J. and Biswas, A. (1987) in *Developments In Oriented Polymers-2*, I.M. Ward (ed.), Elsevier Applied Science Publishers, London and New York, 153–98.
15. DeGennes, P.G. (1974) *The Physics of Liquid Crystals*, Clarendon Press, Oxford.
16. Finkelmann, H. (1983) in *Polymer Liquid Crystals*, A. Ciferri, W.R. Krigbaum and R.B. Meyer (eds),Academic Press, New York, 35–62.
17. Blumstein, A. and Thomas, O. (1982) *Macromol.*, **15**, 1264.
18. Blumstein, A., Thomas, O., Asrar, J. *et al.* (1984) *J. Polym. Sci. Polym. Lett. Ed.*, **22**, 13.
19. Brostow, W. (1988) *Kunststoffe – German Plastics*, **78**, 411.
20. MacDonald, W.A. (1992) in *Liquid Crystal Polymers: From Structures to Applications*, A.A. Collyer (ed.), Elsevier Applied Science Publishers, London and New York, Ch. 8.
21. Huynha-Ba, O. and Cluff, E.F. (1985) in *Polymeric Liquid Crystals*, A. Blumstein (ed.), Plenum Press, New York, 217.
22. Collyer, A.A. (1989) *Mater. Sci. and Technol.*, **5**, 309–21.
23. Cottis, S.G., Economy, J. and Nowak, B.E. (1973) US Patent 3975486, (Carborundum Co.).
24. Calundann, G.W. (1986) in *High Performance Polymers: Their Origin And Development*, R.B. Seymour and G.S. Kirschenbaum (eds), Elsevier, Amsterdam, 235–49.
25. Pletcher, T.C. (1976) US Patent 3991013 and 3991014 (E.I. Du Pont de Nemours & Co.).
26. Payet, C.R. (1978) Ger. Offen, 2751653 (E.I. Du Pont de Nemours & Co.).
27. Zhou, Q.F. and Lenz, R.W. (1983) *Polym. Preprints*, **24** (2), 255.
28. Zhou, Q., Lenz, R.W. and Jin, J.-I. (1984) in *Polymeric Liquid Crystals*, A. Blumstein (ed.), Plenum Press, New York.
29. Antoun, S., Jin, J.-I. and Lenz, R.W. (1981) *J. Polym. Sci., Polym. Chem. Ed.*, **19**, 1901.
30. Jackson, W.J. (1984) in *Contemporary Topics In Polymer Science*, Vol 5, E.J. Vandenberg (ed.), Plenum Press, New York and London.
31. Payet, C.R. (1976) US Patent 4159365 (E.I. Du Pont de Nemours & Co.).
32. Goodman, I., McIntyre, J.E. and Stimpson, J.W. (1962) UK Patent 989552 (ICI).
33. Goodman, I., McIntyre, J.E. and Aldred, D.H. (1962) UK Patent 993272 (ICI).
34. Schaefgen, J.R. *et al.* (1974) US Patent 1507207 (E.I. Du Pont de Nemours & Co.).
35. Van Luyden, D. and Strzelecki, L. (1980) *Eur. Polym. J.*, **16**, 299.
36. Roviello, A. and Sirigu, A. (1982) *Makromol. Chem.*, **183**, 895.
37. Abe, A. (1984) *Macromol.*, **17**, 2280.

38. Aguilera, C., Bartulin, J., Hisgen, B. *et al.* (1983) *Makromol. Chem.*, **184**, 253.
39. Dobb, M.G. and McIntyre, J.E. (1984) *Adv. Polym. Sci.*, **60/61**, 61.
40. Jackson, W.J. and Kuhfuss, H.F. (1976) *J. Polym. Sci. Chem. Ed.*, **14**, 2043.
41. Griffin, B.P. and MacDonald, W.A. (1980) Eur. Patent W5527 (ICI).
42. Pielartzik, H., Dhein, R., Meyer, R.V. *et al.* (1987), Eur. Patent 303935 (Bayer A.G.): Eur. Patent 303931 (Bayer A.G.).
43. Collyer, A.A. (1991) *High Performance Plastics*, Nov., 1–3.
44. Collyer, A.A. (1990) *Mater. Sci. and Technol.*, **6**, 981–92.
45. Kwolek, S.L. (1966) UK Patent 1198081 (I.E. Du Pont de Nemours & Co.).
46. Papkov, S.P., Kulichickhin, V.G., Kalmykovo, V.O. *et al.* (1974), *J. Polym. Sci. Polym. Phys. Ed.*, **12**, 1753.
47. Information Bulletin No. 6E, E.I. Du Pont de Nemours & Co., Wilmington, DE, USA, 1974.
48. Anon (1987) *High Performance Plastics*, **4** (10), 1.
49. Gupta, N. (1980) *Textile Inst. Ind.*, Feb., 39.
50. Wolfe, J.F. (1988) *Encyl. Polym. Sci. Eng.*, **11**, 601–35.
51. Wolfe, J.F., Sybert, P.D. and Sybert, J.R. (1985) US Patent 4533692 (SRI International): 4533693 (SRI International).
52. Northolt, M.G. and Sikkema, D.J. (1992) *Adv. in Polym. Sci.*, **98**, 115–77; also in *Liquid Crystal Polymers: From Structures to Applications*, A.A. Collyer (ed.), Elsevier Applied Science Publishers, London and New York, Ch 6.
53. Matheson, R.R. Jnr and Flory, P.J. (1981) *Macromol.*, **14**, 954.
54. Wong, C.P., Ohnuma, H. and Berry, G.C. (1979) *J. Polym. Sci. Polym. Symp. Ed.*, **65**, 173.
55. Chow, A.W., Penwell, P.E., Bitler, S.P. *et al.* (1987) *Polym. Prepr. Am. Chem. Soc. Div. Polym. Chem.*, **28** (1), 50.
56. Evers, R.C., Arnold, F.E. and Helminiac, T.E. (1981) *Macromol.*, **4**, 925.
57. Evers, R.C. (1982) US Patent 4359567 (US Air Force).
58. Tsai, T.T., Arnold, F.E. and Hwang, W.-F. (1985) *Polym. Prepr. Am. Chem. Soc. Div. Polym. Chem.*, **26** (1), 144.
59. Onsager, L. (1949) *Ann. N.Y. Acad. Sci.*, **51**, 627.
60. Isihara, A. (1951) *J. Chem. Phys.*, **19**, 1142.
61. Flory, P.J. (1956) *Proc. Roy. Soc. London*, **A234**, 73.
62. Flory, P.J. (1956) *Proc. Roy. Soc. London*, **A234**, 66.
63. Flory, P.J. (1956) *Proc. Roy. Soc. London*, **A234**, 60.
64. Oster, G. (1950) *J. Gen. Physiol.*, **34**, 415.
65. Flory, P.J. (1942) *J. Chem. Phys.*, **10**, 51.
66. Flory, P.J. (1953) *Principles of Polymer Chemistry*, Cornell U.P., Ithaca, NY.
67. Huggins, M.L. (1942) *Ann. NY Acad. Sci.*, **41**, 1.
68. Flory, P.J. (1978) *Macromol.*, **11**, 1141.
69. Flory, P.J. and Ronca, G. (1979) *Mol. Cryst. Liq. Cryst.*, **54**, 289–311.
70. Maier, W. and Saupe, A. (1956) *Z. Naturforsh*, **14a**, 882: (1960) *Ibid.*, **15a**, 287.
71. Kwolek, S.L., Morgan, P.W. and Schaefgen, J.R. (1988) *Encyl. Polym. Sci. Eng.*, **9**, 1–61.
72. White, J.L. (1985) *J. App. Polym. Sci. App. Polym. Symp.*, **41**, 3–24.

73. Varshney, S.K. (1986) *JMS Rev. Marcomol. Chem. Phys.*, **C26** (4), 551–650.
74. Picken, S. (1989) *Macromol.*, **22**, 1766: (1990) *Ibid.*, **23**, 464.
75. Jähnig, F. (1979) *J. Chem. Phys.*, **70**, 3279.
76. Tenbosch, A., Maissa, P. and Sixou, P. (1983) *Phys. Let.*, **94A**, 298.
77. Warner, M., Gunn, J.M.F. and Baumgartner, A. (1985) *J. Phys.*, **A18**, 3007.

2

Theoretical aspects of the flow of liquid crystal polymers

G. Marrucci

2.1 INTRODUCTION

It is well known that the flow behaviour of polymeric liquids can be, and often is, much more complicated than that of ordinary fluids. The ultimate reason for the complication is the molecular anisotropy, obviously very large in a polymer chain. Indeed, although most materials are isotropic at rest (because the molecular segments are randomly orientated), the anisotropy reveals itself when the polymeric liquid is set in motion, as the molecules become oriented. The flow-induced molecular orientation, and the consequent material anisotropy, even persists for a while after the flow has been switched off, i.e. during relaxation.

Molecular anisotropy plays a central role in the rheological behaviour of all polymers, yet a profound difference exists between ordinary and liquid crystalline polymers (LCPs). In the case of ordinary polymers, since the system is isotropic at equilibrium, the molecular orientation is entirely determined by the flow process. On the other hand, LCPs are anisotropic materials already at equilibrium. A spontaneous molecular orientation already exists before the flow is switched on. Moreover, that spontaneous orientation generally varies in space, over distances of several microns or less, in the so-called 'polydomain'. The flow behaviour of LCPs is therefore much more complex than for ordinary polymers.

Rheology and Processing of Liquid Crystal Polymers
Edited by D. Acierno and A.A. Collyer
Published in 1996 by Chapman & Hall
ISBN 0 412 59640 7

In this chapter, we first describe the 'local' situation, i.e. how a velocity gradient interacts with the molecular orientation existing at a given point. Unlike the case of ordinary liquids, which are isotropic at equilibrium, we shall find out that the frictional interaction in the anisotropic case cannot be described by a single viscosity. In nematics, as many as six quantities, known as Leslie coefficients, are required. Still at a local level, the nonlinear viscoelasticity of LCPs, arising from the molecular response to fast flows, explains peculiar phenomena such as the negative normal stress effect. Next, we will discuss 'long range' interactions due to Frank elasticity. This is a characteristic elasticity of nematic mesophases, which works in the direction of restoring spatial uniformity of the molecular orientation. Thus, Frank elasticity is expected to play a role in polydomains. Frank elasticity and polymer viscoelasticity should not be confused with one another. Indeed, Frank elasticity was first studied in small molecule nematics (like those used in electronic displays), which are purely viscous rather than viscoelastic, at least in the frequency range relevant to flow.

Before presenting flow effects, it is necessary to describe the quantities which are commonly used to characterize molecular orientation in nematics. In the next section, a few useful concepts, such as 'director' and 'order parameter', are introduced. The important distinction between local (or microscopic) and mesoscopic averages is also illustrated.

2.2 MOLECULAR ORIENTATION

Differently from the case of an ordinary isotropic liquid, in a nematic mesophase (see Chapter 1) the molecules are no longer randomly oriented. Rather, if the molecular segments are seen as rigid rods, they would look as depicted in Fig. 2.1, where most rods are roughly parallel to one another. It is important to remember that the molecules of a liquid are in a highly mobile state. Thus, all the rods in Fig 2.1 should be imagined to undergo thermal motions, rapidly changing their individual positions as well as their orientations in space. Yet, in spite of this thermal agitation, the rod-like molecules forming the nematic phase maintain on average a common orientation, indicated by the unit vector **n** in Fig 2.1. Vector **n** is called the **director**.

The *average* molecular orientation described by the director **n** should not be confused with the individual molecular orientations. Figure 2.1 shows that the latter are distributed around **n**, as a consequence of thermal agitation. The spread in the distribution of molecular orientations is an important thermodynamic variable measured by a scalar quantity S varying from zero to unity. S is called the **order parameter**, $S = 0$ indicating total absence of orientational order (isotropy) and $S = 1$ the perfect nematic order, with all molecules strictly parallel to one another.

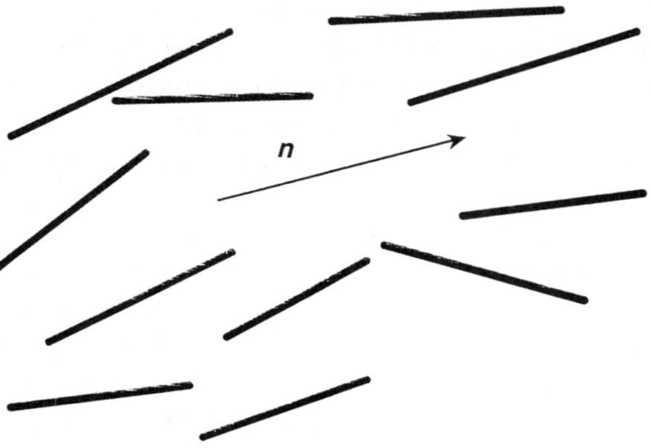

Figure 2.1 Rod-like molecules in the nematic phase. The average molecular orientation is indicated by the unit vector **n**, called 'director'.

The actual value of S in any given case represents a compromise between the ordering effect of the mesogenic interaction and the disordering contribution of temperature.

Notice that, while S is a thermodynamic property of the mesophase (a well-defined function of temperature and concentration), **n** is not. In other words, while temperature and concentration encourage the molecules of a nematic to stay parallel to one another to within a fixed dispersion, the common direction of the molecules is left undetermined. The director **n** is determined by relatively weak forces, such as the effect of a wall, or of magnetic and electric fields or, as we shall presently see, by the action of flow.

Although **n** and S are quite distinct quantities, they both refer to molecular orientation. It can be useful, therefore, to show how they are obtained from the individual orientations of the rod-like molecules. To this end, let us indicate with the unit vector **u** the instantaneous orientation of a single molecule, and consider the ensemble average $\langle \mathbf{uu} \rangle$. This average is a symmetric tensorial quantity $S_{ij} = \langle u_i u_j \rangle$ which can be represented geometrically by an ellipsoid. Under equilibrium conditions, the ellipsoid is axially symmetric, and the symmetry axis defines the director **n** (Fig 2.2). On the other hand, since also the trace of this tensor is fixed $S_{ii} = \langle \mathbf{u} \cdot \mathbf{u} \rangle = 1$, a single scalar quantity is required to specify (completely) the 'shape' of the ellipsoid. We can use the order parameter S to this end. Indeed, if the direction of the symmetry axis is taken to be the z-axis of a cartesian coordinate system (as well as the polar axis of a spherical coordinate system), the square length of the ellipsoid major axis S_{zz} and the order parameter S are related in a simple way:

$$S = \frac{3}{2} S_{zz} - \frac{1}{2}$$
$$S_{zz} = \langle \mathbf{u}_z \mathbf{u}_z \rangle = \langle \cos^2 \vartheta \rangle \tag{2.1}$$

In view of the unit trace condition, the square 'width' of the ellipsoid is given by:

$$S_{xx} = S_{yy} = \frac{1 - S_{zz}}{2}$$

Notice that, if all rods are exactly parallel to one another, then $S = S_{zz} = 1$. At the opposite extreme, isotropy implies $S_{xx} = S_{yy} = S_{zz} = \frac{1}{3}$, hence $S = 0$ and the ellipsoid becomes a sphere. In general, one should also consider the range of negative values of the order parameter, $-\frac{1}{2} < S < 0$ (corresponding to oblate ellipsoids), such that

$$S = -\frac{1}{2} \left(S_{xx} = S_{yy} = \frac{1}{2}, S_{zz} = 0 \right)$$

describes a limiting situation where all rods are parallel to a plane, while randomly oriented within that plane. The negative S range is not important for rod-like molecules, however.

So far, we have tacitly assumed that our nematic phase is spatially

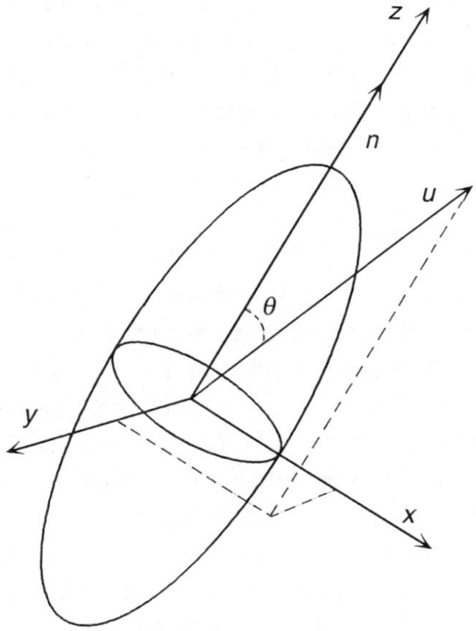

Figure 2.2 The ellipsoid representative of the tensorial average S_{ij}. In the equilibrium, spatially uniform nematic phase, the ellipsoid is uniaxial (and prolate). The symmetry axis defines the director **n**, while the 'slenderness' of the ellipsoid is related to the order parameter S (see equation 2.1). In the isotropic phase, the ellipsoid becomes a sphere, and $S = 0$.

homogeneous, i.e. the sample is an undistorted 'single crystal'. This is hardly obtained in most practical cases, especially for LCPs, where we most often find a so-called polydomain, somehow the liquid analogue of the solid polycrystalline structure. For such a case, the orientational situation so far described, based on molecular averages, still holds true locally, at the micron scale, say. At that level, we still usually find a well-defined director and, consequently, a material anisotropy. At a larger scale, however, we also need to consider that the director varies in space, from one 'domain' to the next. Thus, if we take an average at the scale of 100 μm, say, we should not be surprised that no net molecular orientation is found (in a quiescent sample, for example) as if the material were isotropic. Such averages over domains are called **mesoscopic**. In the previous example of an apparent isotropy, the *mesoscopic order parameter* would be zero, though the *local* order parameter at each 'point' in the sample is nonzero, and is in fact the same everywhere (except for defects), as is appropriate to an equilibrium thermodynamic quantity.

Experimentally, whether we are measuring a local average over molecules, or a mesoscopic average over domains, obviously depends on the technique. Rheological instruments are expected to be sensitive to mesoscopic averages, as the stress is typically obtained from integral values over large boundaries. Conversely, optical and spectroscopic methods can also reveal local properties, though not so easily and usually in very thin samples. In the following two sections, we discuss flow effects on molecular orientation by temporarily ignoring the polydomain complication, i.e. we proceed as if the sample were a monodomain. We go back to the polydomain case in the last sections.

2.3 SLOW FLOWS – THE LINEAR SITUATION

Here we discuss flows such that the shape of the equilibrium ellipsoid of Fig 2.2 remains virtually unaltered in spite of flow. This situation applies as long as the rate of deformation due to flow (the magnitude of the symmetric part of the velocity gradient) remains smaller than the rate by which molecules rearrange their orientational spread by thermal motions (which is measured by a molecular, or 'viscoelastic', relaxation time). For small molecules, which move very fast, this condition is satisfied virtually in all flows. For polymeric nematics, on the contrary, it is obeyed only in slow flows; even then it is with some difficulty, because of the role played by the defects in LCPs, to be discussed later.

If the shape of the ellipsoid does not change with flow, the viscous response of the nematic phase is intrinsically linear in the velocity gradient $\nabla \mathbf{v}$ of the flow. However, such a linearity of the viscous stress cannot be expressed in terms of a simple proportionality constant (i.e. by a single viscosity as in ordinary liquids) because the behaviour still depends on how $\nabla\mathbf{v}$ interacts with the ellipsoid orientation, i.e. with the director $\mathbf{n}$. In other words, even in this relatively simple situation, we have to pay a price to the anisotropy of the material. The price we pay is the complexity of the viscous stress expression in Leslie–Ericksen theory:

$$\sigma = \alpha_1 \mathbf{A} : \mathbf{nnnn} + \alpha_2 \mathbf{nN} + \alpha_3 \mathbf{Nn} + \alpha_4 \mathbf{A} + \alpha_5 \mathbf{nn} \cdot \mathbf{A} + \alpha_6 \mathbf{A} \cdot \mathbf{nn} \tag{2.2}$$

where the six material coefficients α_i have dimensions of a viscosity (but not all of them are necessarily positive in sign), and $\mathbf{A}$ and $\mathbf{N}$ are given by (notice the linearity in velocity gradient):

$$\begin{aligned} \mathbf{A} &= \frac{1}{2}(\nabla\mathbf{v} + \nabla\mathbf{v}^{\mathrm{T}}) \\ \mathbf{N} &= \dot{\mathbf{n}} - \frac{1}{2}(\nabla\mathbf{v} - \nabla\mathbf{v}^{\mathrm{T}}) \cdot \mathbf{n} \end{aligned} \tag{2.3}$$

In the expression for $\mathbf{N}$, the vector $\dot{\mathbf{n}}$ is the rate of rotation of the director.

Equation (2.2) is the most general constitutive equation, linear in $\nabla\mathbf{v}$, for uniaxial nematics. The meaning of equation (2.2) is readily understood in the classical example of the Miesowicz viscosities. Assume that the nematic is subjected to a simple shear flow while holding the director fixed by means, e.g. of a strong magnetic field. Then, although the shear stress will be found to be proportional to the shear rate no matter how the director has been orientated by the magnetic field, the proportionality constant (i.e. the viscosity) will be different depending on the $\mathbf{n}$ orientation in the shear field. The three Miesowicz viscosities correspond to orienting $\mathbf{n}$ either along the shear direction, or along the gradient, or perpendicular to both. By developing the algebra in equations (2.2) and (2.3), one can calculate how these viscosities (all three of them positive, of course) are related to the six Leslie coefficients α_i. Yet other values of the viscosity are calculated for elongational flows, etc.

Experimentally, it is found that the largest Miesowicz viscosity is obtained when $\mathbf{n}$ is along the velocity gradient direction. This is readily understood since, with such a director alignment, the molecules stick out the most in the shear gradient and, therefore, dissipate the most in their relative motion. One should be careful not to generalize too quickly, however, as that result is true only as long as the director is held fixed, i.e. as long as the director rotation $\dot{\mathbf{n}}$ (contained in $\mathbf{N}$, see equation (2.3)) remains zero. Imagine that the magnetic field is suddenly switched off; then the director immediately yields to the shear deformation and starts rotating, thus reducing the dissipation. For a free director rotating in the plane of shear, the maximum dissipation is no longer reached at 90° from the shear direction, but rather in the neighbourhood of 45°, where the relative velocity 'alongside' the unextendible molecules attains the largest value during rotation.

The feature we have just discussed brings up the question. How is the director rotation rate $\dot{\mathbf{n}}$ determined? The answer to this question requires that *torques* be explicitly considered, an aspect which is absent when dealing with isotropic liquids. Within the same linear range where equation (2.2) applies, the viscous torque **T** acting on the director is given by:

$$\mathbf{T} = (\gamma_1 \mathbf{N} + \gamma_2 \mathbf{A} \cdot \mathbf{n}) \times \mathbf{n}$$
$$\gamma_1 = \alpha_3 - \alpha_2 \qquad (2.4)$$
$$\gamma_2 = \alpha_3 + \alpha_2$$

Thus, if no other torques are acting (and since orientational inertia is negligible in all cases) $\dot{\mathbf{n}}$ is determined by the condition $\mathbf{T} = 0$.

Equation (2.4) shows that, among the six Leslie coefficients, α_2 and α_3 play a special role, as they alone control the orientational behaviour of the nematic. What is particularly important is the sign of the ratio α_2/α_3. If it is positive, then a shear flow leads to a stable molecular orientation. The director sets itself in the plane of shear at a particular angle θ (called the **Leslie angle**) from the shear direction, such that $\tan^2\theta = \alpha_3/\alpha_2$. Nematics for which α_2/α_3 is positive are called **flow aligning**. In rod-like nematics, θ is less than 45°, and is expected to be particularly small in the polymeric case, in view of the large molecular anisotropy. If and when a polymeric nematic is flow aligning, the direction of the orientation should virtually coincide with that of shear.

In the opposite case, i.e. if α_2/α_3 is negative, no stable orientation exists. The condition $\mathbf{T} = 0$ can only be satisfied if the director keeps rotating, i.e. if $\dot{\mathbf{n}}$ always remains nonzero (in the absence of other torques, of course, like the magnetic one). Nematics having a negative value of α_2/α_3 are called **tumbling nematics**. Tumbling is a relatively rare occurrence in small-molecule liquid crystals. There are good theoretical indications showing that, for rod-like molecules, the sign of α_2/α_3 depends on how 'fat' is the prolate ellipsoid of Fig. 2.2. If the order parameter S is small, i.e. if the orientational spread is large (a fat ellipsoid), then the nematic is of the flow-aligning category. Conversely, values of S closer to unity (slim ellipsoids) imply a tumbling nematic. This prediction is consistent with the observation that some small-molecule flow-aligning nematics turn to tumbling nematics by decreasing the temperature. Indeed, the order parameter (which, we recall, is a thermodynamic quantity) is expected to increase with decreasing temperature, as the nematogenic interaction increasingly prevails upon thermal agitation.

It is a known fact that LCPs show large values of the order parameter, significantly larger than those of small-molecule nematics. Thus, although

LCP molecules are rarely rod-like in a strict sense, and therefore the above mentioned prediction does not strictly apply, it should perhaps be expected that tumbling is more likely to occur for polymers than for small-molecule nematics. We find confirmation of this expectation in the nonlinear behaviour of LCPs, to be discussed in the next section.

Before concluding here, it should be mentioned that the complete Leslie–Ericksen theory of nemato-dynamics also includes the effect of the *distortions* of the liquid crystal. If the director is not uniform in space (i.e. if the 'crystal' is orientationally distorted), then both an elastic stress and an elastic torque generally arise, to be added to the viscous ones. This kind of elasticity (named after Frank who first developed the theory for the static situation) should not be confused with the typical rubber-like elasticity of polymeric liquids. Indeed, while the latter is measured through elastic moduli (dimensions of force per square length), Frank elasticity is measured through three constants (K_1, K_2 and K_3, one for each of the basic distortions of splay, twist, and bend) which have dimensions of a force. The difference is profound because, while the ratio of a viscosity η to a modulus G gives a relaxation time $\tau = \eta/G$ which is an intrinsic property of a viscoelastic liquid, the ratio of a viscosity to a Frank constant (η/K) cannot generate a characteristic time without invoking the help of some characteristic length, such as the sample thickness. In other words, relaxation times associated with Frank elasticity are size dependent; they are not an intrinsic material property. We shall comment further on this aspect later in the chapter.

Frank elasticity can play an important role in flowing nematics in so far as the elastic torque resulting from a distortion may well immobilize the director (similarly to a strong magnetic field). For example, if the director is 'anchored' at the wall as a consequence of some molecular adsorption process, and a flow is started, then in the neighbourhood of the wall, Frank elasticity counteracts the viscous torque which would carry the director orientation elsewhere. Even if the nematic is of the tumbling type, it cannot be excluded that Frank elasticity might arrest actual tumbling of the director, by virtue of a suitable distortion field permeating the whole sample volume. We shall discuss these concepts again when dealing with polydomains. For more information on Frank elasticity and Leslie–Ericksen theory and, more generally, on the physics of small-molecules nematics, the reader is referred to the monograph by de Gennes [1].

2.4 NONLINEAR BEHAVIOUR–NEGATIVE NORMAL STRESSES

If the velocity gradient of a flow is large enough, polymeric liquids behave nonlinearly. As previously mentioned, this occurs when the rate of molecular relaxation is taken over by the deformation rate due to flow.

Since polymers are sluggish molecules, nonlinearities occur much more easily than for small molecules. With respect to ordinary polymers, however, many LCPs show peculiar nonlinearities. The most striking of them is the behaviour of the normal stress differences, N_1 and N_2, in simple shear flow.

Figure 2.3 reports normal stress results by Magda *et al.* [2] for a lyotropic system in the relevant range of shear rates. The curves show complex nonmonotonic features, and both N_1 and N_2 change sign twice. This behaviour is in sharp contrast to that of ordinary polymers, for which the normal stress differences vary monotonically with shear rate. Historically, the unexpected result of negative values for N_1 was first reported by Kiss and Porter [3] and later confirmed by many others. Since the shear rate where this strange behaviour is found falls outside the linear range, the continuum theory of Leslie and Ericksen cannot be used to interpret it. Fortunately, however, a relatively simple molecular theory convincingly explains it.

The theory is that of rigid rod-like polymers which, for dilute solutions, was laid down in the early 1950s by Kirkwood and Auer [4]. It was then extended to more concentrated solutions by Doi and Edwards [5] and, finally, to nematic phases by Hess [6] and by Doi [7]. Owing to a mathematical approximation, however, the 'mystery' of negative values of N_1 was unravelled only some time later [8]. It is now clear that the physical origin of the strange behaviour reported in Fig. 2.3 is the

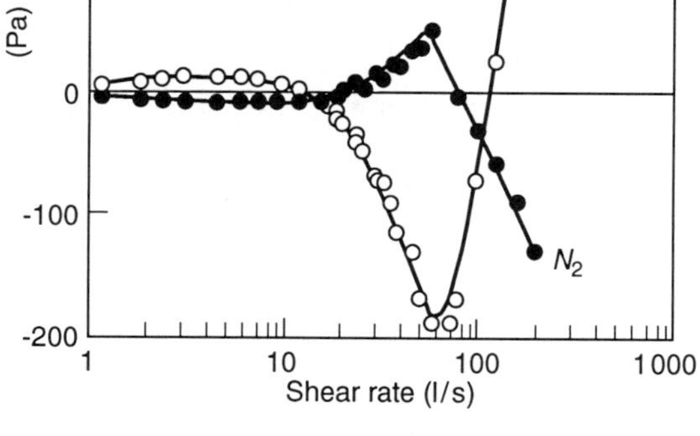

Figure 2.3 The first (N_1) and the second (N_2) normal stress differences in the shear flow of a nematic solution of polybenzylglutamate (PBLG) in m-cresol (12.5% w/w). (Reproduced with permission from Ref. 2.)

tumbling character of the polymeric nematic in the linear range, i.e. at low shear rates. Indeed, in the range of shear rates where N_1 becomes negative (and N_2 positive) a dynamic transition takes place (predicted by the nonlinear theory) from tumbling at low shear rates to flow aligning at high ones. The negative values of N_1 are a manifestation of the rearrangement that takes place in the orientational distribution in the neighbourhood of the transition. With increasing shear rate in that range, the ellipsoid of Fig. 2.2 first becomes 'fatter' than it is at equilibrium, then progressively thins down again, eventually to become even thinner than at equilibrium. (In fact, the ellipsoid also loses the axial symmetry, i.e. the system becomes biaxial during flow.)

Negative values of N_1 have been measured in many lyotropic LCPs. For a few systems, optical evidence of tumbling at the start-up of a shear flow has also been directly obtained by first carefully preparing monodomain samples. It so appears that tumbling at low shear rates is the rule for lyotropic LCPs. Much less clear is the situation for thermotropic main-chain LCPs, for which negative normal stresses are rarely observed, if at all. On the one hand, the molecules of thermotropic LCPs are less rigid than those of lyotropic ones and, as yet, there is no good theory of the dynamics of semi-rigid chains in a nematic phase. There are, however, other indications that, perhaps, also thermotropic LCPs are tumbling nematics. This matter is taken up again in the next section.

2.5 DEFECTS AND POLYDOMAINS

Several kinds of defects can be present in nematics, corresponding to apparent discontinuities of the director field. The most common variety are the **disclination lines**, which give to the nematic phase, observed under the microscope, a characteristic threadlike texture. Owing to defects and distortions, the director is nonuniform in space, and the sample is therefore a polydomain.

Defectless samples are easily obtained in small-molecule nematics. Indeed, defects are energetically unfavoured and Frank elasticity is the driving force which eliminates defects, either by shrinking disclination loops, or by the merging of defects of complementary topology (opposite in sign, according to a conventional classification), etc. On the contrary, it is extremely difficult to rid LCPs of defects, and even when it is done, the result is extremely 'fragile' as a shear flow will soon destroy the monodomain.

One obvious difference is that LCPs are much more viscous than small-molecule nematics, and therefore the dynamics of defect elimination is slowed down considerably. Another possible explanation of the difference between LCPs and small-molecule nematics in this regard could be that defects are more easily generated in LCPs because they are tumbling nematics while small-molecule nematics are mostly flow aligning. Indeed, tumbling induces distortions, and large distortions often relax by generating defects. Some experimental evidence of defect generation during shear flow of LCPs has now been collected. In thermotropic LCPs undergoing oscillatory shear, a new defect-laden texture was first observed by Graziano and Mackley [9], and named **wormlike** by them. The wormlike texture was then studied by De'Neve *et al.* [10] in frozen samples of another thermotropic polymer. Electron micrographs reveal it to be constituted by a large number of defect loops [11]. These observations are hardly reconcilable with the expected behaviour of a flow-aligning nematic, which should be stable in a shear flow, especially in one-directional flow as in Navard's experiments.

The presence of many defects in LCPs, i.e. their polydomain structure, is also invoked to justify why the viscosity curve, differently from ordinary polymers, does not usually show a Newtonian plateau at very low shear rates [12]. Shear thinning at very low shear rates is a very curious nonlinear phenomenon, which cannot be attributed to molecular relaxation mechanisms in the nematic phase. Rather, the nonlinearity might perhaps be explained by looking at the system as if it were a 'suspension' of a more viscous 'defect phase' in the nematic 'solvent'. The defect phase is made up of the material within the defect cores where the nematic order is disrupted (i.e. where the equilibrium ellipsoid of Fig. 2.2 is strongly modified by distortion). The structural evolution of such a defect 'suspension' would generate the shear thinning behaviour at low shear rates [13]. A complete theory is lacking, however.

It should be noted that tumbling is a shear phenomenon which disappears in elongational flows. Thus, in elongational flows, all LCPs are expected to behave in the same way. In particular, they should readily align along the direction (or directions) of stretch. But even in shear, tumbling LCPs might be fairly well aligned because the direction of shear remains privileged even for a tumbling nematic, especially if the molecular anisotropy is large as for LCPs. The main problem of tumbling seems to be the unavoidable large defect density associated with it. More about molecular orientation will be said in the next section. In the case of complex geometries like those encountered in processing, the combination of elongational and shear flows makes it even more difficult to predict the structural evolution. It is noteworthy, however, that the assumption that LCPs are tumbling nematics has allowed Wissbrun [14] to formulate a sensible model to explain the behaviour of LCPs at the entrance of a capillary, thereby interpreting a large body of capillary data in a simple way.

Lacking a complete theory for polydomains, we only present in this section a simple dimensional analysis for the linear case. By linear, we mean that two conditions are simultaneously satisfied:

1. The magnitude of the velocity gradient is smaller than the reciprocal relaxation time of the molecules in the nematic phase, and therefore that the ellipsoid of Fig. 2.2 keeps its equilibrium shape in such a phase. There also follows that the nematic phase obeys the Leslie–Ericksen constitutive equation.
2. Although there are many defects, and they play an important role that we shall presently discuss, yet the contribution of the defect phase to the stress tensor (which would be nonlinear) is in fact negligible.

We believe that such a linear range is observed in those LCPs which exhibit a newtonian plateau in the viscosity curve at some intermediate range of shear rates (Region II according to Onogi and Asada [12]).

To be definite, let us specifically refer to a simple shear flow between two parallel plates and, initially, let us limit ourselves to steady-state conditions. For the purpose of dimensional analysis, we may formally express the dependence of the shear stress σ as

$$\sigma = \sigma(v, L, \alpha, K) \tag{2.5}$$

where v is the velocity of one of the plates (the other being fixed), L is the sample thickness, and α and K stand, respectively, for all six Leslie coefficients (dimensions of viscosity) and all three elastic constants (dimensions of force) of a given nematic. Since $\dot{\gamma} = v/L$ is the apparent shear rate of the flow, and $\eta = \sigma/\dot{\gamma}$ is the apparent viscosity, we may rewrite equation (2.5) in the equivalent, nondimensional form as

$$\frac{\eta}{\alpha} = f\left(\frac{\alpha\dot{\gamma}L^2}{K}\right) \tag{2.6}$$

where $f(\cdot)$ is some material function. A similar dimensional analysis for the capillary flow of a nematic was first made by Ericksen, and $Er = \alpha\dot{\gamma}L^2/K$ is called the **Ericksen number**. Equation (2.6) predicts that the apparent viscosity of a nematic should vary either by varying the shear rate $\dot{\gamma}$ (in spite of linearity! see comments below) or by varying the sample thickness L, both of which are contained in the Ericksen number. The predictions of equation (2.6) are in excellent agreement with various data on *flow-aligning small-molecule nematics*.

The physical meaning of the L and $\dot{\gamma}$ dependence of the apparent viscosity of small-molecule nematics is as follows. Frank elasticity, resulting from the orientational deformation of a 'crystal', is a long-range effect. Thus, the anchoring of the director at the walls may extend its effects well inside the sample where the director wants to align at the Leslie angle. The relative proportion of sample volume oriented in the wall way or in the Leslie way will then depend both on the thickness L and on the flow strength $\dot{\gamma}$. The viscosity, which is sensitive to director orientation (remember the discussion on the Miesowicz viscosities), will consequently vary with both these factors. It should be mentioned, however, that the dependence in equation (2.6) is predicted to *saturate* at high Ericksen numbers. Indeed, if *Er* is large enough, for all purposes the director can be considered as oriented at the Leslie angle throughout the whole sample thickness, and the apparent viscosity then becomes a constant.

Returning to LCPs, the first thing to be mentioned is that the viscosity is *not* found to depend on sample thickness. Should the LCP be a flow-aligning nematic, there would be no problem in interpreting this result. Indeed, since the Ericksen number as defined above is certainly very large in LCPs (they are much more viscous than small-molecule nematics), we could safely assume that the system has reached the above mentioned saturation. Correspondingly, we should also find a uniform director orientation at the Leslie angle. However, this is definitely not the case. In the shear rate range of Region II of the viscosity curve, all exisiting orientation measurements do not indicate a full alignment. Furthermore, there is evidence that the structure 'refines' with increasing shear rate in that range, i.e. the domains become smaller. Clearly, some different interpretation is required.

It is perhaps possible to look at all these facts in the following way. Let us assume that the LCP is a tumbling nematic. Then, in a shear flow, the system soon becomes a polydomian (even if it was not beforehand) with many defects and disclinations. The system is now full of distortions on a short scale, so that the bulk of the material conceivably ignores the orientational influence of the walls. The situation becomes one where the defects act as if they were a sort of internal wall for the bulk nematic. In fact, there are arguments indicating that the director might even become anchored at these internal walls (see next section). In any event, for what concerns dimensional analysis, the important difference is that the externally determined characteristic length *L* must be dropped from the analysis, and replaced by some *internal* characteristic length, which is not predetermined. Rather, it is *self-adjusted*.

If we call *a* this internal characteristic length (*a* can be looked upon as an average distance between neighbouring defects, or else $1/a^2$ as the total length of disclination lines per unit volume), dimensional analysis dictates that [15]

$$\frac{\alpha\dot{\gamma}a^2}{K} = \text{constant} \tag{2.7}$$

Equation (2.7) predicts that the structure refines as $\dot{\gamma}$ increases, since *a* must decrease to keep the group in equation (2.7) a constant. This group might be called an ***internal* Ericksen number**. The important result of dimensional analysis is that such a group stays constant in a given material, a prediction supported by experiments [16]. The prediction that the viscosity also stays constant with varying either sample thickness or shear rate is obtained trivially from the analogue of equation (2.6) where the internal Ericksen number replaces the external one.

Of course, dimensional analysis is incapable of telling us the details of the orientational distribution in the tumbling nematic. It says, however, that the director field must remain self-similar at all shear rates (within the linear range). By varying the shear rate, the 'picture' of the director field only blows up (or down) without changing form. This prediction is confirmed by optical measurements of the mesoscopic orientation, which stays rigorously constant in the linear range [17].

The dimensional analysis can be fruitfully extended to the time-dependent situation encountered in several experiments where the transient response following a stepwise change in the shear rate is monitored (steps up or down, flow reversal, stress relaxation, etc. cf. Chapter 8). In all such experiments, there is an initial value of the shear rate $\dot{\gamma}_{\text{init}}$, a final one $\dot{\gamma}_{\text{final}}$, and, of course, there is the time variable t. On the other hand, we must again cancel the external dimension from the relevant variables because, in the transient case as well as at steady state, LCP experiments do show that the response is independent of the sample thickness.

We write formally for the time-dependent stress

$$\sigma = \sigma(t, \dot{\gamma}_{\text{init}}, \dot{\gamma}_{\text{final}}, \eta, K) \tag{2.8}$$

where the internal dimension a has not been listed because it is self-determined and not externally imposed. We have also conveniently replaced the characteristic Leslie coefficient α with the viscosity η measured at steady state (in the linear range) since the previous analysis has already shown that η, similar to the Leslie coefficients, is a material constant. The nondimensional form of equation (2.8) is then

$$\frac{\sigma}{\eta\dot{\gamma}_{\text{init}}} = f\left(\dot{\gamma}_{init}\, t, \frac{\dot{\gamma}_{\text{final}}}{\dot{\gamma}_{\text{init}}}\right) \tag{2.9}$$

where $f(\cdot)$ is some material function. We have chosen to put $\dot{\gamma}_{\text{init}}$ both in the group on the left side and as $\dot{\gamma}_{\text{init}}\, t$ on the right but, of course, $\dot{\gamma}_{\text{final}}$ would have done equally well (unless $\dot{\gamma}_{\text{final}} = 0$). What is really noteworthy in equation (2.9) is that, although Frank elasticity certainly plays a significant role in determining the polydomain evolution, no group containing K can be accommodated in the dimensional analysis of the stress response, because of the lack of a characteristic length in the problem. Notice further that, in some experiments, the ratio $\dot{\gamma}_{\text{final}}/\dot{\gamma}_{\text{init}}$ is automatically fixed, either because $\dot{\gamma}_{\text{final}}$ is zero (stress relaxation) or because it is equal but opposite in sign to $\dot{\gamma}_{\text{init}}$ (flow reversal). In all such cases, equation (2.9) reduces to

$$\frac{\sigma}{\eta\dot{\gamma}} = f(\dot{\gamma}t) \tag{2.10}$$

where, of course, $f(\cdot)$ is a different function, and $\dot{\gamma}$ is the only shear rate that matters. Equation (2.10), with its own $f(\cdot)$, also applies to recoil experiments. Finally, it should be mentioned that, for the general case of time-dependent flows (which are not necessarily stepwise changes in either the shear rate or the shear stress), Larson and Doi [18] and, for the case of blends, Doi and Otha [19] have formulated a *scaling law* between motions which differ in the time-scale. The law applies to 'complex' systems with a constant viscosity, and equations (2.9) and (2.10) are but special cases of this law.

Equations (2.9) and (2.10) are in excellent agreement with all the experiments of Moldenaers *et al.* [20, 21] in the linear range. The typical transient response shows a sequence of damped oscillations, a feature which further confirms the tumbling nature of most LCPs [16]. As is apparent from the equations, and confirmed by the experiments, no characteristic relaxation time of the material is involved in these scaling laws. The running time t is made dimensionless through the 'strain' units $\dot{\gamma}t$. Indeed, in order to 'feel' the relaxation times of the polymer one must either go to faster flows, approaching the nonlinear range, or to shorter times (the very beginning of the relaxation, for example). Materials which never show a linear range behave in a mixed fashion. For them, the regimes controlled by Frank elasticity and by polymer viscoelasticity are somehow superimposed on one another [22].

Finally, it should be noted that equation (2.10) does not compare favourably with data of shear start up from the quiescent state. Often the start-up transients are not even reproducible. This is due to the fact that the 'structure' of the quiescent state is ill-defined in most cases, as the material at rest continuously evolves by progressively eliminating defects induced by previous motions, a process which becomes increasingly slower as the structure coarsens. A pseudo-relaxation time τ of Frank elasticity can be constructed from dimensional analysis as follows

$$\tau = \frac{\eta a^2}{K} \tag{2.11}$$

Equation (2.11) shows that if the domain size has grown from, say, 1 μm to 100 μm, τ has increased by a factor 10^4.

2.6 FLOW-INDUCED ORIENTATION

As a conclusion to this chapter, we wish to emphasize a few points which, lacking a complete description of the complexities of the polydomain dynamics (not to speak of a reliable constitutive equation which embodies them all), might serve as a qualitative guidance to deal with flow problems, particularly with flow-induced molecular orientation.

The preliminary, and perhaps primary, concept is that main-chain LCPs are much more easily orientable materials than ordinary polymers. This is due to the co-operative nature of a nematic phase where, at least locally, the molecules are already mutually oriented in a roughly parallel fashion. Thus, in order to orient the bunch of them in the desired direction, we have not to fight against thermal agitation, which is the difficult task in the case of ordinary polymers. As previously mentioned, much weaker forces are required to reorient a director. The easier orientability of the nematic phase (at least in principle), together with its reduced viscosity, are among the important factors which make LCPs attractive materials in the first place.

In the case of elongational flows (such as in fibre spinning), a good orientation is more readily obtained. The tumbling nature of LCPs, if present, is anyhow irrelevant because actual tumbling can only occur in flows dominated by shear deformations. A partial obstacle to a good orientation can possibly come from the pre-existing defects, if they do not find the time or the topological conditions to disappear under the influence of flow, which *per se* works in favour of defect elimination.

A word of caution about elongational flows should be said, however. Elongational flows are also encountered in wall-constrained flows, such as in mould filling. They are present, for example, if there is a contraction or an enlargement in the cross-section of the flow. It is important to realize that, for such cases, the spontaneous direction of molecular orientation does not necessarily coincide with the flow direction, as is sometimes erroneously believed. As briefly mentioned in the previous section, an elongational flow orients the molecules (or the director) along the *stretching* directions which, generally, are different from that of the velocity. For example, in the decelerating flow which exists in a section enlargement, the stretching directions are *transversal*, i.e. orthogonal to the velocity. Now, while for the case of ordinary polymers this might simply mean that they lose locally some of the previously acquired molecular stretch in the flow direction, in the case of the readily orientable LCPs, the result could easily be a sudden *flip over* of the orientation in a direction transversal to that of flow.

Passing on to consider shear-dominated flows, another possible misconception could be that tumbling completely destroys orientation, effectively creating an isotropic situation at the mesoscopic level. In the previous section, we have briefly mentioned that such is not the case, but the matter requires a somewhat more detailed discussion. Several distinct arguments, based on quite different assumptions about the tumbling dynamics, lead to conclusions about molecular orientation in shear flows which, though differing in important details, are roughly equivalent in one important feature. The common feature is that the mesoscopic orientation, though not the best possible orientation (which would be that of the monodomain), is roughly *along the shear direction* in spite of tumbling.

In the first, simpler argument, Frank elasticity is neglected altogether, and it is assumed that director tumbling *actually* takes place, i.e. that the director keeps rotating indefinitely, without disturbances of any sort. Let us then consider the projected motion in the plane of shear (the director will generally have an out-of-plane component), i.e. let us calculate how the angle θ between the (projected) director and the shear direction varies with time. From the zero torque condition of equation (2.4):

$$\dot{\theta} = \dot{\gamma} \frac{\sin^2 \theta + \lambda \cos^2 \theta}{1 + \lambda} \tag{2.12}$$

where $\dot{\theta}$ is the director rotation rate, and $\lambda = -\alpha_3/\alpha_2$ is a positive number in tumbling nematics. In rod-like molecule nematics, and particularly in LCPs which are highly anisotropic, λ is a very small number. Thus equation (2.12) shows that $\dot{\theta}$ is very close to the shear rate $\dot{\gamma}$ when θ is close to 90° (gradient direction), while it is much smaller, $\dot{\theta} \approx \lambda\dot{\gamma}$, when $\dot{\theta}$ is close to zero (shear direction). In other words, the director, though continuously rotating, most of the time *stays close to the shear direction*, and only once in a while (so to speak) flips over and quickly performs a half turn. To be precise, we should account for the out-of-plane component and therefore say that, most of the time, the director remains close to the plane made by the shear and by the neutral directions (let us call Π this plane).

With such a picture of the tumbling dynamics, certainly the directors in the polydomain are not randomly distributed. Rather, they are predicted to lay essentially in the Π plane, while the distribution within that plane remains unknown. Yet, if we further assume that the initial orientation in the quiescent sample was random in space (and keep ignoring domain interactions due to Frank elasticity), then it is readily verified that, once the directors have been brought by the shear flow to lay within Π, their distribution in that plane is not uniform: the shear direction is favoured over the neutral direction. The conclusion is reached that the mesoscopic average orientation of the tumbling polydomain is after all in the shear direction, though not 100% so, since some of it is spread over the Π plane, and a tiny bit out of Π. This conclusion, simplistic as the argument may be, is nevertheless roughly in agreement with experiments [17, 23]. (Though it is not so in some revealing detail, e.g. in the fact that the mesoscopic director points slightly *above* Π.)

A more complete, and complex, analysis of the polydomain situation is contained in the constitutive proposal by Larson and Doi [18] of a mesoscopic orientation (and mesoscopic stress tensor). Larson and Doi bring Frank elasticity into the picture, though only in some approximate average way. The prediction of the orientation in steady shear comes out (roughly) similar to that of the previous simple argument, with some improvement (the director is predicted to point slightly above Π, for example), but also with some other disagreement with data [23].

In yet another argument, a great importance is given to defects, specifically to defect cores, which are thought to be capable of *arresting actual tumbling* of the surrounding nematic through an anchoring effect [24]. If and when the defect density is sufficiently large at the given shear rate, the director of the nematic phase would 'hang' to the defect phase, through Frank elasticity, thereby stopping its rotational motion. With such a picture of the tumbling dynamics, again the Π plane is favoured for director orientation. Indeed, a stationary director in the Π plane cannot afford deviating too much from that plane without breaking the balance between the viscous and elastic torques. From equation (2.4), the viscous torque on the stationary director comes out proportional to the same function of θ which appears in equation (2.12), and is therefore approximately constant in the neighbourhood of $\theta = 0$, while growing quadratically with θ as soon as θ becomes larger than *c.* $\sqrt{\lambda}$. Should the director leave the Π plane by more than that, additional defects are generated, a process which takes place whenever the shear rate is further increased, but which should not occur, at least not frequently, at a steady state. The conclusion is reached that, at steady state, the defect density is just such as to keep the director in the neighbourhood of the Π plane.

Unfortunately, this more sophisticated argument has not been developed to the point of predicting a detailed orientational distribution. Thus, the orientation within Π remains undetermined. However, there are recent calculations by Han and Rey [25] showing that, in tumbling nematics constrained between two anchoring walls, a shear flow can induce stable stationary structures for which the distortion is mostly a twist parallel to Π. Both the orientation within Π and the tiny components out of Π are well determined in these solutions, and there again the shear direction comes out favoured over the neutral one in the Π plane, while in one of the solutions (the one called chiral) also the tilt off the Π plane is in the direction indicated by the experiments. It seems interesting to perform more detailed comparisons between the predictions of Rey and the mesoscopic orientation measured by Hongladarom and Burghardt [23].

There remains to be mentioned that shear flows of tumbling LCPs do an even better job of orienting the molecules in the shear direction if the shear rate is so large as to reach the nonlinear range, where the material undergoes a transition from tumbling to flow-aligning behaviour (see a previous section). Very good alignments, close to 100%, at very high shear rates are confirmed by experiments [17].

ACKNOWLEDGEMENTS

This work was jointly supported by the European Community Commission under the HC program, contract No. CHRX-CT93-0200, and by the Italian Ministry for the University, MURST, Rome.

REFERENCES

1. de Gennes, P.G. (1974) *The Physics of Liquid Crystals*, Clarendon Press, Oxford.
2. Magda, J.J., Baek, S.G., de Vries, L. *et al.* (1991) *Macromolecules*, **24**, 4460.
3. Kiss, G. and Porter, R.S. (1978) *J. Polym. Sci.; Polym. Symp.*, **65**, 193.
4. Kirkwood, J.G. and Auer, P.L. (1951) *J. Chem. Phys.*, **19**, 281
5. Doi, M. and Edwards, S.F. (1978) *J. Chem. Soc. Faraday Trans. 2*, **74**, 560.
6. Hess, S. (1976) *Z. Naturforsch.*, **31A**, 1034.
7. Doi, M. (1981) *J. Polym. Sci.; Polym. Phys.*, **19**, 229.
8. Marrucci, G. and Maffettone, P.L. (1989) *Macromolecules*, **22**, 4076.
9. Graziano, D.J. and Mackley, M.R. (1984) *Mol. Cryst. Liq. Cryst.*, **106**, 73.
10. De'Neve, T., Navard, P. and Kléman, M. (1993) *J. Rheol.*, **37**, 515.
11. De'Neve, T. (1993) PhD thesis, Ecole Nationale de Mines de Paris.
12. Onogi, S. and Asada, T. (1980) in *Rheology, Vol. I* (Eds G. Astarita, G. Marrucci and L. Nicolais) Plenum Press, New York, 127.
13. Marrucci, G. and Maffettone, P.L. (1993) in *Liquid Crystalline Polymers* (Ed. C. Carfagna) Pergamon, 127.
14. Wissbrun, K.F. (1993) *J. Rheol.*, **37**, 777.
15. Marrucci, G. (1985) *Pure Appl. Chem.*, **57**, 1545.
16. Burghardt, W.R. and Fuller, G.G. (1990) *J. Rheol.*, **34**, 959.
17. Hongladarom, K., Burghardt, W.R., Baek, S.G. *et al.* (1993) *Macromolecules*, **26**, 772.
18. Larson, R.G. and Doi, M. (1991) *J. Rheol.*, **35**, 539.
19. Doi, M. and Ohta, T. (1991) *J. Chem. Phys.*, **95**, 1242.
20. Moldenaers, P., Fuller, G.G. and Mewis, J. (1989) *Macromolecules*, **22**, 960.
21. Moldenaers, P., Mortier, M. and Mewis, J. (1994) *Chem. Eng. Sci.*, **49**, 699.
22. Sigillo, I. and Grizzuti, N. (1994) *J. Rheol.*, **38**, 589.
23. Hongladarom, K. and Burghardt, W.R. (1994) *Macromolecules*, **27**, 483.
24. Marrucci, G. and Greco, F. (1993) *Adv. Chem. Phys.*, **86**, 331.
25. Han, W.H. and Rey, A.D. (1994) *Phys. Rev. E.*, **49**, 597.

3

Hamiltonian modelling of liquid crystal polymers and blends

M. Grmela and B.Z. Dlugogorski

3.1 INTRODUCTION

Mathematical models of polymeric liquid crystal (PLC) are expected to be useful for developing applications, for gaining an understanding, and for guiding future investigations. The practical usefulness of models is usually measured by the extent to which they provide a setting for organizing results of the experimental observations of our particular interest, by the extent they allow us to extract from the experimental results pertinent material characteristics, and by the extent they can be used to predict flows arising in the course of industrial processing operations. By understanding the physical processes involved, we usually mean establishing relations between microscopic (molecular) properties and the macroscopic properties of our direct interest. It is very unlikely that all the above expectations can be satisfied by a single model. We have probably the best chance to satisfy the expectations with a family of mutually compatible and complementary models, each formulated on a different level of description. The main objective of this chapter is to show how the models in such a family are constructed. Our aim is to present a method – a do-it-yourself prescription. Particular models are introduced only as illustrations. In section 3.2, we present the method and illustrate its use on the well-known models (including Doi's model). In section 3.3, we introduce a new computer-friendly model suitable for

Rheology and Processing of Liquid Crystal Polymers
Edited by D. Acierno and A.A. Collyer
Published in 1996 by Chapman & Hall
ISBN 0 412 59640 7

molecular simulations. In section 3.4, we use the method to model rheological properties of blends.

3.2 FAMILY OF MUTUALLY COMPATIBLE MODELS

Consequences of the models that we are going to introduce are, of course, expected to agree with results of observations. First, we turn our attention to very general observations. In particular, we turn our attention to the observation constituting the basis of equilibrium thermodynamics (i.e. the observation that systems that are left without external disturbances reach a state – called an equilibrium state – at which their behaviour is well described by equilibrium thermodynamics). We focus first on this observation, not because we are particularly interested in it (there are more specific observations that we want to understand with the help of our models) but because we expect that an analysis of this universal observation will reveal a universal structure – a grammar, if we borrow the terminology used in linguistics – that will be common to all models in the family. Particular models in the family are then introduced by filling the skeleton provided by the general structure with the physical insight collected in the course of our experimental observations and in the course of our reflections about the physical processes involved. The models introduced in this way are then automatically guaranteed to be mutually compatible.

3.2.1 Nonlinear Onsager–Casimir equation

The structure that has been identified as the structure expressing mathematically the compatibility of a dynamical theory with the equilibrium thermodynamics has been collected in [1–4] in an abstract equation,

$$\frac{\partial f}{\partial t} = L\frac{\partial \Phi}{\partial f} - \frac{\partial \Psi}{\partial(\partial \Phi/\partial f)} \tag{3.1}$$

called a nonlinear Onsager–Casimir equation (NOC equation – see section 3.2.5 for explanation of the terminology). By f we denote the chosen state variables (section 3.2.2), $\Phi(f)$ is the thermodynamical potential (section 3.2.3), L is an operator expressing mathematically hamiltonian kinematics (section 3.2.4) and $\Psi(\partial\Phi/\partial f)$ is a dissipative potential expressing mathematically dissipative kinematics (section 3.2.4). To introduce a specific model means to introduce a particular realization of the abstract NOC equation (3.1). This then means that the state variables f are chosen (in other words a level of description is chosen) and the physical insight collected from our investigations is expressed in the specification of Φ, L and Ψ. In the rest of this section, we shall discuss the physical meaning of these quantities and show how to specify them. As an illustration, we introduce Doi's model.

3.2.2 State variables

Anticipated applications of the model together with the experience collected in observations constitute a basis for choosing the state variables. Models with state variables that are able to express molecular details will very likely be more useful for gaining understanding than for developing applications. This is because the more microscopic are the state variables the more difficult it will be, in general, to solve the governing equations of the model. The experience collected in observations is needed in order to choose state variables whose time evolution can be decoupled, with some accuracy, from the time evolution of the rest of the microscopic details. The state variables for which this is the case will be called **dynamically closed**. For example, the five hydrodynamic fields

$$f \equiv (\rho(\mathbf{r}), \mathbf{u}(\mathbf{r}), s(\mathbf{r})) \tag{3.2}$$

denoting mass density, momentum density and entropy density respectively with $\mathbf{r}$ standing for the position coordinate, are found to be dynamically closed if the fluids under consideration are simple fluids (e.g. water), but not if the fluids that are studied are complex fluids. Any subset of (3.2) does not constitute a set of dynamically closed state variables even for simple fluids. Some fields in (3.2) can, however, be omitted if the system under consideration is controlled from outside. For example, in this chapter, we shall limit ourselves to isothermal processes and consequently the field of entropy $s(\mathbf{r})$ will be omitted.

If complex fluids are investigated, some state variables describing the internal (molecular) structure have to be added to the set of hydrodynamic state variables. We shall introduce some examples.

Following Ericksen [5] and Leslie [6], we can choose to characterize the molecular structure by a unit vector field $\mathbf{n}(\mathbf{r})$ (called the **director** vector field). We thus introduce the state variables

$$f \equiv (\rho(\mathbf{r}), \mathbf{u}(\mathbf{r}), \mathbf{n}(\mathbf{r})) \tag{3.3}$$

a more detailed description of the molecular structure is achieved by [7],

$$f \equiv (\rho(\mathbf{r}), \mathbf{u}(\mathbf{r}), \mathbf{c}(\mathbf{r})) \tag{3.4}$$

where $\mathbf{c}(\mathbf{r})$, called a conformation tensor field, expresses an internal (molecular) strain. In terms of $\mathbf{c}$, the unit vector $\mathbf{n}$ appearing in (3.3) can be seen as the principal eigenvector of $\mathbf{c}$ normalized to unity. Still more detailed characterization of the molecular structure can be achieved by [8, 9] the configuration space distribution function $\psi(\mathbf{r}, \mathbf{R})$, where $\mathbf{R}$ is the end-to-end vector of a model macromolecule. We thus introduce

$$f \equiv (\rho(\mathbf{r}), \mathbf{u}(\mathbf{r}), \psi(\mathbf{r}, \mathbf{R})) \tag{3.5}$$

The conformation tensor **c** can be seen as being the second moment of ψ,

$$c_{\alpha\beta} = \int d\mathbf{R} R_\alpha R_\beta \psi(\mathbf{r}, \mathbf{R}) \tag{3.6}$$

where $\alpha, \beta = 1, 2, 3$. If we would model macromolecules as chains, then $\psi(\mathbf{r}, \mathbf{R}_1, \mathbf{R}_2, \ldots)$, where $\mathbf{R}_1, \mathbf{R}_2, \ldots$ are end-to-end vectors of the chain links, would replace $\psi(\mathbf{r}, \mathbf{R})$. The last example of state variables f, which we mention in this chapter, is the phase space distribution function

$$f \equiv f_2(\mathbf{r}_1, \mathbf{P}_1, \mathbf{r}_2, \mathbf{P}_2) \tag{3.7}$$

where $\mathbf{r}_1, \mathbf{r}_2$ are position vectors and $\mathbf{P}_1, \mathbf{P}_2$ momenta of two beads of a 'dumb-bell' that is regarded as a model of macromolecules. The state variables (3.5) can be expressed in terms of (3.7) as follows [8, 9]:

$$\begin{aligned}
\rho(\mathbf{r}) &= \int d\mathbf{r}_1 \int d\mathbf{r}_2 \int d\mathbf{P}_1 \int d\mathbf{P}_2 [m\delta(\mathbf{r} - \mathbf{r}_1) \\
&\quad + m\delta(\mathbf{r} - \mathbf{r}_2)] f_2(\mathbf{r}_1, \mathbf{P}_1, \mathbf{r}_2, \mathbf{P}_2) \\
\mathbf{u}(\mathbf{r}) &= \int d\mathbf{r}_1 \int d\mathbf{r}_2 \int d\mathbf{P}_1 \int d\mathbf{P}_2 [\mathbf{P}_1 \delta(\mathbf{r} - \mathbf{r}_1) \\
&\quad + \mathbf{P}_2 \delta(\mathbf{r} - \mathbf{r}_2)] f_2(\mathbf{r}_1, \mathbf{P}_1, \mathbf{r}_2, \mathbf{P}_2) \\
\psi(\mathbf{r}, \mathbf{R}) &= \int d\mathbf{r}_1 \int d\mathbf{r}_2 \int d\mathbf{P}_1 \int d\mathbf{P}_2 \delta\left(\mathbf{r} - \frac{\mathbf{r}_2 - \mathbf{r}_1}{2}\right) \\
&\quad \times \delta(\mathbf{R} - \mathbf{r}_2 + \mathbf{r}_1) f_2(\mathbf{r}_1, \mathbf{P}_1, \mathbf{r}_2, \mathbf{P}_2)
\end{aligned} \tag{3.8}$$

By m, we denote the mass of one bead.

Experience collected in [5–9] indicates that the state variables (3.3), (3.4), (3.5) and (3.7) are dynamically closed. Dynamical closeness of state variables can also be studied by analysing solutions of fully microscopic dynamical theory (quantum mechanics or classical mechanics of the molecules composing the fluid under consideration). Dynamically closed state variables appear in this analysis as quantities whose time evolution is slower and sufficiently separated from the time evolution of the rest of microscopic state variables. Such analysis is, however, rarely available.

3.2.3 Thermodynamic potential

The next task is to express equilibrium thermodynamics in terms of the chosen state variables f. For this purpose we introduce the thermodynamic potential

$$\Phi(f;\, T, \mu) = E(f) - k_B T S(f) - \mu N(f) \tag{3.9}$$

where $E(f)$ is the total energy, $S(f)$ the total entropy and $N(f)$ the total number of moles, T is the temperature, μ is the chemical potential and k_B is the Boltzmann constant. Let us assume for a moment that $S(f)$, $E(f)$, $N(f)$ have been specified. How do we then deduce from (3.9) the fundamental thermodynamic relation $P = P(\mu, T)$? Following [1–4] we look first for equilibrium states (we denote them by f_{eq}) that we define to be solutions of

$$\frac{\partial \Phi}{\partial f} = 0 \tag{3.10}$$

The fundamental thermodynamic relation $P = P(\mu, T)$ is then

$$-PV = \Phi(f_{eq};\, T, \mu) \tag{3.11}$$

where V is the volume of the region in which the fluid under consideration is confined.

Now we turn our attention to the specification of $S(f)$, $E(f)$ and $N(f)$. The physical meaning of the state variables included in (3.7) implies that

$$N(f) = \int d\mathbf{r}_1 \int d\mathbf{r}_2 \int d\mathbf{P}_1 \int d\mathbf{P}_2 f_2(\mathbf{r}_1, \mathbf{P}_1, \mathbf{r}_2, \mathbf{P}_2)$$

or in terms of the state variables (3.3)–(3.5)

$$N(f) = \frac{1}{M_O} \int d\mathbf{r} \rho(\mathbf{r}) \tag{3.12}$$

where M_O is the molar mass.

To specify the energy $E(f)$, we have to specify the intra- and intermolecular interactions in the fluid under consideration. For example, if the molecules are regarded as mutually non-interacting Hookean dumb-bells then

$$E(f) = \int d\mathbf{r} d\mathbf{R} \frac{1}{2} H R^2 \psi(\mathbf{r}, \mathbf{R})$$

where H is the dumb-bell (spring) constant. Many other examples of $E(f)$ can be found [9, 3]. A particularly interesting intermolecular interaction corresponding physically to the inextensible (rigid) dumb-bell, can be expressed by introducing the constraint

$$|\mathbf{R}| = \text{constant} \tag{3.13}$$

if the state variables (3.5) are used, or the constraint

$$\text{tr}\mathbf{c} = \text{constant} \tag{3.14}$$

if the state variables (3.4) are used.

The entropy $S(f)$ can be specified with the help of Gibbs' equilibrium statistical mechanics [10–12]. Here, we shall only mention a few examples. The simplest expression for $S(f)$ is the Boltzmann entropy, that, in terms of the state variables (3.7), reads

$$S_B(f) = -k_B \int d\mathbf{r}_1 \int d\mathbf{r}_2 \int d\mathbf{P}_1 \int d\mathbf{P}_2 f_2(\mathbf{r}_1, \mathbf{P}_1, \mathbf{r}_2, \mathbf{P}_2) \ln f_2(\mathbf{r}_1, \mathbf{P}_1, \mathbf{r}_2, \mathbf{P}_2) \tag{3.15}$$

and in terms of the state variables (3.5)

$$S_B(f) = -k_B \int d\mathbf{r} \int d\mathbf{R} \psi(\mathbf{r}, \mathbf{R}) \ln \psi(\mathbf{r}, \mathbf{R}) \tag{3.16}$$

where k_B is the Boltzmann constant. If the distribution function in (3.15) be an N-molecule distribution function (where N is Avogadro's number) then the expression (3.15) would be exact. Since f in (3.15) is one dumb-bell distribution function, the expression (3.15) is only approximate.

A modification of (3.16)

$$S_O(f) = S_B(f) - \frac{1}{2} k_B K \int d\mathbf{r} \int d\mathbf{R} \int d\mathbf{R}' |\mathbf{R} \wedge \mathbf{R}'| \psi(\mathbf{r}, \mathbf{R}) \psi(\mathbf{r}, \mathbf{R}') \tag{3.17}$$

has been suggested by Onsager [13]; the symbol $\wedge$ denotes the vector product. The second term on the right-hand side of (3.17) takes into account orientation dependent topological constraints among molecules. The parameter K is proportional to the excluded volume and to the concentration of the macromolecules if the fluid under consideration is a macromolecular solution. We note that the second term on the right-hand side of (3.17) reaches a maximum if $\mathbf{R}$ and the end-to-end vector $\mathbf{R}'$ of the partner molecule are parallel. This term thus expresses the preference for parallelism. One can easily show [14, 15] that the Boltzmann entropy becomes, in terms of the state variables (3.4),

$$S_B(f) = \frac{1}{2} k_B \int d\mathbf{r} \ln \det \mathbf{c} \tag{3.18}$$

If, in the second term in the right-hand side of (3.17), we replace $|\mathbf{R} \wedge \mathbf{R}|$ by $(\mathbf{R} \wedge \mathbf{R})^2$ we obtain,

$$S_{MS}(f) = \frac{1}{2} k_B \int d\mathbf{r} \ln \det \mathbf{c} - \frac{1}{2} k_B K \int d\mathbf{r} ((\mathrm{tr}\mathbf{c})^2 - \mathrm{tr}\mathbf{c}\mathbf{c}) \tag{3.19}$$

This expression has been used by Maier–Saupe [16]. Useful modifications of the entropy $S(f)$ arising due to the flexibility of macromolecular chains have been introduced in [10, 11]; see [17] for these modifications formulated in terms of the state variables (3.4).

Equilibrium theory of PLCs consists of three steps: In the first step, the thermodynamic potential Φ is specified. In the second step, equilibrium states, i.e. solutions of (3.10), are sought. Isotropy-anisotropy phase transitions appear as bifurcations of solutions of (3.10). In the classical Onsager theory of lyotropic liquid crystals, Φ is specified by (3.12), (3.13) and (3.17). The bifurcation parameter is K. The isotropy-anisotropy transition appears as a pitchfork bifurcation. This can be seen particularly easily with the state variables (3.4). To simplify the calculations, we shall limit ourselves only to homogeneous fluids (i.e. the state variables are independent of $\mathbf{r}$). The thermodynamic potential that we choose is (3.12), (3.14), (3.19). The constraint (3.14) (we use a constant of 1) is taken into account by using the Lagrangian multiplier. Without loss of generality, we can assume that $\mathbf{c}$ is diagonal ($c_{11} = c_1, c_{22} = c_2, c_{33} = 1 - c_1 - c_2$). Equation (3.10) becomes

$$\frac{\partial}{\partial c_\alpha}\left(-\frac{1}{2}\ln \det\mathbf{c} + \frac{1}{2}K(1 - \mathrm{tr}\mathbf{c}\mathbf{c}) + \lambda(\mathrm{tr}\mathbf{c} - 1)\right) = 0$$

λ is the Lagrangian multiplier. This equation implies $c_1 = c_2 = c$ and

$$\left(c - \frac{1}{3}\right)(4Kc^2 - 2Kc + 1) = 0$$

determining c.

If $K < K_{\mathrm{crit}} = 4$, then there is only one solution, $c = \frac{1}{3}$ representing isotropic phase. If $K \geq K_{\mathrm{crit}}$, then in addition to the isotropic phase with $c = \frac{1}{3}$ there appears also an anisotropic phase with c that is a solution of

$$4Kc^2 - 2Kc + 1 = 0$$

Quantitatively, the pitchfork bifurcation appearing in Fig. 3.1 represents also solutions of (3.10) with state variables (3.5) (see [13, 18]).

3.2.4 Kinematics

Gradient $\partial\Phi/\partial f$ of the thermodynamic potential Φ is a driving force of the time evolution. By letting this force act in a way that is compatible with kinematics, the state variable f evolves in time. In mathematical terms, kinematics is represented by an operator L that transforms the covector $\partial\Phi/\partial f$ into a vector that is then equated with $\partial f/\partial t$. From the physical point of view, kinematics is determined by the physical nature of the state variables and by the requirement that this nature is preserved during the time evolution.

For example, in the classical mechanics of N particles, the state variables f are position coordinates $\mathbf{r}_1, \ldots, \mathbf{r}_N$ and the momenta $\mathbf{P}_1, \ldots, \mathbf{P}_N$.

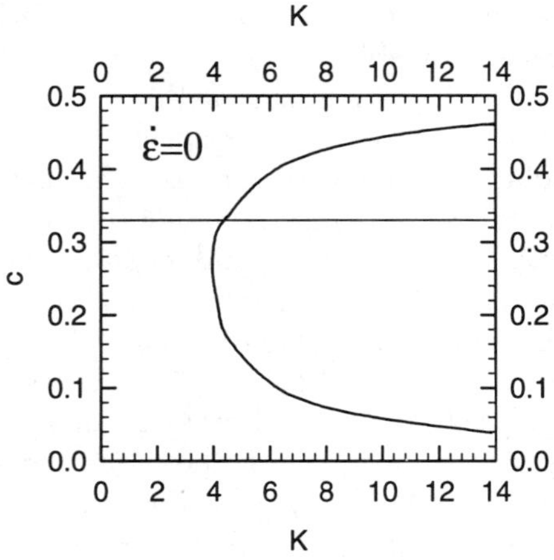

Figure 3.1 Isotropy–anisotropy phase transition at equilibrium.

We shall use the shorthand notation: $x_i \equiv (\mathbf{r}_i, \mathbf{P}_i)$, $i = 1, \ldots, N$. This physical interpretation of the state variable x implies that x is an element of the cotangent bundle and consequently the operator L, denoted $L^{(\mathrm{p})}$ ('p' stands for 'particle'), is the co-symplectic matrix

$$L^{(\mathrm{p})} = \begin{pmatrix} 0 & \mathbf{I} \\ -\mathbf{I} & 0 \end{pmatrix} \tag{3.20}$$

$\mathbf{I}$ is the unit matrix in $\mathscr{R}^{3N}$, representing the natural structure (symplectic structure) of the cotangent bundle. An alternative way of introducing the operator $L^{(\mathrm{p})}$ is by introducing a bracket

$$\{A, B\}^{(\mathrm{p})} =< A_x, L^{(\mathrm{p})} B_x > \tag{3.21}$$

where A, B, are sufficiently regular functions of $x \equiv (x_1, \ldots, x_N)$ $A_x = \partial A/\partial x$, and $<, >$ is the euclidian scalar product. It can be easily verified that (3.21) is a Poisson bracket; i.e. $\{A, B\}^{(\mathrm{p})}$ is a linear function of A_x, B_x, $\{A, B\}^{(\mathrm{p})} = -\{A, B\}^{(\mathrm{p})}$ and the Jacobi identity

$$\{A, \{B, C\}^{(\mathrm{p})}\}^{(\mathrm{p})} + \{B, \{C, A\}^{(\mathrm{p})}\}^{(\mathrm{p})} + \{C, \{A, B\}^{(\mathrm{p})}\}^{(\mathrm{p})} = 0$$

holds.

If instead of choosing $f \equiv x \equiv (x_1, \ldots, x_N)$ as the state variable, we choose the N-particle distribution function $f_N(x_1, \ldots, x_N)$ as the state variable, (i.e. $f = f_N(x_1, \ldots, x_N)$), then [19] instead of the Poisson bracket (3.21) we obtain the Poisson bracket

$$\{A, B\} = \int \mathrm{d}x\, f_N \{A_{f_N}, B_{f_N}\}^{(\mathrm{p})} \tag{3.22}$$

where A, B are sufficiently regular (real valued) functions of f_N, where

$$A_{f_N} \equiv \frac{\partial A}{\partial f_N(x_1, \ldots, x_N)}$$

The operator L corresponding to (3.22) is found by rewriting (3.22) in the form

$$\{A, B\} = \int \mathrm{d}x A_{f_N} L B_{f_N}$$

In the calculations involved, we use the boundary conditions that guarantee that all integrals over the boundaries, arising in integration by parts, equal zero. It can be shown, by a direct verification or by using the analysis developed in [19], that (3.22) is a Poisson bracket. It is useful to note that if f_N in (3.22) is the Klimontovich distribution function, i.e.

$$f_N(x_1, \ldots, x_N) = \delta(x_1 - x_1(t)) \cdots \delta(x_N - x_N(t))$$

where $x_1(t), \ldots, x_N(t)$ are particle coordinates and momenta, and (3.22) is restricted to functions A, B that are linear as functions of f_N, then (3.22) reduces to the standard Poisson bracket (3.21) of classical mechanics.

In the case of a general state variable f, the operator L characterizing the hamiltonian kinematics is conveniently introduced by introducing the Poisson bracket

$$\{A, B\} =< A_f, LB_f > \tag{3.23}$$

where $A \equiv \partial A/\partial f$ and $<, >$ is a scalar product. We note that if (3.23) is a Poisson bracket then the time evolution equation $\partial f/\partial t = L\Phi_f$ implies $\mathrm{d}\Phi/\mathrm{d}t = 0$.

If N be Avogadro's number then the operator L introduced in (3.23) would represent the complete kinematics. If, however, N is smaller than Avogadro's number, then due to interactions with the state variables that are ignored, the state variables f (or at least a part of the state variables f) will relax. To reproduce the experience constituting the basis of the equilibrium thermodynamics, the relaxation has to be such that Φ does not increase. In other words, in the course of the time evolution the inequality

$$\frac{\mathrm{d}\Phi}{\mathrm{d}t} \leq 0 \tag{3.24}$$

holds. This is because if (3.24) holds then the thermodynamic potential Φ can be interpreted as a Lyapunov function associated with the approach (as $t \to \infty$) of f to equilibrium states, i.e. solutions to (3.10). Introduction of this relaxation-type dynamics amounts to the introduction of the second kinematics. A general way of introducing such kinematics is to introduce a potential Ψ, called a dissipative potential, satisfying the following properties:

1. Ψ is a function of Φ_f
2. $\Psi = 0$ for $\Phi_f = 0$
3. Ψ reaches its minimum if $\Phi_f = 0$ (3.25)
4. Ψ is a convex function in a neighbourhood of $\Phi_f = 0$

The covector Φ_f is transformed with the help of the dissipative potential Ψ into a vector $\partial\Psi/\partial\Phi_f$.

In the rest of this section, we shall introduce some examples of the operator L (in addition to the examples introduced already in (3.21), (3.22)) and examples of the dissipative potential Ψ.

We begin by establishing the nondissipative kinematics (i.e. operator L or alternatively the Poisson brackets) for the state variables (3.7)–(3.8). We shall see that the Poisson brackets for these state variables can be obtained rigorously (i.e. without using any approximations as, for example, closure approximations) from the Poisson brackets (3.22). The derivation proceeds as follows. First, we have to know the relation between f_N and the chosen state variable. Relations among state variables (3.3), (3.4), (3.5), (3.7) have been established in section 3.2.2. The only relation that remains to be specified is the relationship between f_N and f_2. Following the well-known concept of the reduced distribution functions, we propose:

$$f_2(\mathbf{r}_1, \mathbf{P}_1, \mathbf{r}_2, \mathbf{P}_2) = \text{constant} \int d\mathbf{r}_3 \int d\mathbf{P}_3 \cdots \int d\mathbf{r}_N \int d\mathbf{P}_N f_N \qquad (3.26)$$

As the second step in the derivation of the Poisson bracket, we restrict (3.22) to functions A, B that depend on $f_N(x)$ only through their dependence on the chosen state variables (we denote them by the symbol $f(y)$). This then means that we replace $\partial/\partial f_N$ in (3.22) by

$$\int dy \frac{\partial}{\partial f(y)} \frac{\partial f(y)}{\partial f_N(x)}$$

i.e. we use the chain rule. Consequently, we obtain in this way a bracket for the variable $f(y)$. The bracket is a Poisson bracket since it has been obtained from the Poisson bracket (3.22) by restricting the class of admissible functions A, B. In calculations leading to the brackets, we often have to use integration by parts. The boundary conditions that we assume are those that make all the integrals over the boundary, appearing in the calculations, equal zero. Here, we shall only list, leaving the details of the calculations to the reader, the Poisson brackets for some of the state variables.

At this point, we make a couple of remarks. The procedure described above does not always lead to a bracket involving only the state variable $f(y)$. However, in the case of the state variables (3.3)–(3.8), it does. Also, we should point out that the derivation of the Poisson bracket sketched above is not the only way the hamiltonian kinematics can be studied. For other models see [19].

The Poisson bracket expressing hamiltonian kinematics of the state variables (3.4) is

$$\{A, B\} = \int d\mathbf{r}\left[\rho\frac{\partial}{\partial r_\alpha}(A_\rho)B_{u_\alpha} + u_\gamma\frac{\partial}{\partial r_\alpha}(A_{u_\gamma})B_{u_\alpha} + c_{\gamma\beta}\frac{\partial}{\partial r_\alpha}(A_{c_{\gamma\beta}})B_{u_\alpha}\right.$$
$$\left.+ c_{\gamma\beta}A_{c_{\alpha\beta}}\frac{\partial}{\partial r_\gamma}B_{u_\alpha} + c_{\gamma\beta}A_{c_{\alpha\gamma}}\frac{\partial}{\partial r_\beta}(B_{u_\alpha}) - (A \leftrightarrow B)\right] \tag{3.27}$$

where $(A \leftrightarrow B)$ stands for the sum of the same terms as above except that the role of A and B in these terms is interchanged. We use hereafter the summation convention $\alpha, \beta, \gamma \ldots = 1, 2, 3$. If the conformation tensor $\mathbf{c}$ is restricted by the constraint $\mathrm{tr}\mathbf{c} = 1$, the third term on the right-hand side of (3.27) is replaced by

$$-\frac{\partial}{\partial \mathbf{r}_\alpha}(c_{\gamma\beta})A_{\gamma\beta}B_{u_\alpha}$$

and an additional term

$$2\int d\mathbf{r}\left[c_{\alpha\beta}c_{\gamma\varepsilon}\frac{\partial}{\partial r_\alpha}(A_{u_\beta})B_{c_{\gamma\varepsilon}} - (A \leftrightarrow B)\right] \tag{3.28}$$

appears. The Poisson bracket expressing hamiltonian kinematics of the state variables (3.5) is

$$\{A, B\} = \int d\mathbf{r}\left[\rho\frac{\partial}{\partial r_\alpha}(A_\rho)B_{u_\alpha} + u_\gamma\frac{\partial}{\partial r_\alpha}(A_{u_\gamma})B_{u_\alpha} - (A \leftrightarrow B)\right]$$
$$+ \int d\mathbf{r}\int d\mathbf{R}\psi\left[\frac{\partial}{\partial r_\alpha}(A_\psi)B_{u_\alpha} + R_\beta\frac{\partial}{\partial R_\gamma}(A_\psi)\frac{\partial}{\partial r_\beta}(B_{u_\gamma})\right.$$
$$\left. - (A \leftrightarrow B)\right] \tag{3.29}$$

If the vector $\mathbf{R}$ is restricted by the constraint $|\mathbf{R}| = 1$, the last term on the right-hand side of (3.29) is replaced by

$$\int \mathrm{d}\mathbf{r} \int \mathrm{d}\mathbf{R}\left[\psi(R_\beta \delta_{\alpha\gamma} - R_\gamma R_\alpha R_\beta)\frac{\partial}{\partial R_\alpha}(A_\psi)\frac{\partial}{\partial r_\beta}(B_{u_\gamma}) - (A \leftrightarrow B)\right] \tag{3.30}$$

Other useful examples of Poisson brackets can be found in [19, 3, 4].

Next, we present some examples of the dissipative potential Ψ. We recall that Ψ is a function of Φ_f. But $\Phi_f = 0$ at equilibrium, so that if our interest is focused on states that are not too far from equilibrium, we can consider Φ_f to be small and consequently limit ourselves to Ψ that is a quadratic function of Φ_f. The quadratic dissipative potentials are indeed the most often used dissipative potentials. For example, in order to recast the Navier–Stokes equations into the form (3.1) we have to use

$$\Psi = \int \mathrm{d}\mathbf{r}\left(\frac{\partial u_\alpha}{\partial r_\beta} + \frac{\partial u_\beta}{\partial r_\alpha}\right)\frac{1}{8}\eta\left(\frac{\partial u_\alpha}{\partial r_\beta} + \frac{\partial u_\beta}{\partial r_\alpha}\right) \tag{3.31}$$

where η is the viscosity coefficient. Quadratic dissipative potentials will be used later in the next subsection where Doi's model is presented, and also in section 3.3 where dynamical equations suitable for molecular simulations are presented. More general dissipative potentials (but still satisfying the properties (3.25)) have to be used in chemical kinetics and kinetic theory; see [20].

3.2.5 Examples of the nonlinear Onsager–Casimir equation (3.1)

We have now collected all the ingredients of the abstract equation (3.1) and are thus in the position to construct its realizations. Before presenting some of the realizations expressing PLCs, we make a few observations about properties of solutions of (3.1) and the terminology.

First, we note that if $\Psi \equiv 0$ then (3.1) represents a hamiltonian dynamical system [19]. This feature of (3.1) gives the name hamiltonian modelling to the modelling that uses (3.1). If, on the other hand, $L \equiv 0$ the (3.1) becomes the governing equation of the gradient relaxation dynamics of the Cahn–Hilliard and Ginzburg–Landau type [21, 22]. Let (3.1) be linearized about an equilibrium state, i.e. about a solution of (3.10). It is easy to see (e.g. [20]) that the structure of the NOC equations presents itself as the Onsager–Casimir reciprocity relation [23, 24]. The NOC equation (3.1) can be thus seen as a nonlinear extension of the Onsager–Casimir relations; thus we call (3.1) a nonlinear Onsager–Casimir equation.

Having specified the Poisson bracket, the dissipative potential and the thermodynamic potential, we can write down the corresponding specific form of (3.1). The calculations involved are straightforward except, perhaps, for the passage from the Poisson bracket to the operator L. The easiest way to identify L is to write (3.1) with $\Psi \equiv 0$ in the form

$$\frac{\mathrm{d}A}{\mathrm{d}t} = \{A, \Phi\} \tag{3.32}$$

The requirement that (3.32) holds for all sufficiently regular functions A implies that $\partial f/\partial t = L\partial\Phi/\partial t$. Indeed, the left-hand side of (3.32) is $\int \mathrm{d}x A_{f(x)} \partial f(x)/\partial t$, the right-hand side, after integration by parts, $\int \mathrm{d}x A_{f(x)}[\ldots]$. The expression in $[\ldots]$ is then identified with $L\partial\Phi/\partial f$.

Well known and well tested with experimental observations equations governing the time evolution of macroscopic systems have been cast into the unifying form (3.1) in [19, 20, 1] (Navier–Stokes–Fourier hydrodynamic equations), [1, 3, 4, 20] (Boltzmann kinetic equation), [25, 26] (Ericksen–Leslie theory) etc. Here, we present only the governing equations of Doi's model [27] of PLCs. As the state variables, we choose (3.5) or alternatively (3.4). The molecules are assumed to be inextensible so that (3.13) or alternatively (3.14) constrains the admissible state space. The hamiltonian kinematics is thus specified in (3.29), (3.30) or alternatively in (3.27), (3.28). We choose the quadratic dissipative potential

$$\Psi = \int \mathrm{d}\mathbf{r} \int \mathrm{d}\mathbf{R} \frac{\partial}{\partial R_\alpha}(\Phi_\psi)\Psi D_{\alpha\beta}\frac{\partial}{\partial R_\beta}(\Phi_\psi) \tag{3.33}$$

where $\mathbf{D}$ is a positive definite tensor, or alternatively

$$\Psi = \int \mathrm{d}\mathbf{r} \int \mathrm{d}\mathbf{R} \Phi_{c_{\alpha\beta}} \Lambda_{\alpha\beta\gamma\delta} \Phi_{c_{\gamma\delta}} \tag{3.34}$$

where $\boldsymbol{\Lambda}$ is a positive definite tensor.

The NOC equation (3.1) becomes

$$\frac{\partial\rho}{\partial t} = -\frac{\partial}{\partial r_\gamma}(\rho\Phi_{u_\gamma}) \tag{3.35a}$$

$$\frac{\partial u_\alpha}{\partial t} = -\frac{\partial}{\partial r_\gamma}(u_\alpha \Phi_{u_\gamma}) - \frac{\partial}{\partial r_\alpha} p - \frac{\partial}{\partial r_\gamma}\tau_{\alpha\gamma} \tag{3.35b}$$

$$\begin{aligned}\frac{\partial\psi}{\partial t} = &-\frac{\partial}{\partial r_\gamma}(\psi\Phi_{u_\gamma}) - \frac{\partial}{\partial R_\gamma}\left(R_\alpha \frac{\partial}{\partial r_\alpha}\Phi_{u_\gamma}\right) - \left(R_\gamma R_\alpha R_\beta \frac{\partial}{\partial r_\alpha}(\Phi_{u_\beta})\right)\\ &+\frac{\partial}{\partial R_\alpha}\left(D_{\alpha\beta}\psi \frac{\partial}{\partial R_\beta}(\Phi_\psi)\right)\end{aligned} \tag{3.35c}$$

where the scalar pressure

$$p = -\varphi + \rho\Phi_\rho + u_\gamma \Phi_{u_\gamma} + \int \mathrm{d}\mathbf{R}\psi\Phi_\psi \tag{3.36}$$

and the extra stress tensor

$$\tau_{\beta\gamma} = \int d\mathbf{R}\psi(R_\beta\delta_{\alpha\gamma} - R_\beta R_\alpha R_\gamma)\frac{\partial}{\partial R_\alpha}(\Phi_\psi) \tag{3.37}$$

or alternatively, if the state variables (3.4) are chosen,

$$\frac{\partial\rho}{\partial t} = -\frac{\partial}{\partial r_\gamma}(\rho\Phi_{u_\gamma}) \tag{3.38a}$$

$$\frac{\partial u_\alpha}{\partial t} = -\frac{\partial}{\partial r_\gamma}(u_\alpha\Phi_{u_\gamma}) - \frac{\partial}{\partial r_\alpha}p - \frac{\partial}{\partial r_\gamma}\tau_{\alpha\gamma} \tag{3.38b}$$

$$\frac{\partial c_{\alpha\beta}}{\partial t} = -\frac{\partial}{\partial r_\gamma}(c_{\alpha\beta}\Phi_{u_\gamma}) + c_{\gamma\beta}\frac{\partial}{\partial r_\gamma}(\Phi_{u_\alpha}) + c_{\gamma\alpha}\frac{\partial}{\partial r_\gamma}(\Phi_{u_\beta})$$
$$- 2c_{\alpha\beta}c_{\gamma\varepsilon}\frac{\partial}{\partial r_\gamma}\Phi_{u_\varepsilon} - \Lambda_{\alpha\beta\gamma\varepsilon}\Phi_{c_{\gamma\varepsilon}} \tag{3.38c}$$

where the scalar pressure

$$p = -\varphi + \rho\Phi_\rho + u_\gamma\Phi_{u_\gamma} + c_{\alpha\beta}\Phi_{c_{\alpha\beta}} \tag{3.39}$$

and the extra stress tensor

$$\tau_{\beta\gamma} = -2(c_{\beta\varepsilon}\Phi_{c_{\varepsilon\gamma}} - c_{\beta\gamma}c_{\varepsilon\delta}\Phi_{c_{\varepsilon\delta}}) \tag{3.40}$$

By φ we denote the density of Φ, i.e. $\Phi = \int d\mathbf{r}\varphi(\mathbf{r})$

The time evolution equations (3.35)–(3.40) involve the thermodynamic potential Φ (see (3.9)) that still remains undetermined. By choosing $E \equiv 0$ (the intramolecular energy is taken into account by the constraints (3.13) and (3.14) and the intermolecular energy is neglected) and the entropy $S = S_B + S_O$ (see (3.16), (3.17)) equations (3.35)–(3.37) become the governing equations of Doi's model of PLCs. We end this section by making a few remarks.

Remark 1

One of the advantages of presenting the time evolution equations as realizations of the NOC equation (3.1) is that the expressions for the scalar pressure p (generalizing the expression obtained in the standard presentation of classical hydrodynamics from the assumption of local equilibrium) and the extra stress tensor $\boldsymbol{\tau}$ are obtained automatically, together with the equation governing the time evolution of the extra state variable. In the traditional derivations, the time evolution equations and the expressions for the extra stress tensor are derived independently. This carries a risk that these two terms of the time evolution are mutually incompatible; for an example see [28] and section 3.4.

Remark 2

Another advantage of the hamiltonian modelling is that the simpler version of Doi's model, i.e. the version that uses the conformation tensor **c** rather than the configuration space distribution function ψ as the state variables characterizing molecules, is obtained directly without the need of closure approximation. The standard derivation of (3.38) from (3.35) proceeds as follows. First, we apply the projection (3.6) on (3.35c) (i.e. we multiply (3.35c) by $R_\alpha R_\beta$ and integrate over $\mathbf{R}$). The resulting time evolution equation (called a projected equation) is the equation governing the time evolution of **c**. However, on its right-hand side we still, in general, find ψ. The equation has to be closed (i.e. ψ appearing in the equation has to be expressed in terms of **c**). Equation (3.38c) can be regarded as being such a closed equation. The closure has been chosen in such a way that both (3.35) and (3.38) are realizations of the NOC equation (3.1). We recall that from the physical point of view this means that both (3.35) and (3.38) are compatible with equilibrium thermodynamics. This requirement can thus be seen as serving to select appropriate closure approximations; see also section 3.4.

Remark 3

Having written the governing equations of Doi's model in the form (3.35) or (3.38), we can discuss the problem of their adaptation to varying physical situations.

For example, let the molecules comprising our fluid be not strictly rigid, as Doi assumes, but semiflexible [10, 11]. To adapt the models (3.35), (3.38) to this situation, one has to choose appropriately the potentials Φ and Ψ. The thermodynamic potential Φ expressing semiflexible macromolecules is discussed extensively in [10, 11]. We can thus use this Lifshitz–Khokhlov–Semenov Φ in (3.35) and (3.38); see [17]. It appears to be more difficult to find an appropriate modification of the dissipative potential Ψ. This type of investigation probably has to follow and extend the analysis presented in [29].

Another physical situation that occurs very often but is not taken into account in Doi's theory is the situation in which the fluid under consideration is strongly inhomogeneous. To adapt Doi's model to this situation, we have to include in Φ and Ψ terms depending on derivatives with respect to the position coordinate **r**; see [17, 26]. The new terms that arise in the extra stress tensor are the Frank–Eriksen stresses [30, 5] formulated with the state variables (3.5) and (3.4); originally, these stresses have been introduced with the state variables (3.3) [30, 5].

Remark 4

The main advantage of the model that uses the state variable **c** is that its governing equations are easier to solve. We have already appreciated this in the analysis of equilibrium solutions; see section 3.2.3. We can now ask how the pitchfork bifurcation expressing the isotropy–anisotropy phase transition at equilibrium is modified if the fluid is subjected to a flow. To answer this question we have to solve (3.38c) with given

$$\kappa_{\alpha\beta} = \frac{\partial}{\partial r_\alpha}(\Phi_{u_\beta})$$

We shall consider two types of flow, the simple elongation

$$\boldsymbol{\kappa} = \dot{\varepsilon}\begin{pmatrix} \frac{1}{2} & 0 & 0 \\ 0 & \frac{1}{2} & 0 \\ 0 & 0 & -1 \end{pmatrix} \tag{3.41}$$

and the simple shear

$$\boldsymbol{\kappa} = \dot{\gamma}\begin{pmatrix} 0 & 0 & 0 \\ 1 & 0 & 0 \\ 0 & 0 & 0 \end{pmatrix} \tag{3.42}$$

$\dot{\varepsilon}$ and $\dot{\gamma}$ are considered to be constant parameters. In the case of the flow (3.41), we look for **c** in the form,

$$\mathbf{c} = \begin{pmatrix} c & 0 & 0 \\ 0 & c & 0 \\ 0 & 0 & 1-2c \end{pmatrix} \tag{3.43}$$

in the case of the flow (3.42), *c* is searched in the form

$$\mathbf{c} = \begin{pmatrix} c_1 & c_2 & 0 \\ c_2 & c_3 & 0 \\ 0 & 0 & 1-c_1-c_3 \end{pmatrix} \tag{3.44}$$

If we use in (3.38c)

$$\Lambda_{\alpha\beta\gamma\varepsilon} = \frac{1}{2}\Lambda(c_{\alpha\gamma}\delta_{\varepsilon\beta} + c_{\beta\gamma}\delta_{\varepsilon\alpha})$$

and dimensionless variables $t \to k_B T\Lambda t$, $\dot{\varepsilon} \to \dot{\varepsilon}/k_B T\Lambda$, $\dot{\gamma} \to \dot{\gamma}/k_B T\Lambda$, we obtain solutions shown in Figs. 3.2 and 3.3; for more details see [17].

In the equilibrium analysis (see section 3.2.3), we noted that the pitchfork bifurcation that we have found to represent the isotropy–anisotropy transition in the description that uses **c** as the state variables remains qualitatively unchanged if **c** is replaced by ψ. We expect that this is also true for results shown in Figs. 3.2 and 3.3. It has to be pointed out, however, that there certainly are some phenomena that appear in solutions of (3.35c) but are absent in solutions of (3.38c). An example of such a property is the tumbling and negative first normal stress observed

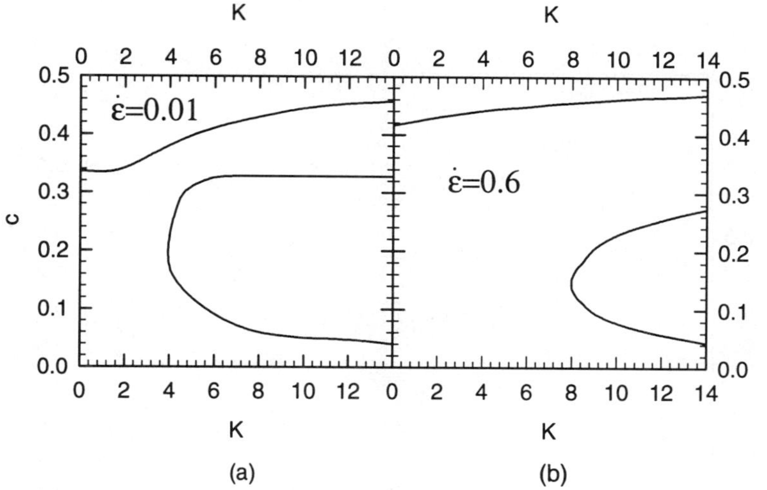

Figure 3.2 Isotropy–anisotropy phase transition under flow – elongation flow.

by Marrucci and Maffetone [31] in solutions of [3.35c]. Solutions of (3.38c) do not show these properties. The question arises whether there exists an intermediate level of description (i.e. the level of description that uses state variables that are more macroscopic than ψ but less macroscopic than **c**) on which the tumbling and the negative first normal stress could be observed in solutions of the governing equations. Preliminary studies indicate that an example of such a level of description is the level of description that uses two conformation tensors, one is **c** and the other is the time derivative of **c**. In this way, the time evolution includes inertia that then implies the effect of tumbling.

3.3 MOLECULAR SIMULATIONS

At this point, we are in position to follow the strategy described above and formulate a model or a family of models best suited to our needs. We continue this chapter by illustrating the whole process of model construction, starting with the formulation of objectives and ending with the derivation of model predictions; for details concerning this illustration see [32, 33].

Step 1: Objectives

We assume that we have at our disposal a modest state-of-the-art computer equipment (e.g. one or several RISC stations). We look for a model that would be formulated in such a way that its governing equations

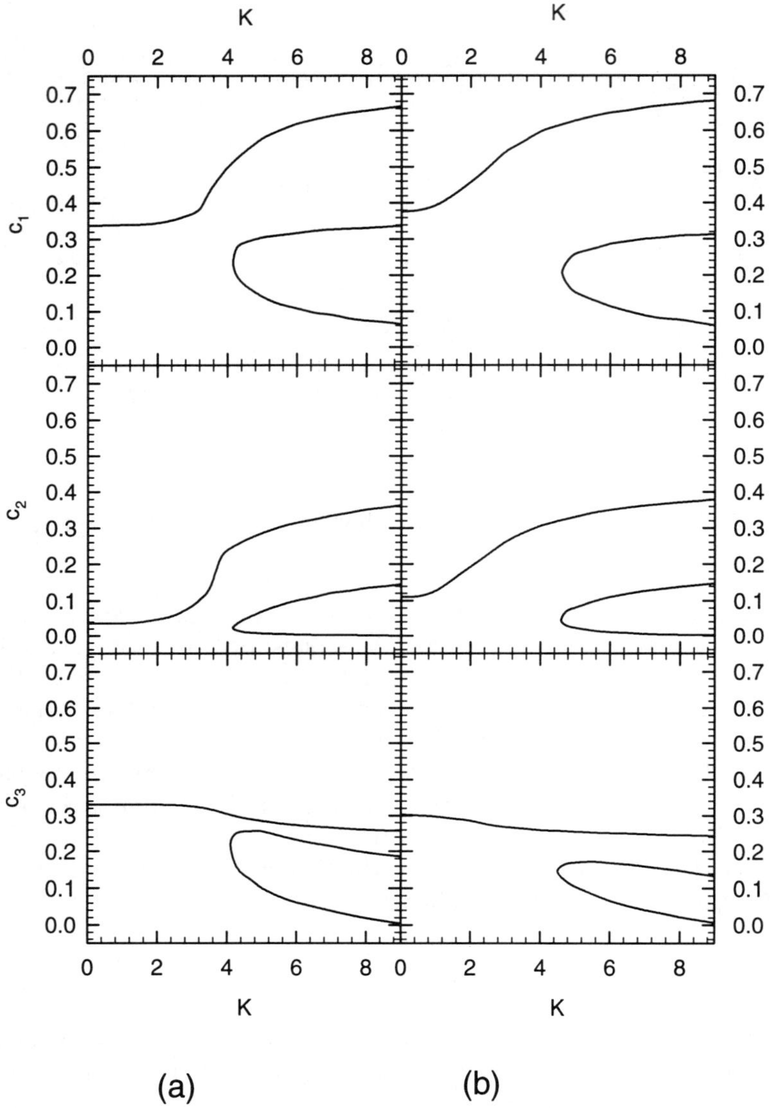

Figure 3.3 Isotropy–anisotropy phase transition under flow – shear flow: (a) $\dot{\gamma} = 0.1$; (b) $\dot{\gamma} = 0.4$.

can be directly, or almost directly, implemented into computers. With such a model (we shall call it a computer friendly model) we shall be able to exploit fully our tool.

We make a few observations. First, we note that the modelling that uses the state variables (3.3)–(3.7) is not computer friendly. The governing equations (typically these equations are partial differential equations) that have to be solved in order to derive model predictions, can be solved with the assistance of computer only after they have been discretized. The discretization (preparation for computer) is a very complex process. The questions that arise (e.g. how do we discretize, how do we know that solutions to the discretized equations approximate well solutions to the original equations?) are very difficult to answer and, moreover, they have nothing to do with the physics of the process involved. We want therefore to avoid such questions as much as possible. As the second observation, we note that ordinary differential equations (i.e. time evolution equations in a finite dimensional space) can be rather directly transformed into equations entering computers. In order to formulate computer friendly models, we therefore look for state variables that are elements of a finite dimensional space; note that the state variables (3.3)–(3.7) are elements of infinite dimensional spaces. One possible physical picture that leads to this type of state variable is the picture in which the fluids under consideration are regarded as composed of macromolecules whose states are characterized by a finite dimensional vector. The governing equations of the model are the equations (ordinary differential equations) governing that time evolution of a finite number of macromolecules. If we follow this line of thinking [34–36], however, we meet several difficulties. We mention three of them:

1. The number of molecules in real fluids is proportional to 10^{23}. Even with the most advanced computers we cannot follow more than, say, 10^6 molecules. It means that the molecules that we follow have to be representative of many real molecules. How do we specify such molecules?
2. Usually, we are not interested in molecular details but in macroscopic properties. We have to know, therefore, how to extract the macroscopic properties from the microscopic information obtained as the output of our calculations.
3. The fluids under consideration are controlled by macroscopic forces (e.g. they are subjected to a given flow or a given temperature gradient). We have to know how to express the boundary conditions and how the macroscopic forces act on a single molecule.

We shall see below that the modelling strategy described in the first two sections of this chapter is useful in dealing with these questions.

Step 2: State variables, kinematics

First, we have to decide how to model a single macromolecule. This decision depends, of course, on the specific nature of the fluid under consideration. Let us assume that a rigid body, in particular a spheroid, represents well the macromolecules composing our fluid. States of a spheroid are described by

$$x_1 = (\mathbf{r}, \mathbf{P}, \mathbf{p}, \mathbf{m})$$

where $\mathbf{r}$, $\mathbf{P}$, denote the position coordinate vectors and momenta of the centre of mass of the molecules, $\mathbf{p}$ is the unit director vector and $\mathbf{m}$ is the angular momentum (in the coordinate system rotating with the molecule). Following [37], the hamiltonian kinematics of x_1 is described by the Poisson bracket; see (3.21)

$$\{a, b\}_1^{(\mathrm{p})} = \{a, b\}_1^{(\mathrm{p},1)} + \{a, b\}_1^{(\mathrm{p},2)} \tag{3.45}$$

where $\{a, b\}_1^{(\mathrm{p},1)}$ is the Poisson bracket introduced in (3.21) (in (3.21) the bracket is denoted by the symbol $\{,\}^{(\mathrm{p})}$) and

$$\{a, b\}_1^{(\mathrm{p},2)} = \mathbf{m}\left(\frac{\partial a}{\partial \mathbf{m}} \wedge \frac{\partial b}{\partial \mathbf{m}}\right) + \mathbf{p}\left[\left(\frac{\partial a}{\partial \mathbf{m}} \wedge \frac{\partial b}{\partial \mathbf{p}}\right) - \left(\frac{\partial b}{\partial \mathbf{m}} \wedge \frac{\partial a}{\partial \mathbf{p}}\right)\right] \tag{3.46}$$

the symbol $\wedge$ denotes the vector product.

Now we consider N molecules (spheroids). The state variable is $x \equiv (x_1, \ldots, x_N)$, the Poisson bracket is

$$\{a, b\}_N^{(\mathrm{p})} = \sum_{i=1}^{N} \{a, b\}_i^{(\mathrm{p})}$$

If we pass from the particle description to the field description in which the N-molecule distribution function $f_N(x, t)$ serves as the state variables, then the hamiltonian kinematics of $f_N(x, t)$ is expressed by the Poisson bracket (3.22) in which $\{,\}^{(\mathrm{p})}$ is replaced by $\{,\}_N^{(\mathrm{p})}$; see section 3.2.4.

Next, we introduce state variables that are moments of the N-molecule distribution function $f_N(x)$. In particular, we introduce two types of moments,

$$f_n(x_n) = \int \mathrm{d}x_N \cdots \int \mathrm{d}x_{N-n+1} f_N(x_N) \tag{3.47}$$

$n = 1, \ldots, N-1$, and,

$$f^{(1)}(\mathbf{r}) = \begin{cases} \mathbf{u}(\mathbf{r}) & = \int \mathrm{d}\mathbf{p} \int \mathrm{d}\mathbf{P} \int \mathrm{d}\mathbf{m} \mathbf{P} f_1(x_1) \\ \boldsymbol{\mu}(\mathbf{r}) & = \int \mathrm{d}\mathbf{p} \int \mathrm{d}\mathbf{P} \int \mathrm{d}\mathbf{m} \mathbf{m} f_1(x_1) \\ \rho(\mathbf{r}) & = \int \mathrm{d}\mathbf{p} \int \mathrm{d}\mathbf{P} \int \mathrm{d}\mathbf{m} f_1(x_1) \\ c_{\alpha\beta}(\mathbf{r}) & = \int \mathrm{d}\mathbf{p} \int \mathrm{d}\mathbf{P} \int \mathrm{d}\mathbf{m} p_\alpha p_\beta f_1(x_1) \end{cases} \tag{3.48}$$

$\alpha, \beta = 1, 2, 3$. The distribution functions $f_N(x_n)$, $n = 1, \ldots, N-1$ are the reduced distribution functions, the fields $\mathbf{u}$ and ρ are the classical hydrodynamic fields of momentum and mass density; see the state variables (3.3)–(3.5). The new fields are the angular momentum field $\boldsymbol{\mu}(r)$ and the tensor field $\mathbf{c}(\mathbf{r})$ that expresses the orientation; compare with $\mathbf{c}$ introduced in (3.4), (3.6). It is an easy matter to show that the hamiltonian kinematics of f_n and $f^{(1)}$ can be obtained directly and rigorously (i.e without using any approximations as for example closure relations) from the hamiltonian kinematics of f_N. It suffices to restrict the bracket (3.22) (with $x \equiv (x_1, \ldots, x_N)$ and $\{,\}^{(\mathrm{p})} \equiv \{,\}_N^{\mathrm{P}}$) to functionals A, B, that depend on f_N only through their dependence on f_n (or $f^{(1)}$). We thus replace $\partial/\partial f_N$ in (3.23) by $\partial/\partial f_n$ (or $\partial/\partial\rho + p_\alpha p_\beta \partial/\partial c_{\alpha\beta} + P_\alpha \partial/\partial u_\alpha$ $+ m_\alpha \partial/\partial u_\alpha$). Straightforward calculations then lead to

$$\{A, B\}_n = \int \mathrm{d}x_1 \cdots \int \mathrm{d}x_n\, f_n \left\{\frac{\partial A}{\partial f_n}, \frac{\partial B}{\partial f_n}\right\}_n^{(\mathrm{p})} \tag{3.49}$$

and

$$\begin{aligned}\{A, B\}^{(1)} = \int \mathrm{d}\mathbf{r}\Bigg[&\rho\left(\frac{\partial}{\partial r_\alpha}(A_\rho)B_{u_\alpha} - \frac{\partial}{\partial r_\alpha}(B_\rho)A_{u_\alpha}\right)\\ &+ u_\beta\left(\frac{\partial}{\partial r_\alpha}(A_{u_\beta})B_{u_\alpha} - \frac{\partial}{\partial r_\alpha}(B_{u_\beta})A_{u_\alpha}\right)\\ &+ \mu_\beta\left(\frac{\partial}{\partial r_\alpha}(A_{\mu_\beta})B_{u_\alpha} - \frac{\partial}{\partial r_\alpha}(B_{\mu_\beta})A_{u_\alpha}\right)\\ &+ \varepsilon_{\alpha\beta\gamma}\mu_\alpha A_{\mu_\beta} B_{\mu_\gamma}\\ &+ (\varepsilon_{\alpha\beta\gamma}c_{\alpha\delta} + \varepsilon_{\alpha\beta\delta}c_{\alpha\gamma})(A_{\mu_\beta}B_{c_{\gamma\delta}} - B_{\mu_\beta}A_{c_{\gamma\delta}})\Bigg]\end{aligned} \tag{3.50}$$

the tensor $\boldsymbol{\varepsilon}$ is the alternating tensor.

To introduce the description that is suitable for molecular simulations, we combine the descriptions that use f_n, and $f^{(1)}$ (see (3.47), (3.48)) as state variables. (We follow the strategy presented in [38]). We include new state variables $\hat{f}_n$, $\hat{f}^{(1)}$ by

$$\hat{f}_n = f_n$$

$$\hat{f}^{(1)}(\mathbf{r})$$

$$= \begin{cases} \hat{\mathbf{u}}(\mathbf{r}) & = \mathbf{u}(1) + n\int \mathrm{d}2 \ldots \int \mathrm{d}n \int \mathrm{d}\mathbf{p}_1 \int \mathrm{d}\mathbf{P}_1 \int \mathrm{d}\mathbf{m}_1 \mathbf{P}_1 f_n(x_n)\\ \hat{\boldsymbol{\mu}}(\mathbf{r}) & = \boldsymbol{\mu}(1) + n\int \mathrm{d}2 \ldots \int \mathrm{d}n \int \mathrm{d}\mathbf{p}_1 \int \mathrm{d}\mathbf{P}_1 \int \mathrm{d}\mathbf{m}_1 \mathbf{m}_1 f_n(x_n)\\ \hat{\rho}(\mathbf{r}) & = \rho(1) + n\int \mathrm{d}2 \ldots \int \mathrm{d}n \int \mathrm{d}\mathbf{p}_1 \int \mathrm{d}\mathbf{P}_1 \int \mathrm{d}\mathbf{m}_1 f_n(x_n)\\ \hat{c}_{\alpha\beta}(\mathbf{r}) & = c_{\alpha\beta}(1) + n\int \mathrm{d}2 \ldots \int \mathrm{d}n \int \mathrm{d}\mathbf{p}_1 \int \mathrm{d}\mathbf{P}_1 \int \mathrm{d}\mathbf{m}_1 p_{1\alpha} p_{1\beta} f_n(x_n) \end{cases} \tag{3.51}$$

The new fields $\hat{f}^{(1)}$ represent now the global hydrodynamic fields and

$\hat{f}_n$, represents the extra microscopic information. Since the transformation $(f_n, f^{(1)}) \to (\hat{f}_n, \hat{f}^{(1)})$ given by (3.51) is one-to-one, the hamiltonian kinematics (3.49), (3.50) transforms into another hamiltonian kinematics specified by the Poisson bracket (to simplify the notation we use hereafter $f_n, f^{(1)}$ instead of $\hat{f}_n, \hat{f}^{(1)}$)

$$\{A, B\} = \{A, B\}_n + \{A, B\}^{(1)} + \{A, B\}_n^{(1)} \tag{3.52}$$

where

$$\begin{aligned}
\{A, B\}_n^{(1)} = \int d1 \ldots \int dnf_n \sum_{I=1}^{n} \Bigg[\Bigg(& P_{I\gamma} \frac{\partial}{\partial r_{I\gamma}} (A_{u_\gamma(\mathbf{r}_I)}) + \frac{\partial}{\partial r_{I\alpha}} (A_{\rho(\mathbf{r}_I)}) \\
& + p_{Ik} p_{Il} \frac{\partial}{\partial r_{I\alpha}} (A_{c_{kl}(\mathbf{r}_I)}) + m_{I\gamma} \frac{\partial}{\partial r_{I\alpha}} (A_{\mu_\gamma(\mathbf{r}_I)}) \Bigg) \frac{\partial}{\partial P_{I\alpha}} (B_{f_n(1,\ldots,n)}) \\
& + \frac{\partial}{\partial r_{I\alpha}} (A_{f_n(1,\ldots,n)}) B_{u_\alpha(\mathbf{r}_I)} + \varepsilon_{ijk} m_{Ii} A_{\mu_j(\mathbf{r}_I)} \frac{\partial}{\partial m_{Ik}} (B_{f_n(1,\ldots,n)}) \\
& + \varepsilon_{ijk} p_{Ii} \left(A_{\mu_j(\mathbf{r}_I)} \frac{\partial}{\partial p_{Ik}} (B_{f_n(1,\ldots,n)}) \right) \\
& + \varepsilon_{ijk} p_{Ii} \left(\frac{\partial}{\partial m_{Ij}} (A_{f_n(1,\ldots,n)}) \right) \Bigg(P_{I\gamma} \frac{\partial}{\partial p_{Ik}} (B_{u_\gamma(\mathbf{r}_I)}) \\
& + m_{I\gamma} \frac{\partial}{\partial p_{Ik}} (B_{u_\gamma(\mathbf{r}_I)}) \\
& + \frac{\partial}{\partial p_{Ik}} (B_{\rho(\mathbf{r}_I)}) + \frac{\partial}{\partial p_{Ik}} (p_{I\alpha} p_{I\beta} B_{c_{\alpha\beta}(\mathbf{r}_I)}) \Bigg) \Bigg) - (A \leftrightarrow B) \Bigg]
\end{aligned} \tag{3.53}$$

and $\{A, B\}_n$ and $\{A, B\}^{(1)}$ are the brackets (3.49) and (3.50). Now, we recall an observation made after (3.22). If we restrict ourselves to A, B that are linear functionals of f_n and consider f_n to be the Klimontovich distribution function then f_n as well as its kinematics reduces to the particle state variables and their kinematics. The hamiltonian time evolution equation (i.e. (3.1) with $\Psi \equiv 0$) implied by (3.52), (3.53) are:

$$\frac{\partial \rho}{\partial t} = -\frac{\partial}{\partial r_\gamma} (\rho \Phi_{u_\gamma}) - \frac{\partial}{\partial r_k} \left(\sum_{I \in V(\mathbf{r})} \frac{\partial}{\partial P_{Ik}} \Phi_{f_n} \right) \tag{3.54a}$$

$$\begin{aligned}
\frac{\partial u_\alpha}{\partial t} = & -\frac{\partial}{\partial r_\gamma} (u_\alpha \Phi_{u_\gamma}) - \frac{\partial}{\partial r_\alpha} p - \frac{\partial}{\partial r_k} \Bigg(\sum_{I \in V(\mathbf{r})} P_{I\alpha} \frac{\partial}{\partial P_{Ik}} \Phi_{f_n} \\
& - \frac{1}{2} \sum_{J \in V(\mathbf{r})} (r_{J\alpha} - r_{I\alpha}) \frac{\partial}{\partial (r_{J\alpha} - r_{I\alpha})} \Phi_{f_n} \Bigg)
\end{aligned} \tag{3.54b}$$

$$\frac{\partial c_{\alpha\beta}}{\partial t} = -\frac{\partial}{\partial r_\gamma}(c_{\alpha\beta}\Phi_{u_\gamma}) - \varepsilon_{\alpha\gamma\delta}c_{\gamma\beta}\Phi_{\mu_\delta} - \varepsilon_{\beta\gamma\delta}c_{\gamma\alpha}\Phi_{\mu_\delta}$$
$$-\frac{\partial}{\partial r_k}\left(\sum_{I\in V(\mathbf{r})} p_{I\alpha}p_{I\beta}\frac{\partial}{\partial P_{Ik}}\Phi_{f_n}\right)$$
$$-\sum_{I\in V(\mathbf{r})} p_{I\alpha}p_{I\beta}\frac{\partial}{\partial p_{Ik}}\left(\varepsilon_{ijk}p_{Ii}\frac{\partial}{\partial m_{Ij}}\Phi_{f_n}\right) \quad (3.54c)$$

$$\frac{\partial \mu_\alpha}{\partial t} = -\frac{\partial}{\partial r_\gamma}(\mu_\alpha\Phi_{u_\gamma}) - \varepsilon_{\alpha\beta\gamma}\mu_\beta\Phi_{\mu_\gamma} - \varepsilon_{\alpha\beta\gamma}c_{\beta\delta}\Phi_{c_{\gamma\delta}} - \varepsilon_{\alpha\beta\delta}c_{\beta\gamma}\Phi_{c_{\gamma\delta}}$$
$$-\frac{\partial}{\partial r_k}\left(\sum_{I\in V(\mathbf{r})} m_{I\alpha}\frac{\partial}{\partial P_{Ik}}\Phi_{f_n}\right) - \sum_{I\in V(\mathbf{r})}\varepsilon_{\alpha ik}m_{Ii}\frac{\partial}{\partial m_{Ik}}\Phi_{f_n} \quad (3.54d)$$

$$\frac{\mathrm{d}r_{I\alpha}}{\mathrm{d}t} = \frac{\partial}{\partial P_{I\alpha}}\Phi_{f_n} + \Phi_{u_\alpha(\mathbf{r}_I)} \quad (3.54e)$$

$$\frac{\mathrm{d}P_{I\alpha}}{\mathrm{d}t} = -\frac{\partial}{\partial r_{I\alpha}}\Phi_{f_n} - \frac{\partial}{\partial r_{I\alpha}}\Phi_{\rho(\mathbf{r}_I)} - P_{I\gamma}\frac{\partial}{\partial r_{I\alpha}}\Phi_{u_\gamma(\mathbf{r}_I)}$$
$$- m_{I\gamma}\frac{\partial}{\partial r_{I\alpha}}\Phi_{\mu_\gamma(\mathbf{r}_I)} \quad (3.54f)$$

$$\frac{\mathrm{d}p_{I\alpha}}{\mathrm{d}t} = -\varepsilon_{\alpha ij}p_{Ii}\frac{\partial}{\partial m_{Ij}}\Phi_{f_n} - \varepsilon_{\alpha ij}p_{Ii}\frac{\partial}{\partial m_{Ij}}\Phi_{\mu_j(\mathbf{r}_I)} \quad (3.54g)$$

$$\frac{\mathrm{d}m_{I\alpha}}{\mathrm{d}t} = -\varepsilon_{\alpha ij}m_{Ii}\frac{\partial}{\partial m_{Ii}}\Phi_{f_n} - \varepsilon_{\alpha ij}p_{Ii}\frac{\partial}{\partial p_{Ij}}\Phi_{f_n} - \varepsilon_{\alpha ij}m_{Ii}\Phi_{\mu_\gamma(\mathbf{r}_I)}$$
$$- \varepsilon_{\alpha ij}p_{Ii}P_{I\gamma}\frac{\partial}{\partial p_{Ij}}\Phi_{\mu_\gamma(\mathbf{r}_I)} - \varepsilon_{\alpha ij}p_{Ii}\frac{\partial}{\partial p_{Ij}}\Phi_{\rho(\mathbf{r}_I)} - \varepsilon_{\alpha ij}p_{Ii}m_{i\gamma}\frac{\partial}{\partial p_{Ij}}\Phi_{\mu_j(\mathbf{r}_I)}$$
$$- \varepsilon_{\alpha ij}p_{Ii}\frac{\partial}{\partial p_{Ij}}(p_{Ik}p_{Il}\Phi_{c_{kl}(\mathbf{r}_I)}) \quad (3.54h)$$

By $V(\mathbf{r})$ we denote particles in a neighbourhood of $\mathbf{r}$. The thermodynamic potential Φ is left in (3.54) unspecified. The scalar pressure p in (3.54b) is given by

$$p = -\varphi + \rho\Phi_\rho + u_\alpha\Phi_{u_\alpha} + \mu_\alpha\Phi_{\mu_\alpha} + c_{\alpha\beta}\Phi_{c_{\alpha\beta}} \quad (3.55)$$

where φ is the density of the thermodynamic potential Φ, i.e.

$$\Phi = \int \mathrm{d}\mathbf{r}\varphi$$

We note three particular cases of equations (3.54), (3.55). First, we consider only the fields ρ, $\mathbf{u}$, $\mathbf{c}$, and $\boldsymbol{\mu}$ as state variables (i.e. we put $\mathbf{r}$, $\mathbf{P}$, $\mathbf{p}$, $\mathbf{m}$ equal to zero and consider Φ independent of f_n). We remain then only with (3.54a)–(3.54d) describing hamiltonian hydrodynamics that includes the additional fields **c** and **μ**. The second particular case are equations describing the time evolution of n spheroids (we consider only **r**, **P**, **p**, **m** as state variables, the fields ρ, **u**, **c** and **μ** are put equal to zero). These are the standard time evolution equations of n rigid bodies. The third particular case are the governing equations of SLLOD dynamics [35, 38] that are very often used in molecular simulations. Equations (3.54) reduce to the governing equations of SLLOD dynamics if only the state variables $(\mathbf{u}(\mathbf{r}), \mathbf{r}_1, \ldots, \mathbf{r}_n, \mathbf{p}_1, \ldots, \mathbf{p}_n)$ are kept (i.e. we omit all terms involving $\rho(\mathbf{r}), \boldsymbol{\mu}(\mathbf{r}), \mathbf{c}(\mathbf{r}), \mathbf{m}_1, \ldots, \mathbf{m}_n, \mathbf{P}_1, \ldots, \mathbf{P}_n$). From the physical point of view, this reduction of the state variables means that the fluid under consideration is modelled as composed of spheres rather than ellipsoids.

Next we turn our attention to dissipative kinematics. We recall that to specify the dissipative kinematics means to specify the dissipative potential Ψ. We shall assume that the state variables that dissipate fastest are the angular momentum field $\boldsymbol{\mu}(\mathbf{r})$ and the angular-momentum-like fields that can be constructed from the gradient of the momentum field $\mathbf{u}(\mathbf{r})$. We thus introduce three vectors

$$\begin{aligned} &\Phi_{\mu_\alpha} \\ &\zeta_\alpha = \varepsilon_{\alpha\beta\gamma}\frac{\partial}{\partial r_\beta}\Phi_{u_\gamma} \\ &\xi_\alpha = \frac{1}{2}\varepsilon_{\alpha\beta\gamma}c_{\delta\beta}\left(\frac{\partial}{\partial r_\gamma}\Phi_{u_\delta} + \frac{\partial}{\partial r_\delta}\Phi_{u_\gamma}\right) \end{aligned} \tag{3.56}$$

and a quadratic dissipative potential

$$\Psi = \int d\mathbf{r} D^{-1}(\boldsymbol{\Phi}_{\boldsymbol{\mu}}, \boldsymbol{\zeta}, \boldsymbol{\xi})\mathbf{K}\begin{pmatrix}\boldsymbol{\Phi}_{\boldsymbol{\mu}}\\ \boldsymbol{\zeta}\\ \boldsymbol{\xi}\end{pmatrix} \tag{3.57}$$

where

$$\mathbf{K} \equiv \begin{pmatrix} \boldsymbol{\delta} & \lambda^{(4)}\boldsymbol{\delta} & \frac{1}{2}\lambda^{(1)}\boldsymbol{\delta} \\ \lambda^{(4)}\boldsymbol{\delta} & \lambda^{(2)}\boldsymbol{\delta} & 0 \\ \frac{1}{2}\lambda^{(1)}\boldsymbol{\delta} & 0 & \lambda^{(3)}\boldsymbol{\delta} \end{pmatrix} \tag{3.58}$$

$\boldsymbol{\delta} \equiv$ Kronecker delta, $D, \lambda^{(1)}, \lambda^{(2)}, \lambda^{(3)}, \lambda^{(4)}$ are constant parameters. We can see easily that the dissipative potential (3.57) satisfies the properties (3.25) provided $D > 0, \lambda^{(2)} > (\lambda^{(4)})^2$ and $\lambda^{(2)}(\lambda^{(3)} - \frac{1}{4}(\lambda^{(1)})^2) > (\lambda^{(4)})^2\lambda^{(3)}$. The complete NOC equation (3.1) corresponding to the kinematics introduced above is thus (3.54) with modified (3.54b), (3.54d) and (3.55). The modification of (3.54b) and (3.54d) consists of adding to their right-hand sides the terms $-\Psi_{u_\alpha}$ and $-\Psi_{\mu_\alpha}$ respectively. The term $-\Psi_{u_\alpha}$ can be recast into the form $-\partial\tau_{\alpha\beta}/\partial_{r\beta}$. In this way, we obtain the extra stress tensor $\boldsymbol{\tau}$. It can be shown [20] that in order to recover, as a particular case of (3.54), the time evolution equation of suspension dynamics introduced by Jeffery [39] we have to put $\lambda^{(4)} = 1$. The thermodynamic potential Φ as well as the coefficients D, $\lambda^{(1)}$, $\lambda^{(2)}$, $\lambda^{(3)}$, appearing in the dissipative potential remains still unspecified. These quantities will be specified as the last step before implementing the governing equations into computers.

Step 3: Solutions of the governing equations

The NOC equation that appeared in Step 2 is not yet computer friendly since it still involves partial differential equations. Our next task is, therefore, to eliminate the partial differential equations. Our strategy will be to solve the partial differential equations (3.54a)–(3.54d) and insert the solutions into the ordinary differential equations (3.54e)–(3.54h). These ordinary differential equations will then be the computer friendly equations that we introduce into computers.

To simplify the task of solving the partial differential equations (3.54a)–(3.54d), we shall limit ourselves in this illustration only to calculating rheological data. This means that the momentum field $\mathbf{u}(\mathbf{r})$ is given ($\mathbf{u}(\mathbf{r})$ is the flow field produced in and controlled by the rheometric apparatus). The remaining fields $\rho(\mathbf{r})$, $\mathbf{c}(\mathbf{r})$, and $\boldsymbol{\mu}(\mathbf{r})$ are specified as follows. We shall assume that the fluid under consideration is incompressible and thus put $\rho(\mathbf{r}) = $ constant. As for the field $\mathbf{c}(\mathbf{r})$, we note that this field influences only the time evolution of $\mathbf{m}(t)$; see (3.54h). Moreover, we see that only the off-diagonal terms of $\mathbf{c}(\mathbf{r})$ contribute to the right-hand side of (3.54h). In this illustration, we shall assume that the off-diagonal elements of $\mathbf{c}(\mathbf{r})$ are small and we thus omit the last term on the right-hand side of (3.54h). In (3.54d), we shall assume that the dissipative term $-\Psi_{\Phi_\mu}$ is dominant. This then means that the field $\boldsymbol{\mu}(\mathbf{r})$ evolves in time faster than the other fields. Consequently, we can put in (3.54d) $\partial\mu_\alpha/\partial t = 0$ and solve it approximately. As a result, we obtain

$$\Phi_{\mu_\alpha} \approx \frac{1}{2}\lambda^{(4)}\zeta_\alpha \tag{3.59}$$

where the vorticity ζ is known since the momentum field is known. With (3.59) we conclude the task of solving the partial differential equations (3.54a)–(3.54d). We have limited ourselves in this chapter to isothermal systems. This allowed us to ignore the hydrodynamical fields of entropy (or internal energy). With state variables x and the time evolution equations (3.54e)–(3.54h), the requirement of the constant temperature is not guaranteed. The problem of imposing the temperature in molecular simulations is discussed in spirit of the NOC equations in [38]. In this illustration, we shall guarantee the constant temperature by constraining the kinetic energy; see more in [35].

What remains now is to specify the thermodynamic potential Φ (see (3.9)) and the parameter $\lambda^{(4)}$. Since we have already eliminated the fields ρ, $\mathbf{u}$, $\boldsymbol{\mu}$, $\mathbf{c}$, it is sufficient to specify only the dependence of Φ on x. Both entropy S and the number of moles N are independent of x. Only the energy E depends on x. To specify E, we have to specify the interactions among spheroids. This question has been extensively discussed in [40]. Following Gay and Berne we choose (for two interacting spheroids I and J)

$$E(x) = E_{IJ}(\mathbf{p}_I, \mathbf{p}_J, \mathbf{r}_{IJ}) = 4\epsilon_{IJ}(\mathbf{p}_I, \mathbf{p}_J, \mathbf{r}_{IJ})\left[\left(\frac{1}{\mathbf{r}_{IJ} - \sigma_{IJ}(\mathbf{p}_I, \mathbf{p}_J, \mathbf{r}_{IJ}) + 1}\right)^{12} - \left(\frac{1}{\mathbf{r}_{IJ} - \sigma_{IJ}(\mathbf{p}_I, \mathbf{p}_J, \mathbf{r}_{IJ}) + 1}\right)^{6}\right] \tag{3.60}$$

where the direction dependent collision radius and the depth of the potential well are defined as

$$\begin{aligned}\epsilon_{IJ}(\mathbf{p}_I, \mathbf{p}_J, \mathbf{r}_{IJ}) &= \left(\frac{1-\kappa^2}{1-\kappa^2(\mathbf{p}_I\cdot\mathbf{p}_J)^2}\right)^{\frac{1}{2}}\left\{1 - \frac{K}{2}\left[\frac{(\mathbf{r}_{IJ}\cdot(\mathbf{p}_I+\mathbf{p}_J))^2}{1+K(\mathbf{p}_I\cdot\mathbf{p}_J)}\right.\right.\\ &\quad\left.\left.+\frac{(\mathbf{r}_{IJ}\cdot(\mathbf{p}_I-\mathbf{p}_J))^2}{1-K(\mathbf{p}_I\cdot\mathbf{p}_J)}\right]\right\}^2\\ \sigma_{IJ}(\mathbf{p}_I, \mathbf{p}_J, \mathbf{r}_{IJ}) &= \left\{1 - \frac{\kappa}{2}\left[\frac{(\mathbf{r}_{IJ}\cdot(\mathbf{p}_I+\mathbf{p}_J))^2}{1+\kappa(\mathbf{p}_I\cdot\mathbf{p}_J)} + \frac{(\mathbf{r}_{IJ}\cdot(\mathbf{p}_I-\mathbf{p}_J))^2}{1-\kappa(\mathbf{p}_I\cdot\mathbf{p}_J)}\right]\right\}^{-\frac{1}{2}}\end{aligned} \tag{3.61}$$

K and κ are the anisotropy parameters: κ relates to the ratio of long to short axes of the spheroids (taken here as 1.9) according to $(\sqrt{1.9}-1)/(\sqrt{1.9}+1)$, and K is an adjustable parameter which corrects for the depth of the potential well in the side-by-side configuration. Here, K is calculated from $(\sqrt{2.5}-1)/(\sqrt{2.5}+1)$; 2.5 is the ratio of depths of the potential wells in side-by-side and end-to-end configurations. Non-dimensional variables are used in (3.60), (3.61), and in subsequent equations of this section.

Finally, the time evolution equations that we shall solve with the assistance of computers are ($\boldsymbol{\omega}$ stands for angular velocity):

$$
\begin{aligned}
\frac{d\mathbf{r}_I}{dt} &= \frac{\mathbf{P}_I}{m_I} + \mathbf{r}_I \cdot \boldsymbol{\kappa} \\
\frac{d\mathbf{P}_I}{dt} &= \mathbf{F}_{rI} - \mathbf{P}_I \cdot \boldsymbol{\kappa} - \alpha_1 \mathbf{P}_I \\
\frac{d\mathbf{p}_I}{dt} &= -\mathbf{p}_I \wedge \boldsymbol{\omega}_I - \mathbf{p}_I \wedge \boldsymbol{\zeta} - \lambda_1 \mathbf{p}_I \\
\frac{d\mathbf{m}_I}{dt} &= -\mathbf{m}_I \wedge \boldsymbol{\omega}_I + \mathbf{p}_I \wedge \mathbf{F}_{pI} - \mathbf{m}_I \wedge \boldsymbol{\zeta} - \alpha_2 \mathbf{m}_I
\end{aligned}
\tag{3.62}
$$

The Lagrangian multipliers α_1 and α_2 guarantee constant total translational and rotational kinetic energies. A set of Lagrangian multipliers (λ_I) ensures rigidity of each spheroid,

$$
\begin{aligned}
\alpha_1 &= \frac{\sum_{I=1}^{N} \mathbf{P}_I \cdot (\mathbf{F}_{rI} - \mathbf{P}_I \cdot \boldsymbol{\kappa})}{\sum_{I=1}^{N} \mathbf{P}_I \cdot \mathbf{P}_I} \\
\alpha_2 &= \frac{\sum_{I=1}^{N} \boldsymbol{\omega}_I \cdot (\mathbf{p}_I \wedge \mathbf{F}_{pI} - \mathbf{m}_I \wedge \boldsymbol{\omega}_I - \mathbf{m}_I \wedge \boldsymbol{\zeta})}{\sum_{I=1}^{N} \boldsymbol{\omega}_I \cdot \mathbf{m}_I} \\
\lambda_I &= -\frac{\mathbf{p}_I \cdot (\mathbf{p}_I \wedge \boldsymbol{\omega}_I + \mathbf{p}_I \wedge \boldsymbol{\zeta})}{\mathbf{p}_I \cdot \mathbf{p}_I}
\end{aligned}
\tag{3.63}
$$

The forces are obtained from the following expressions:

$$
\begin{aligned}
\mathbf{F}_{rI} &= -\frac{\partial}{\partial \mathbf{r}_I} \sum_{I \neq J, J=1}^{N} E_{IJ} = -\mathbf{F}_{rJ} \\
\mathbf{F}_{pI} &= -\frac{\partial}{\partial \mathbf{p}_I} \sum_{I \neq J, J=1}^{N} E_{IJ}
\end{aligned}
\tag{3.64}
$$

Equations (3.62) together with (3.60, 3.61, 3.63, 3.64) are solved for 256 spheroids by a means of Gear five-value predictor-corrector method, assuming simple shear flow ($\dot{\gamma}$ = constant). The molecular dynamics simulations are performed in conjunction with orthogonal and Lees–Edwards shearing blocks [41] (in the direction of the velocity gradient) periodic boundary conditions. Calculations are done for two fluids having the particle number densities (ρ) of 0.25 and 0.40. At equilibrium (no shear) these fluids remain isotropic.

Step 4: Results

From the snapshots of molecular configurations, such as presented in Fig. 3.4, we observe that for low shear rates ($\dot{\gamma} < 0.1$) the fluid is isotropic. However, once the shear rate exceeds unity, the spheroids become aligned in the direction of flow. Upon further increase of $\dot{\gamma}$, this ordered

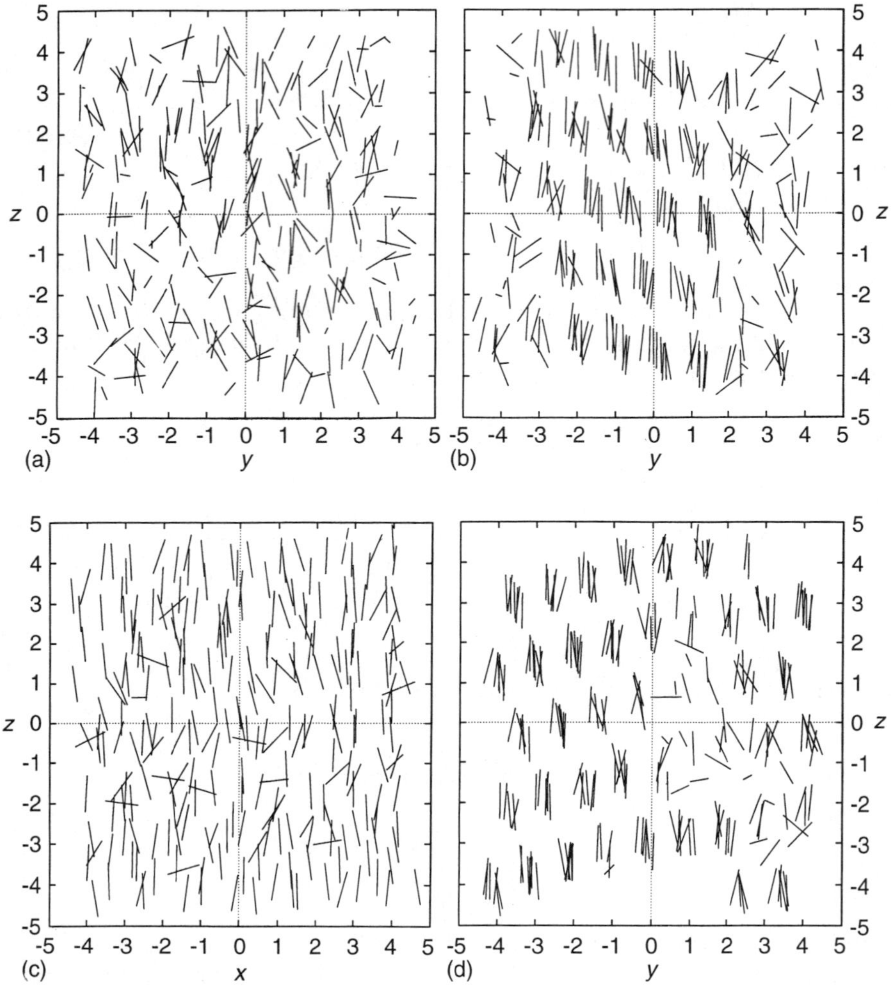

Figure 3.4 Instantaneous configuration of 256 spheroids ($\rho = 0.4$). The line segments reflect the orientation of long axes of the spheroids; $x =$ direction of flow, $y =$ direction of the velocity gradient, $z =$ neutral direction. (a) $\dot{\gamma} = 5$; (b) $\dot{\gamma} = 7$; (c) $\dot{\gamma} = 9$; (d) $\dot{\gamma} = 9$.

structure breaks down, and the fluid regains its isotropic property; see Fig. 3.4(a). Finally, the imposition of even higher shear rates leads to the formation of highly organized fluid structure, characterized by spheroids aligned, on average, parallel to the direction of vorticity, i.e. perpendicular to the direction of flow. There is an experimental evidence, in the case of viscoelastic fluids, indicating shear-related alignment of suspended rigid rods in the direction of vorticity.

This shear induced transformation in the organization of fluids composed of spheroids is well reflected in the order parameter (S); see Fig. 3.5. The order parameter measures the dispersion in the orientation of all spheroids around the director (**n**) and is defined as,

$$S(\dot{\gamma}) = \frac{3}{2N}\left\langle \sum_{I=1}^{N}(\mathbf{n}\cdot\mathbf{p}_I)\right\rangle - \frac{1}{2} \tag{3.65}$$

where $\langle \ldots \rangle_t$ refers to the time average. The ordering in the direction of flow manifests itself by peaks in the order parameter, for shear rates around 2.5–3; the lower density fluid exhibits the peak at the higher shear rate. Since this ordering is destroyed with increasing shear rates, the peaks are followed by troughs for shear rate around 4.5. A sudden increase in S after this point corresponds to the alignment of spheroids in the direction of vorticity, i.e. in the neutral direction z.

Since the total stress tensor, calculated from the expression

$$\sigma_{\alpha\beta}(\dot{\gamma}) = -\frac{1}{V}\left\langle \sum_{I=1}^{N}\frac{P_{I\alpha}P_{I\beta}}{m_I} + \sum_{I=1}^{N-1}\sum_{J>I}^{N} r_{IJ\alpha}F_{rIJ\alpha}\right\rangle_t \tag{3.66}$$

is in general not symmetric, but in the equilibrium limit, two viscosity coefficients are introduced

$$\eta_1(\dot{\gamma}) = \frac{\sigma_{12}(\dot{\gamma})}{\dot{\gamma}}$$
$$\eta_2(\dot{\gamma}) = \frac{\sigma_{21}(\dot{\gamma})}{\dot{\gamma}} \tag{3.67}$$

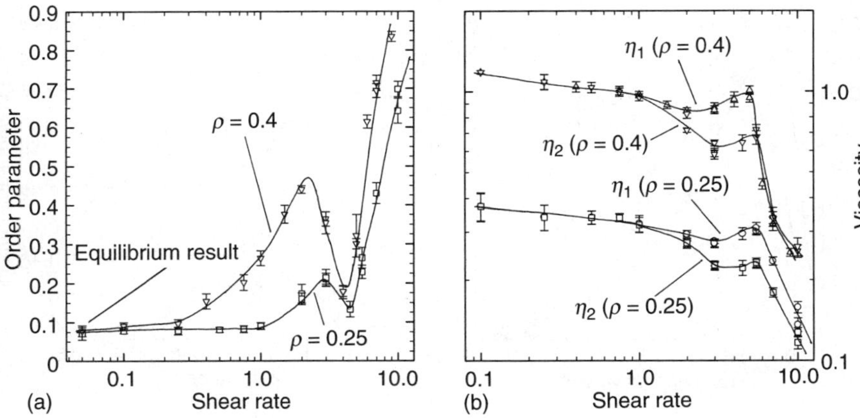

Figure 3.5 Effect of applied shear rate on (a) the order parameter and (b) viscosity coefficients ($\rho = 0.4$).

The trends in η_1 and η_2 closely follow those displayed by the order parameter. One readily notices that, in the newtonian limit, η_1 and η_2 collapse into one curve and the fluids display shear thinning for $\dot{\gamma} > 5$ (Fig. 3.5).

One of the attractive features of molecular dynamics simulations is the fact that results can be presented at any level of detail. This gives a valuable insight into phenomena which are too coarse to be seen in the snapshots but still too fine to be observed at the fully macroscopic level. For example, Fig. 3.6 demonstrates that distribution of the translational velocities of spheroids tends to deviate from the classical maxwellian curve. This deviation increases with the imposed shear rate. On the other hand, the tensorial pair radial distribution function,

$$g_{\alpha\beta}(r, \dot{\gamma}) = \frac{2}{N\rho} \frac{\left\langle \sum_{I=1}^{N-1} \sum_{J>I}^{N} \frac{r_{IJ\alpha} r_{IJ\beta}}{r_{IJ} r_{IJ}} \right\rangle_t}{4\pi r^2 \Delta r} \tag{3.68}$$

where $4\pi r^2 \Delta r$ denotes volume of the spherical shell), generalizes the structural information, derived from very many single configurations, into a single curve. We note in Fig. 3.6 that g_{33} remains anisotropic for $\dot{\gamma} = 1.5$ and $\dot{\gamma} = 9$, but is isotropic for $\dot{\gamma} = 5$. This is, of course, consistent with the other observations.

3.4 BLENDS

Finally, we illustrate the hamiltonian modelling in the context of blends of immiscible simple fluids. The first question that we ask is, as always:

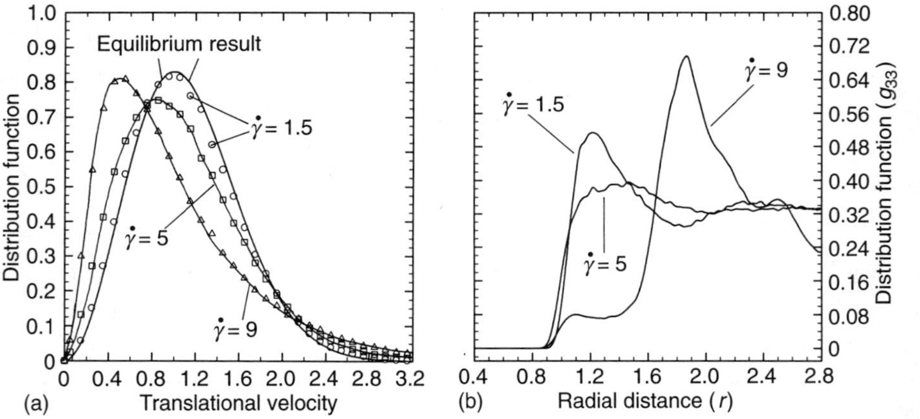

Figure 3.6 Dependence of reduced distribution functions on shear rates; on the left – distribution of (centre-of-mass) translational velocities, on the right – diagonal element of the pair radial distribution function ($\rho = 0.4$).

How do we characterize the internal structure of blends? Following [42], we focus our attention on the interface and characterize it by a trace-free symmetric tensor **q** (characterizing the orientation of the normal vector to the interface) and by a scalar Q (characterizing the total interfacial area). Our next task, if we follow the hamiltonian modelling, is to find kinematics of the fields $(\rho(\mathbf{r}), u(\mathbf{r}), \mathbf{q}(\mathbf{r}), Q(\mathbf{r}))$. This becomes easy if we realize that the tensor $(Qq_{\alpha\beta} + \frac{1}{3}Q^2\delta_{\alpha\beta})$ has the same kinematics as $c^{-1}_{\alpha\beta}$, i.e. the inverse of the conformation tensor **c** introduced in (3.4). This is because an element of the interface is characterized by a vector orthogonal to the interface. We thus have a one-to-one transformation between **c** and $(\mathbf{q}, Q)$. Under this transformation, the Poisson bracket (3.27) transforms into another Poisson bracket (we write it here only for incompressible fluids – the field $\rho(\mathbf{r})$ is omitted – and homogeneous fluids – $(\mathbf{q}, \mathbf{Q})$ are assumed to be independent of **r**)

$$
\begin{aligned}
\{A, B\} = \int \mathrm{d}\mathbf{r}\Big[& u_\gamma \frac{\partial}{\partial r_\alpha}(A_{u_\gamma})B_{u_\alpha} - q_{\gamma\beta}Aq_{\alpha\beta}\frac{\partial}{\partial r_\gamma}(B_{u_\alpha}) \\
& - q_{\gamma\beta}Aq_{\alpha\gamma}\frac{\partial}{\partial r_\beta}(B_{u_\alpha}) + \frac{2}{Q}q_{\gamma\alpha}q_{\varepsilon\delta}Aq_{\varepsilon\delta}\frac{\partial}{\partial r_\gamma}(B_{u_\alpha}) \\
& - \frac{1}{3}QAq_{\alpha\gamma}\frac{\partial}{\partial r_\gamma}(B_{u_\alpha}) - \frac{1}{3}QAq_{\alpha\gamma}\frac{\partial}{\partial r_\alpha}(B_{u_\gamma}) \\
& + \frac{2}{3}\delta_{\varepsilon\delta}q_{\gamma\alpha}Aq_{\varepsilon\delta}\frac{\partial}{\partial r_\gamma}(B_{u_\alpha}) - \frac{1}{2}q_{\gamma\alpha}A_Q\frac{\partial}{\partial r_\gamma}(B_{u_\alpha}) \\
& - \frac{1}{2}q_{\gamma\alpha}A_Q\frac{\partial}{\partial r_\alpha}(B_{u_\gamma}) - (A \leftrightarrow B)\Big]
\end{aligned} \tag{3.69}
$$

The NOC equation (3.1) becomes thus (3.38b)

$$
\begin{aligned}
\frac{\mathrm{d}q_{\alpha\beta}}{\mathrm{d}t} = & -q_{\gamma\beta}\frac{\partial}{\partial r_\gamma}(\Phi_{u_\alpha}) - q_{\gamma\alpha}\frac{\partial}{\partial r_\gamma}(\Phi_{u_\beta}) + \frac{2}{3}\delta_{\alpha\beta}q_{\mu\gamma}\frac{\partial}{\partial r_\gamma}(\Phi_{u_\mu}) \\
& - \frac{Q}{3}\left(\frac{\partial}{\partial r_\alpha}(\Phi_{u_\beta}) + \frac{\partial}{\partial r_\beta}(\Phi_{u_\alpha})\right) + \frac{2}{Q}q_{\gamma\varepsilon}q_{\alpha\beta}\frac{\partial}{\partial r_\gamma}(\Phi_{u_\varepsilon}) \\
& - \frac{\partial\Psi}{\partial(\Phi_{q_{\alpha\beta}})} \\
\frac{\mathrm{d}Q}{\mathrm{d}t} = & -q_{\alpha\beta}\frac{\partial}{\partial r_\alpha}(\Phi_{u_\beta}) - \frac{\partial\Psi}{\partial(\Phi_Q)}
\end{aligned} \tag{3.70}
$$

where the extra stress tensor

$$
\tau_{\alpha\beta} = q_{\alpha\gamma}\Phi_Q + \frac{2}{3}Q\Phi_{q_{\alpha\gamma}} + 2q_{\gamma\beta}\Phi_{q_{\beta\alpha}} - \frac{2}{Q}q_{\gamma\alpha}q_{\varepsilon\delta}\Phi_{q_{\varepsilon\delta}} \tag{3.71}
$$

The thermodynamic potential Φ and the dissipative potential Ψ are left unspecified.

If we compare (3.70) and (3.71) with the governing equations derived in [42] we note the following

1. If Φ and Ψ are appropriately chosen then (3.70) is identical with the governing equations in [42] except for the factor 2 in the fifth term (that is quadratic in $\mathbf{q}$) on the right-hand side of (3.70). We could arrive at (3.70) by following the derivation in [42] if we chose a slightly modified closure approximation (that is chosen in an *ad hoc* manner in [42]).
2. The two last terms on the right-hand side of the expression (3.71) for the extra stress tensor are missing in [42]. This is again, because the stress tensor is derived separately from the derivation of the time evolution equations. We note that the terms missing in [42] are quadratic and higher order in $\mathbf{q}$.
3. Equations (3.70) and (3.71) involve explicitly the thermodynamic potential Φ. By choosing it appropriately, we can adapt (3.70) and (3.71) to varying physical conditions. A more detailed analysis of blends that follows the approach sketched above can be found in [43].

To study thermodynamics and rheology of blends of PLCs, the state variables characterizing the interface have also to include state variables characterizing the internal orientation of the interfacial layer. An example of a static (thermodynamic) analysis of surfaces and interfaces of liquid crystals can be found in [44, 45].

3.5 CONCLUDING REMARKS

Classical hydrodynamics provides a unifying setting for modelling (of engineering interest) of flows of simple fluids (e.g. water). This setting has to be abandoned if the simple fluids are replaced by complex fluids (e.g. liquid crystals). This is because the time evolution of the internal structure of the complex fluids cannot be decoupled from the macroscopic (hydrodynamic) fluid evolution. Since the unifying setting has been found so useful in the context of simple fluids, the question arises as to whether there exists a unifying setting for modelling of flows of both simple and complex fluids. Such setting is provided by the nonlinear Onsager–Casimir equation (3.1). The modelling conducted inside this setting (called here hamiltonian modelling) proceeds as follows. First, the state variables characterizing the fluid under consideration are chosen. The choice depends on the nature of the fluid and also on our interests and intended applications of the model. As the second step, kinematics, both non-dissipative (expressed in Poisson bracket) and dissipative (expressed in dissipative potential), of the chosen state variables is identified. For many standard choices of state variables, this is just a matter of looking it up in the literature. The third step consists of the specification of the thermodynamic potential. This is the place where our insight into the microscopic interactions that take place in our fluid is expressed. Having made the above three steps, the governing equation of the model is the nonlinear Onsager–Casimir equation (3.1).

Advantages of this type of modelling, illustrated in this Chapter in several examples, are the following:

1. Solutions of the governing equations are guaranteed to agree with results of certain basic observations.
2. The separate search for kinematics and thermodynamic potential enlarges our possibility to express our physical insight in governing equations without using *ad hoc* assumptions. For example, nondissipative kinematics can often be derived from the microscopic nondissipative kinematics rigorously, without the need for an *ad hoc* closure approximation.
3. The governing equations are easily adaptable to varying physical conditions.
4. The governing equations always come together with an expression for the extra stress tensor. The compatibility of the governing equations and the formula for the extra stress tensor is guaranteed.

Finally, we would like to emphasize importance of the model presented in section 3.3. If we anticipate using digital computers to solve governing equations of our model, we should certainly think about expressing our physical insight in the way that is directly acessible to the computers. The strategy of the model building provided by the hamiltonian modelling is shown to be useful in formulating such models.

3.6 LIST OF SYMBOLS

Latin symbols

A, B	sufficiently regular functions of state variables
$\mathbf{c}(\mathbf{r})$	conformation tensor
c, c_1, c_2, c_3	elements of $\mathbf{c}$
D	constant in dissipative potential, (3.57)
$\mathbf{D}$	positive definite tensor
$E(f), E(x)$	total (inter and intramolecular) energy, Gay–Berne potential (3.60)
f	state variable or variables
f_1	one-particle distribution function
$f^{(1)}$	hydrodynamic fields defined in (3.48)

f_{eq}	values assumed by f at thermodynamic equilibrium
f_N	N-particle distribution function
f_n	n-particle distribution function, $n < N$
$\hat{f}_{\mathrm{n}}, \hat{f}^{(1)}$	distribution function and hydrodynamic fields defined in (3.51)
$\mathbf{g}$	tensorial pair radial distribution function
H	spring constant
K	constant, introduced in (3.17); bifurcation parameter
K_{crit}	critical value of the bifurcation parameter
k_{B}	Boltzmann constant
L	kinematics (Poisson) operator
$L^{(\mathrm{p})}$	co-symplectic matrix (classical kinematics operator)
M_{O}	molar mass
m	mass of a bead, or spheroid
$\mathbf{m}$	angular momentum coordinate
N	number of particles, Avogadro's number
$N(f)$	total number of mols
$\mathbf{n}(\mathbf{r})$	director vector
p	pressure
$\mathbf{p}$	unit director coordinate
$\mathbf{P}$	momentum coordinate
$\mathbf{q}$	orientation of the interface (section 3.4)
Q	total interfacial area (section 3.4)
$\mathbf{R}, \mathbf{R}'$	end-to-end vectors of a molecule
$\mathbf{r}, r$	position coordinate, $\lvert\mathbf{r}\rvert$
$\mathbf{r}_{IJ}, r_{IJ}$	$\mathbf{r}_I - \mathbf{r}_J, \lvert\mathbf{r}_I - \mathbf{r}_J\rvert$
$\mathbf{u}(\mathbf{r})$	momentum density
S	order parameter
$S(f)$	total entropy
$S_{\mathrm{B}}(f)$	Boltzmann entropy
$S_{\mathrm{O}}(f)$	Onsager entropy
$S_{\mathrm{MS}}(f)$	Maier–Saupe entropy
$s(\mathbf{r})$	entropy density
T	absolute temperature
t	time
V	volume
$V(\mathbf{r})$	neighbourhood of $\mathbf{r}$
x	position and momentum coordinates $(\mathbf{r}, \mathbf{P})$
$\{,\}$	Poisson bracket
$\{,\}_N^{(\mathrm{p})}$	N-particle Poisson bracket
$\{,\}_n, \{,\}^{(1)}, \{,\}_n^{(1)}$	Poisson brackets introduced in (3.49), (3.50) and (3.53)
$\{,\}_1^{(\mathrm{p},1)}, \{,\}_1^{(\mathrm{p},2)}$	one-particle Poisson brackets relating to $\mathbf{r}$, $\mathbf{P}$ and $\mathbf{p}$, $\mathbf{m}$ coordinates; see (3.45)

Greek symbols

α_1, α_2	Lagrangian multipliers (defined in (3.63))
δ	Dirac delta
$\boldsymbol{\delta}$	Kronecker delta
$\boldsymbol{\varepsilon}$	Levi–Civita (alternating) tensor
$\dot{\varepsilon}$	elongation rate
$\epsilon_{IJ}(\mathbf{p}_I, \mathbf{p}_J, \mathbf{r}_{IJ})$	depth of the potential well (3.60)
$\Phi(f)$	thermodynamic potential
$\dot{\gamma}$	shear rate
η, η_1, η_2	viscosity coefficients
$\mathbf{K}$	tensor defined in (3.58)
$\boldsymbol{\kappa}$	rate of strain tensor
K, κ	anisotropy parameters in Gay–Berne potential (section 3.3)
$\boldsymbol{\Lambda}$	positive definite tensor
λ	Lagrangian multiplier
$\lambda^{(1)} - \lambda^{(4)}$	constants in dissipative potential, (3.58)
μ	chemical potential
$\boldsymbol{\mu}$	angular momentum field
ρ	particle number density
$\rho(\mathbf{r})$	mass density
$\sigma_{IJ}(\mathbf{p}_I, \mathbf{p}_J, \mathbf{r}_{IJ})$	collision diameter (3.60)
$\boldsymbol{\sigma}$	total stress tensor
$\boldsymbol{\tau}$	extra stress tensor
$\boldsymbol{\omega}$	angular velocity
$\Psi(\Phi_f)$	dissipative potential
$\psi(\mathbf{r}, \mathbf{R})$	configuration space distribution function
$\boldsymbol{\xi}$	momentum field defined in (3.56)
$\boldsymbol{\zeta}$	vorticity

Subscripts

- denote components of vectors and tensors, e.g. u_α, $c_{\alpha\beta}$
- signify partial derivatives, e.g. $A_x = \frac{\partial A}{\partial x}$
- denotes indices, e.g. $\mathbf{r}_1$–position coordinates of the first particle.

REFERENCES

1. Grmela, M. (1986) *Physica D*, **21**, 179.
2. Grmela, M. (1988) *Phys. Lett. A*, **130**, 81.
3. Grmela, M. (1991) in 'Rheological Modelling: Thermodynamical and Statistical Approaches'. (Eds. J. Casas-Vasquez and D. Jou) *Lecture Notes in Physics*, Springer-Verlag, **381**, 99.
4. Grmela, M. and Jou, D. (1993) *J. Math. Phys.*, **34**, 2290.

5. Ericksen, J.L. (1960) *Arch. Rat. Mech. Anal.*, **4**, 231.
6. Leslie, F.M. (1968) *Arch. Rat. Mech. Anal.*, **28**, 265; in *Advances Liquid Crystals*, G.H. Brown (ed.), (1979) Academic Press, **4**(1).
7. Hand, C.L. (1962) *J. Fluid Mech.*, **13**, 33.
8. Kirkwood, J.G. (1967) in *Documents in Modern Physics*, P.L. Auer (ed.), Gordon and Breach, N.Y.
9. Bird, R.B., Hassager, O., Armstrong, R.C. and Curtiss, C.F. (1987) *Dynamics of Polymeric Fluids*, Wiley, N.Y. vol. 2.
10. Lifschitz, I.M. (1968) *Zh. Eksp. Theor. Fiz.*, **55**, 2408.
11. Khokhlov, A.R. and Semenov, A.N. (1985) *J. Stat. Phys.*, **38**, 161.
12. Raynes, E.P. and Borden, N. (eds), (1993) *Royal Soc. Trans., Phys. Sci. Eng.*, **344**, 305–440.
13. Onsager, L. (1949) *Ann. N.Y. Acad. Sci.*, **51**, 627.
14. Sarti, G.C. and Marrucci, G. (1973) *Chem. Eng. Sci.*, **28**, 1053.
15. Grmela, M. (1986) *J. Rheol.*, **30**, 707.
16. Maier, W. and Saupe, A. (1959) *Z. Naturforschung A*, **14**, 882; (1960) **15**, 287.
17. Grmela, M. and Chhon Ly, (1987) *Phys. Lett. A*, **120**, 281; Grmela, M. and Chhon Ly, *Spatial Nonuniformities in Lyotropic Liquid Crystals*, Preprint, École Polytechnique de Montréal, Jan. 1988.
18. Keyser, R.F. Jr and Raveché, H.P. (1978) *Phys. Rev. A*, **17**, 2067.
19. Marsden, J.E. and Weinstein, A. (1982) *Physica D*, **4**, 394.
20. Grmela, M. (1993) *Phys. Rev. E*, **47**, 351; **48**, 919.
21. Cahn, J.W. (1961) *Acta Metal.*, **9**, 795.
22. Landau, L.D. (1965) *Collected Papers of L.D. Landau*, D. Ter Haar (ed.), Pergamon, Oxford.
23. Onsager, L. (1931) *Phys. Rev.*, **37**, 405; **38**, 2265.
24. Casimir, H.G. (1945) *Rev. Mod. Phys.*, **17**, 343.
25. Dzyaloshinskii, I.E. and Volovick, G.E. (1980) *Ann. Phys. (N.Y.)*, **125**, 67.
26. Grmela, M. (1989) *Phys. Lett. A*, **137**, 342.
27. Doi, M. (1981) *J. Polym. Sci. Polym. Phys.*, **19**, 229.
28. Grmela, M. (1989) *J. Rheol.*, **33**, 207.
29. Yamakawa, H. (1984) *Ann. Rev. Chem.*, **35**, 23; Yamakawa, H. and Fujii, M. (1973) *Macromolecules*, **6**, 407.
30. Frank, R.C. (1958) *Discuss. Faraday Soc.*, **25**, 19.
31. Marrucci, G. and Maffetone, P.L. (1989) *Macromolecules*, **22**, 4076.
32. Grmela, M. (1993) *Mesoscopic Hydrodynamics of Suspensions*, Preprint, École Polytechnique de Montréal, Jan.
33. Dlugogorski, B.Z., Grmela, M., Carreau, P.J. and Lebon, J. (1994) *J. Non-Newt. Fluid Mech.*, **53**, 25.
34. Allen, M.P. and Tildesley, D.J. (1987) *Computer Simulation of Liquids*, Clarendon Press, Oxford.
35. Evans, D.J. (1987) in *Molecular-Dynamics Simulation of Statistical-Mechanical Systems*, G. Ciccotti and W.G. Hoover (eds), Clarendon Press, Oxford.
36. Hoover, W.G. (1986) in 'Molecular Dynamics', *Lecture Notes in Physics*, Springer-Verlag, **258**.
37. Sudarshan, E.C.G. and Mukunda, N. (1974) *Classical Mechanics – Modern Perspectives*, Wiley, N.Y.
38. Grmela, M. (1993) *Phys. Letters A.*, **174**, 59; **182**, 270.
39. Jeffery, G.B. (1922) *Proc. Roy. Soc. A*, **102**, 161.

40. Gay, J.G. and Berne, B.J. (1981) *J. Chem. Phys.*, **74**, 3316.
41. Lees, A. and Edwards, S. (1972) *J. Phys. C: Solid State. Phys.*, **5**, 1921.
42. Doi, M. and Ohta, T. (1991) *J. Chem. Phys.*, **95**, 1242.
43. Grmela, M. and Aït-Kadi, A. (1994) *J. Non-Newt. Fluid Mech.*, **55**, 191.
44. Osipov, M.A. and Hess, S. (1993) *J. Chem. Phys.*, **99**, 4181.
45. Evangelista, L.R. and Barbero, G. (1993) *Phys. Rev. E*, **48**, 1163.

4

Rheology and processing of liquid crystal polymer melts

F.N. Cogswell and K.F. Wissbrun

4.1 INTRODUCTION

Typical conventional thermoplastic polymers, such as polypropylene and polystyrene, are based on long, linear chain, molecules. In commercial plastics, those molecules have an aspect ratio, of chain length to chain diameter, of the order 1000 or more [1]. In such polymers, the links between individual units along the backbone are angled, typically at about 120° and, at temperatures where the polymer melts, they have a high degree of rotational freedom. Thus, in a melt state at rest, those molecules tend to adopt a random coil configuration entangling with their neighbours. It is that level of entanglement which gives rise to the elastic response of the melt, since the chains can be deformed from their rest state between such entanglement points but, once the stress causing deformation is removed, those chains revert to their random coil configuration. Viscous deformation takes place when the stress is applied for a sufficiently long time that the entanglements start to slip, so that one chain is translated to another position in the matrix of its fellows. The total number of entanglement points in a certain volume of polymer is independent of the lengths of the chains of which it is comprised; thus the elastic response of a melt is largely independent of the molecular weight of the polymer. For the usual case of polydispersity of chain lengths, the effectiveness of the entanglements also depends strongly on the breadth of the chain length distribution: the longer, more entangled, chains will

Rheology and Processing of Liquid Crystal Polymers
Edited by D. Acierno and A.A. Collyer
Published in 1996 by Chapman & Hall
ISBN 0 412 59640 7

form a network which absorbs most of the stress. The resistance to viscous deformation, where the position of a single chain is altered relative to its neighbours, depends very much on the degree of entanglement, and so molecular weight, of that chain: in general, a factor of two difference in molecular weight will result in an order of magnitude change in the resistance to viscous deformation. In practical shaping processes, there is a beneficial interaction between the elastic and viscous response of the melt. Under high stress levels, the chains become oriented, and the number and effectiveness of the entanglements are reduced, so that the chains can slip more readily past one another. Without such 'shear thinning' behaviour, few of the major plastics processing operations, such as injection moulding, would be practicable. The role of such rheological response of polymer melts in plastics processing is considered in more detail elsewhere [2]. As a first approximation, ease of processing for such polymers depends on the flexibility of the backbone chain and the number of entanglements per chain.

Service performance of a polymer, in plastics applications, usually demands stiffness and strength. Stiffness is largely determined by the freedom of rotation of the linkages between the units comprising the backbone chain. Strength, and in particular toughness, depends on the ability of individual chains to share their loads with their neighbours. That ability to transfer stress between molecules is achieved through the entanglements between chains; thus longer chains, with more entanglements, give rise to tougher products.

We therefore have a conflict of interest: between ease of processing and service performance. Good service performance requires high chain stiffness and a high level of entanglement, the very factors that tend to decrease the ease of processing, Fig. 4.1.

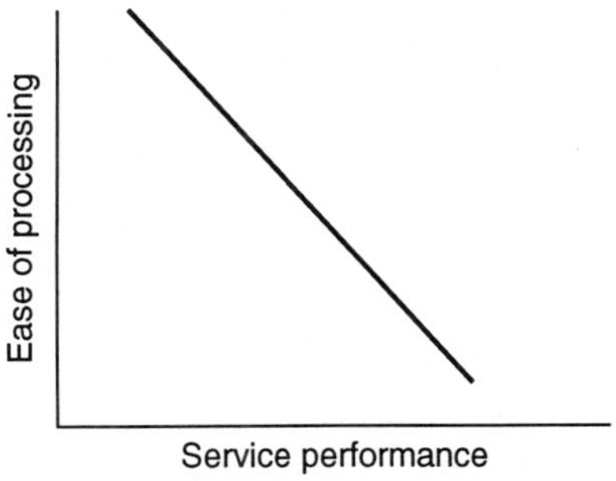

Figure 4.1 Relationship between ease of processing and service properties for conventional polymers.

Figure 4.1 represents a rule of common experience; it is not a law of polymer science. Every polymer chemist knows this to be true: the history of the development of plastics is studded with new polymers that are very difficult to process and have indifferent service performance. The line in Fig. 4.1 is simply the boundary of experience of what are useful materials. It is a boundary that is sensitive to innovative processing technology and new concepts in polymer engineering. Laws must be respected; boundaries can be advanced.

One way of altering the balance between ease of processing and service performance is to alter the structure of the polymer.

4.1.1 Structure in polymer melts

There are many ways in which the structure of a polymer can be tailored to ease processing or enhance service performance.

The most widely used variant is molecular weight distribution; more sophisticated is the introduction of branches on the chain. Polyethylene is one polymer which can readily be branched. While linear polyethylene melts have a well-defined viscosity, it has been noted that the viscosity of branched polyethylene, and so its ease of processing, can be dramatically influenced by the previous mechanical history to which the polymer has been subjected [3]. There is a tendency for high pre-shearing of the material to produce a melt of lower viscosity, which, over a period of time, gradually reverts to the higher viscosity state. It would appear that some different level of molecular organization, a transitory structure, which eases processing, can be induced in branched materials.

The molecules of many polymers can crystallize. Such crystallization enhances many of the service properties, in particular resistance to long-term deformation and to attack by hostile environments. Usually, all such local order must be melted out before processing, but some polymers can be processed with a small amount of crystallinity present. Of commercial materials, polyvinylchloride (PVC) is the best known example of this. Although this polymer has only a low level of crystallinity, it degrades before that crystallinity melts. Plasticizers can depress the melting temperature of the crystallinity, and thereby cause a drastic decrease of viscosity, and improved processability [4]. PVC is processed by a particulate flow mechanism [5], the nature of which has been examined in detail [6, 7]. One of the key features of such supramolecular structure is that it is extremely sensitive to thermo-mechanical history; intense shearing at low temperature leads to easier processing, and heat treatments at high temperature cause a reduction in mobility.

A different form of supramolecular structure is sometimes observed in ABA block copolymer systems. Most notably this is exploited in the family of thermoplastic elastomers, of which styrene-butadiene-styrene copolymers are a significant commercial example. Surface tension forces drive the compatible end groups to associate as amphiphiles forming star shaped accretions [8, 9]. The structure, and so rheology, of such materials can be modified by thermo-mechanical history [10], which can be controlled to produce defined morphologies in the final product [11].

The most dramatic changes in the balance between ease of processing and service performance is achieved by increasing the rigidity of the chain. Superficially, it is easy to see that increasing the rigidity of the molecule will increase the inherent service properties of the polymer, but the experience demonstrated in Fig. 4.1 leads to the expectation of very difficult processing. However, if the chain is made not only rigid but also straight, it ceases to behave like a random-coil molecule and becomes, instead, an extended-chain molecule. When this occurs, a sudden transition to an easy flow 'nematic' state is observed. The change in viscosity can be more than an order of magnitude. For the case of polymer melts, this transition was first reported in detail for the addition of para-hydroxybenzoic acid, a rigid rod-like entity, into polyethylene-terephthalate [12, 13]. As the rigid rod-like elements were added into the chain so, as expected, the viscosity increased until, above a critical concentration of rigid rod-like sections in the chain, the melt became highly mobile with a persistent birefringence indicating local order, even when the melt was at rest – Fig. 4.2.

If too much rigid rod-like material is included in the chain, then the polymer becomes highly crystalline and intractable to normal processing. The optimum is to have as many rigid rod-like entities as possible, to maximize service performance, but still consistent with achieving easy processing. Such rigid rod-like entities are usually provided by aromatic ring structures. Several ways of controlling the liquid crystal phase have

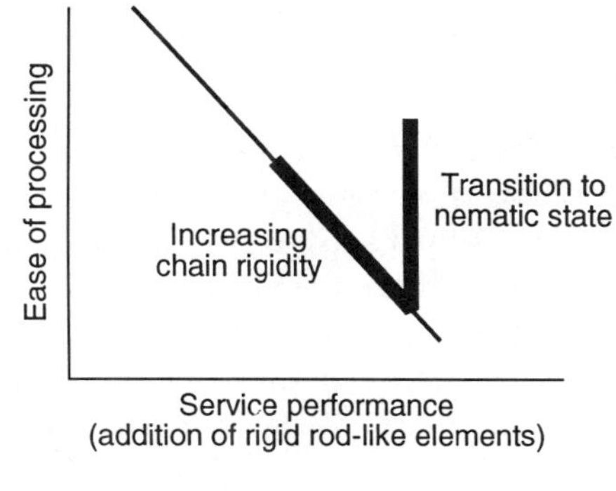

Figure 4.2 Effect of the addition of rigid rod-like elements in the backbone causing a transition in the balance between ease of processing and service performance.

been explored. These include:

- the uses of flexible spacers between the rigid rod-like blocks;
- the introduction of kinks along the rigid chain backbone;
- the incorporation of side group disrupters to prevent crystallization; and
- the inclusion of different aromatic ring structures, which maintain the long-range order in the molecule but encourage local, rotational, movement of the chain.

The various chemistry approaches to this end have been widely reviewed [14–19]. The greatest interest has been in wholly aromatic copolyester systems in which para-hydroxybenzoic acid is a major component. One system includes a naphthalene linkage to prevent regular sequencing between neighbouring chains and permit easier rotation within a chain; these polymers have been commercialized under the 'Vectra' trademark. In another system, marketed as 'Xydar', a longitudinal mismatch of adjacent chains is created by copolymerization with biphenol and a corresponding amount of terephthalic acid, see Chapter 1.

Although the main interest in thermotropic nematic LCPs has been focused on the aromatic polyesters, such behaviour is not confined to that family of polymers. Aside from more exotic monomer systems, Cox [14] notes reports of thermotropic polycarbonates, polyurethanes and cellulose derivatives. The case of hydroxypropyl cellulose is of particular interest. Although this is relatively well known as a lyotropic liquid crystal, at about 50% concentration in water, it is less well known that the bulk polymer has a small thermotropic nematic range at about 200°C [20]. While, in this study, we shall concentrate on the behaviour of thermotropic polyesters, such behaviour may, in the future, be designed into an increasingly broad family of polymers of commercial interest.

4.1.2 Liquid crystal phenomena and the rheology of fibre filled polymers

Thermotropic LCP melts allow a change in the rules which govern the balance between ease of processing and service performance, Fig. 4.2. This is not the exception which proves the rule. An analogous change is found in the field of fibre reinforced materials.

Short fibre reinforcement is a well-known method of enhancing the service performance of polymers. If short fibres are added to a polymer melt, they increase the viscosity and so make processing more difficult. The system thus conforms to the expectation of Fig. 4.1. Changing that short fibre reinforcement to long fibres further enhances service performance, especially in aspects like toughness and dimensional stability but, surprisingly, there is no corresponding increase in the resistance to flow.

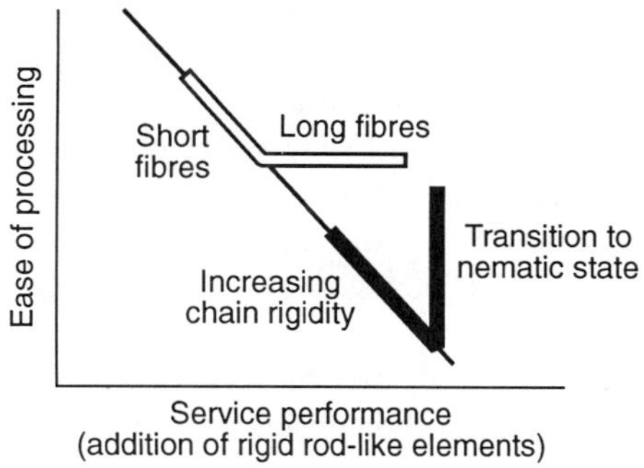

Figure 4.3 Comparison of the advantages of liquid crystal and long fibre reinforced.

Long fibre reinforced materials thus provide a similar advantage to LCPs: they break away from the limitations of conventional materials in the balance between service performance and ease of processing – Fig. 4.3.

That this transition can be achieved is not, in retrospect, entirely surprising. A suspension of fibres in a polymer melt is superficially similar to a solution of rod-like molecules, which was the theoretical scenario postulated by Flory for the formation of a liquid crystal state [21]. In that theoretical study, the importance of the interaction between aspect ratio of the suspended particle and its concentration is particularly noted. High loadings, typically up to 40% by volume, of long fibres, typically with an aspect ratio of the order 200–400, have little option other than to form ordered domain structures. In the development of long fibre injection moulding materials, such as 'Verton', it seems at least probable that the processing advantage of those materials have a common theoretical ancestor with thermotropic LCPs.

There are many similarities between these two families of materials, both in respect of processing and service behaviour, that are explored elsewhere [22]. At the present time, long fibre reinforced thermoplastic materials are not generally perceived as LCPs. This may be no more than a matter of scale: the diameter of reinforcing fibres being about 20 μm in comparison with that of molecules at about 0.5 nm, there is a scale factor difference of the order 40 000 between the two systems. From a purist standpoint, it may be more correct to compare long fibre reinforced thermoplastics with lyotropic, rather than thermotropic, systems; but there are so many similarities between the two that this may be irrelevant. The difference in scale should, of course, be irrelevant to dimensionless theoretical considerations. If the analogy is accepted, it may be possible to learn a considerable amount about the organization of molecules in domain structures by looking at what is apparent with long fibre reinforced materials. What is certain, from a practical standpoint, is that much useful knowledge can be translated from the experience of one type of material to the other in respect of processing 'know-how' and component design for injection moulding applications.

It is thus not surprising to find that short fibre reinforced LCP melts are of considerable commercial interest. Such materials are considered by Kulichikhin in Chapter 5. It should be noted that, rather than further enhancing orientation in LCP melts, the addition of fibres tends to lead to more isotropic mouldings suggesting some disruption of the orientation patterns associated with the polymer molecules by the suspended fibres.

It is of interest to note also that unfilled thermotropic melts and short fibre reinforced isotropic polymer melts are similar in other respects. For example, injection moulded articles are composed of multiple laminae, in each of which the orientation of the director, in one, and of the glass fibres of the other, are different [23, 24, 25]. Another similarity is their difficulty in producing strong weld lines [26].

Thermotropic melts can be used in combination with other polymers as blends. The liquid crystal phase gives such blends low viscosity in the melt and fibrous reinforcement to the solid phase. Such blends, once again, offer another route to advancing the frontier of useful plastics. This area is discussed by Collyer in Chapter 6.

4.1.3 Outline characteristics of LCP melt rheology

The main characteristic of thermotropic LCP melts is the persistence of order in the material even when the stresses causing deformation are removed. When viewed through a microscope a mass of thread-like textures (nemata) are seen. These are related to the disclination lines between various domains of different orientation in the material.

At rest, although locally ordered, the various domains tend to be randomly organized. That state of organization has a high resistance to flow but, as the stress level is increased, it progressively yields. After this, the material flows with an approximately constant viscosity. At higher stress levels, pseudoplastic, 'shear thinning', behaviour becomes the dominant response. When combined together, these give rise to the characteristic 'three region' flow curve associated with liquid crystal behaviour. It should be noted, however, that not all of the regions are always observable in the range of shear rates normally accessed by capillary rheometers. For example, 'Vectra' A900 displays only power-law flow behaviour over a wide range of shear rates and temperatures [27]. Only by combining capillary data with slit and rotational rheometer results was it possible to detect a small 'Region II' plateau [28]. On the other hand, the shape of the flow curve of another LCP varies remarkably with temperature, showing clearly the 'three-region' behaviour at an intermediate temperature [29].

A second feature, which sets LCPs aside from most other thermoplastics, is their unusual manifestation of melt elastic response. It is evidenced by partial recovery from shear flow [27, 30], but not by post-extrusion swelling, which is generally small and may even be negative [31, 32, 33].

As these are thermotropic materials, temperature is an important variable. The temperature sensitivity of LCP melts, especially at low melt temperatures, when they are just becoming mobile, is considerably greater than that of most other polymer melts. In part, this can be associated with the final melting of some residual 'solid' crystallinity.

The fourth significant rheological feature is the sensitivity of the material to previous thermo-mechanical history. Intense working promotes the formation of a highly mobile state, while a prolonged period of rest leads to less tractable material. This characteristic, which is common to other structured melts, such as PVC, branched polythene and ABA block copolymers, means that the rheology of the material cannot be defined accurately without reference to its history. In Chapter 8, Moldenaers makes a particular study of such time-dependent response for simple shearing flows of lyotropic systems. Practical plastics processing involves highly complex flow histories, and this is not all bad news. It is only because of pre-shearing, adventitiously carried out in the barrel of a screw extruder, that some LCPs can be processed at all [34].

If the chemical composition of the polymer is close to that where the transition from normal to liquid crystal behaviour occurs, the rheology of the material is very sensitive to the chemical structure of the chain. All of the commercially important thermotropic LCPs are copolymers; the size and sequencing of the blocks plays a role in determining the detail of the flow. In thermotropic liquid crystal polyesters, it is possible that ester interchange, and so chemical structure changes, could occur at melt temperatures.

Finally, most commercially important thermotropic polymers are made by condensation reactions. In such polymers, it is essential to take every precaution to avoid exposure to water if meaningful data are to be obtained.

4.1.4 Thermotropic aromatic polyesters

From the standpoint of commercial interest, it is the family of aromatic polyesters which is of particular importance. A high level of aromatic rings ensures that the polymer will give service at high temperature and have good fire resistance. The inextensibility of the molecules leads to high values of stiffness, even when they are not highly oriented macroscopically. This advantage is gained without the weight penalty encountered when using fibre reinforcement. Low elastic response means the ability to make precise mouldings with good dimensional stability. This is enhanced by the very low thermal expansion coefficient – one of their outstanding features associated with extended chain molecules. Modest levels of crystallinity at service temperatures give the materials good resistance to solvents. These service performance advantages are all properties which are highly valued by the user. However, the property that has, above all others, been important in gaining a place in the market for these materials is their ease of moulding, especially of complex parts with long, thin-walled flow paths, and being able to do so with minimal flash. Even so, the sophistication necessary in processing and design means that these are still specialist engineering materials sold in relatively low volumes but commanding a high price. The price of thermotropic aromatic polyesters is not necessarily expensive in the long term. Some of the main building blocks, from which they are constructed, are relatively low cost monomer systems.

4.2 SOME CHARACTERISTICS OF THERMOTROPIC POLYESTERS

4.2.1 Molecular structure and mobility

Amongst the most widely used monomers for thermotropic aromatic polyesters are para-hydroxybenzoic acid, and equimolar mixtures of hydroquinone and terephthalic acid (Fig. 4.4(a)).

Polymers based on the simple addition of these monomer units tend to be intractable. The problem is that they are highly crystalline and decompose before they finally melt. The main tailoring of the material is to disrupt that solid crystallinity.

One method of disrupting the crystallinity is by the introduction of flexible aliphatic linkages into the chain (Fig. 4.4(b)). This is only achieved at the expense of high temperature service performance.

A second approach is to incorporate bulky side groups to prevent regular association of the aligned chains (Fig. 4.4(c)).

A third approach (Fig. 4.4(d)) introduces kinks into the chain by using meta or even ortho linkages. These prevent that chain crystallizing in long sequences, but also reduce the level of order in the material and so the intensity of the liquid crystal behaviour.

A fourth approach (Fig. 4.4(e)) is to replace some of the benzene rings by naphthalene or by biphenyl linkages. In this approach, as shown in Fig. 4.4(f), the naphthalene ring and the biphenyl link alter the spacing along the chain, so effectively disrupting crystallization.

Naphthalene rings can introduce an extra degree of rotational freedom. In an aromatic polyester linked by benzene rings, rotational movement of one linkage relative to its neighbour through the ether links can only

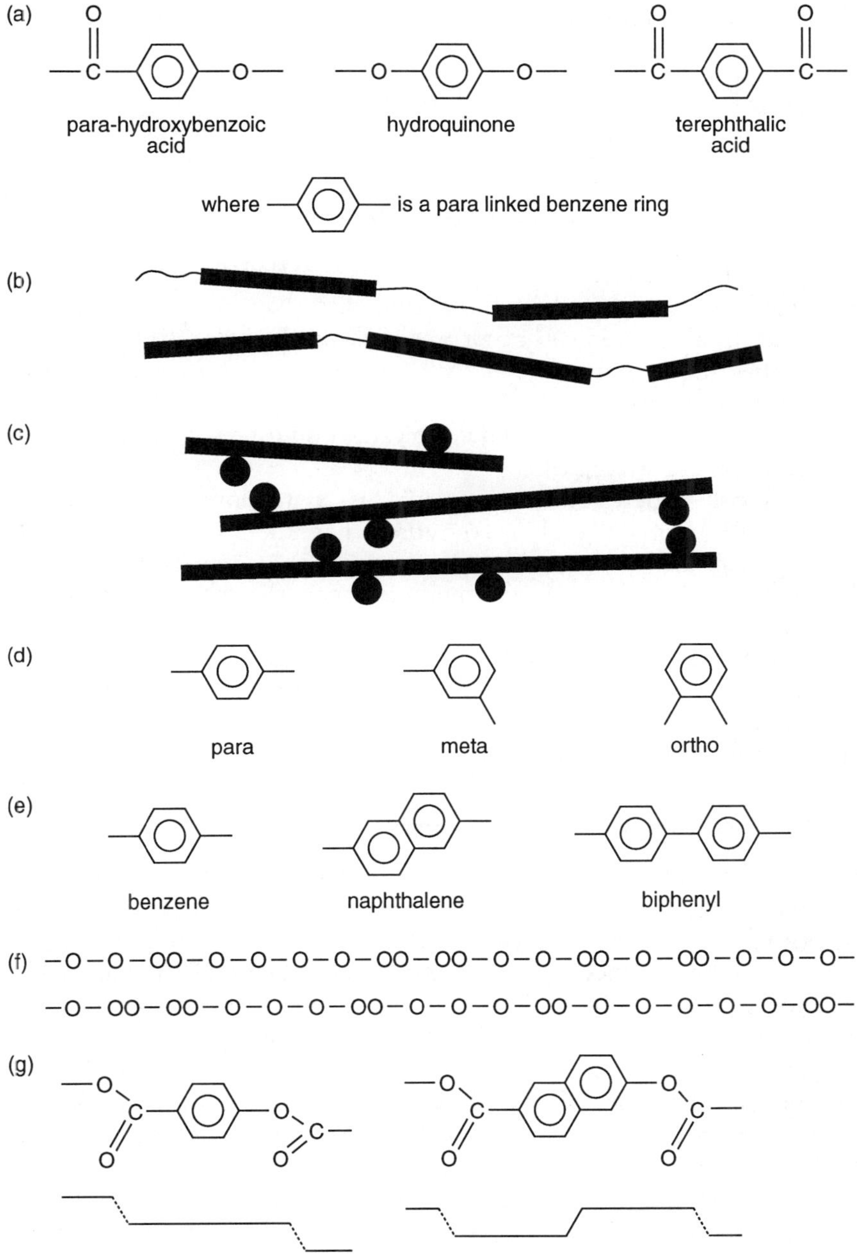

Figure 4.4 (a) Common monomer constituents in thermotropic aromatic polyesters; (b) mixture of rigid aromatic with flexible aliphatic linkages; (c) incorporation of bulky side groups; (d) para, meta and ortho linkages; (e) benzene ring, naphthalene, and biphenyl; (f) benzene and naphthalene, or biphenyl rings, randomly arrayed in neighbouring chains; (g) rotation of benzene and naphthalene rings through the ether linkages.

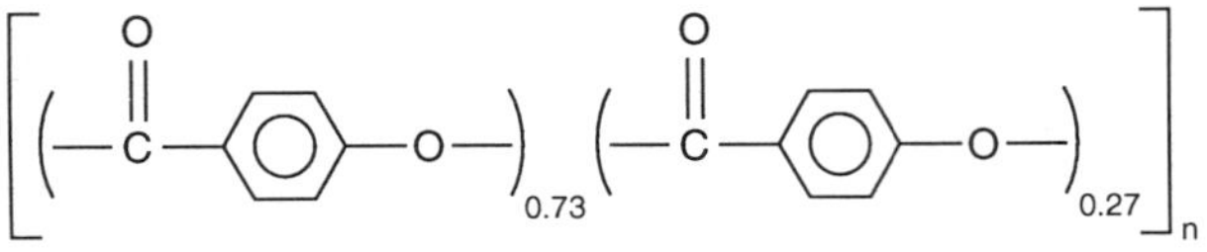

Figure 4.5 73/27 HBA/HNA.

take place by a crank-shaft motion. The naphthalene ring is itself a crank-shaft molecule, so that those two disturbances can cancel each other out, thus rotation can occur freely about the axis, see Fig. 4.4(g). This makes for a more mobile melt, at the expense of some loss in mechanical performance.

Commercial thermotropic liquid crystal polyesters may include an amalgam of these strategies.

The most widely studied material is a random copolymer of 73 mol %HBA with 27 mol % 2,6-hydroxynaphthoic acid (HNA). The structure, Fig. 4.5, is more conveniently written 73/27 HBA/HNA. The chemistry of this class of polymer has been described in detail by Calundann [35]. The 73/27 HBA/HNA copolymer is sold commercially under the trade name 'Vectra' A900.

Another variant, 'Vectra' B950, is also a copolymer of HNA, but instead of HBA the other comonomer is an equimolar mixture of para-aminophenol and terephthalic acid; that polymer is a copolyesteramide.

4.2.2 Molecular weight and distribution

Because the polymers are not readily soluble, definitive values of molecular weight and weight distribution are difficult to determine. Using the normal techniques of polyester polymerization, a range of molecular weights can be obtained. Qualitatively, as molecular weight increases so does the viscosity of the polymer and its service properties, particularly in respect of toughness. However, as with other condensation polymers, a practical limit is set on the useful maximum molecular weight that can be obtained because of the difficulty of eliminating condensation products from the system down to the very low levels necessary. For 73/27 HBA/HNA, the weight average molecular weight is believed to be about 30 000, and the ratio of weight to number average molecular weight is thought to be about 2, in common with other condensation polymers. Elsewhere [27], it has been established that, at such a molecular weight, the polymer is capable of repeated processing without any significant degradation, provided that proper precautions are taken to eliminate water and oxygen.

A chain of 73/27 HBA/HNA of molecular weight 30 000 would have a length of about 150 nm and an effective diameter of approximately 0.5 nm. As such, it has an aspect ratio of about 300. This is of the same order as 6 mm long reinforcing fibres in a long fibre moulding compound. The molecules have an extended chain configuration; at rest, they will naturally form nearly linear rod-like entities. Although the molecules are extremely stiff, it would not necessarily be correct to assume that they must always be straight. Carbon fibres are also extremely stiff, but they are so fine that they can be readily bent, or even tied into knots. Inextensible, but with silk-like flexibility, would be the correct mental picture for the molecules in a thermotropic liquid crystal.

It is also possible to introduce branching into the chain [36]. This does not necessarily destroy the liquid crystal response but, as with conventional polyesters, it does lead to significantly higher melt viscosities. Such materials have special experimental interest. The higher viscosity means that time dependent phenomena are more protracted, and it is therefore possible to inspect transition phenomena more closely in such materials.

4.2.3 Domain structure

After prolonged periods of rest in the melt, the molecules organize themselves into 'domain' structures [37]. The size scale of these domains can be of the order 100 μm, or greater, in some cases; in others, such as 'Vectra' A900, the domain texture is very dense, with a size scale on the order of 1 μm. To overcome this structure, a certain stress level must be exceeded. Once the material starts to flow, very much smaller domains, with orientation, are generated [37, 38]. Microscopic examination of fractured mouldings usually suggests smaller and smaller levels of microstructure, one within the other. Sawyer and Jaffe [39] detect three distinct microfibrillar structures at 5 μm, 0.5 μm and 0.05 μm diameter. At 0.05 μm, or 50 nm, the domain diameter is small relative to the length of an extended chain, so that all the chains must be approximately oriented along the axis of the microfibril. The cross-section of a 50 nm microfibril would, however, still contain some 10 000 individual molecules.

As evidenced by the dependence on thermo-mechanical history, the rheological behaviour of the melt is closely linked to the microstructure. Working mostly with lyotropic systems, Onogi and Asada [40] link the 'three region flow curve' to structure by considering the gradual translation of the plastic flow of piled domains into a monodomain continuous phase structure at very high shear rate. Wissbrun [41] confirms that the observations on thermotropic systems are consistent with that picture although, in such materials, there is also an interaction with the 'solid' crystallinity of the polymer. Marrucci, in Chapter 2, provides an overview of the theories describing the flow of LCPs and its dependence on domain structure.

4.2.4 Crystallinity

Most of the thermotropic LCPs of commercial interest are semi-crystalline materials at ambient temperature. Amorphous liquid crystalline materials have also been prepared. One advantage of amorphous, in comparison to semi-crystalline, polymers is that they can be processed closer to their glass transition temperature, T_g. A rule of experience suggests processing amorphous materials about 100°C above T_g, whereas semi-crystalline polymers must be processed 200°C above their T_g. In practice, since maximum processing temperature is limited by the onset of thermal degradation, this means that it is possible to produce mouldable amorphous materials with a higher glass transition temperature. By contrast, semi-crystalline polymers are generally perceived as having superior environment resistance combined with the ability to be used, especially if reinforced with rigid fillers, up to just below the melting point. The selection of amorphous or crystalline polymer is a matter for consideration in the context of projected service application. Both amorphous and semi-crystalline thermotropic LCPs have similar melt rheology [42].

Although the disruption in the sequencing of the monomer units along the chain backbone only allows a low level of crystallinity to be developed, Wissbrun [29] emphasizes that the presence of residual 'solid' crystallinity plays an important role in determining the rheological response near to the melting point. Only by taking the temperature to at least 20°C above the final melting point detectable by differential scanning calorimetry can well-defined data be obtained. However, once the final residues of crystallinity have been removed in that way, it is then possible to supercool the material [43].

In contrast to the effect of rapid heating followed by supercooling to eliminate the crystalline structure, it has been demonstrated [44, 45] that, by annealing close to the melting point, the final melting point of the thermotropic polymers can be driven up by as much as 50°C. Such annealing may occur inadvertently during processing operations with a broad distribution of residence times, such as in stagnant corners of manifolds or dies. The viscosity increase, or even solidification, that results from such annealing can have a deleterious effect on the operation, and the process design should take this possibility into account when dealing with these materials.

These interactions between crystallinity and processing behaviour make it essential to consider both morphology and rheology in parallel.

4.2.5 Thermal and thermodynamic properties

The thermal and thermodynamic properties of thermotropic aromatic polyesters are similar to those for conventional polymers [27].

73/27 HBA/HNA has a melting point of about 280°C. For this polymer at 300°C:

- melt density is 1270 kg m^{-3};
- bulk modulus is 1.9 GN m^{-2};
- specific heat is 1.9 kJ (kg °C)$^{-1}$;
- and, thermal diffusivity is 1.8×10^{-7} m^2 s^{-1}.

The value of thermal diffusivity is nearly double that of normal polymer melts. This value is for a sample of material at rest, where the domains are randomly oriented. It would be reasonable to assume that thermal diffusivity will be increased along the direction of order in a highly oriented material and reduced in a direction perpendicular to that orientation. In most practical processes, the major heat flow will tend to be across the primary direction of orientation.

4.2.6 Precautions in use

Although thermotropic aromatic polyesters, such as 73/27 HBA/HNA, are not as sensitive to water as conventional thermoplastic polyesters, such as polyethyleneterephthalate, it is necessary to dry the materials before use. It is also desirable to exclude oxygen during the study, in order to enhance the reproducibility of measurements. Typical drying conditions are recommended as four hours under vacuum at 120°C, followed by testing under an argon blanket [27].

4.3 RHEOLOGY

The rheological properties of interest range from measurements at small strain amplitude in oscillatory flow, which can be particularly illuminating of the structure of a material, to large-strain steady-state, shearing flows, such as are encountered in extrusion or moulding processes. As well as shear deformations, extensional flows are important, since the very high orientation that can easily be obtained in these materials makes them of considerable interest as drawn fibres or tapes.

4.3.1 Small amplitude oscillatory shear

Dynamic measurements have the advantage of obtaining a wide spectrum of the viscoelastic response from a small sample of material. As well as being useful data to illuminate the small-strain response of the material, such measurements are frequently used as a method of gauging the steady flow response. The Cox–Merz rule, equating the modulus of the complex viscosity as a function of angular velocity to the steady shear viscosity as a function of shear rate, is an empirical relationship which is widely acceptable for conventional thermoplastic materials [2].

In respect of thermotropic aromatic polyesters, although the general shape of the flow curve is the same in dynamic and steady flow measurements, the Cox–Merz rule is usually not satisfied quantitatively. There is evidence that the dynamic response depends strongly on the amplitude of deformation [29], with the complex viscosity reducing as strain amplitude is increased. Further, for isotropic melts, the approximate equality of N_1, the first normal stress difference, with twice the elastic storage modulus G' is commonly observed. For one thermotropic polymer, N_1 is about ten times G', even when the Cox–Merz rule for viscosity holds. That the correlation between small amplitude dynamic measurements and large strain, steady flow studies is not generally followed for thermotropic materials is not particularly surprising when the differences in domain structure between those two conditions are considered.

The quantitative failure of the Cox–Merz rule for LCPs does not mean that small amplitude oscillatory measurements are not useful for them. They still represent a valid tool for comparative measurements of different samples of a polymer, for temperature sweeps to detect transitions, and to diagnose changes after imposition of a mechanical or thermal history [46–48].

A dependence of dynamic viscoelastic response on strain amplitude has also been reported for conventional thermoplastic melts reinforced with continuous [49] and short [50] fibres. Such materials also show high elastic response in small amplitude dynamic measurements, but there is virtually no elastic response to large deformation steady flows. In such systems, the dynamic results can be related to the steady flow response by plotting the apparent Maxwell viscosity as a function of the maximum shear rate in the material, the product of strain amplitude and angular velocity [51].

For 73/27 HBA/HNA, Guskey and Winter [52] deduced that the linear, strain amplitude independent region extended only to strain amplitudes of 0.02 at low frequency and 0.005 at high frequency; see Fig. 4.6.

The inference from these results is that, for heavily structured melts, the dynamic response depends on the strain history of the material, especially in respect of the maximum strain rate, or stress, to which the melt is exposed [46, 48].

4.3.2 Torsional flow transient behaviour and normal force measurements

The simplest rheological measurements are made under steady flow conditions in a cone and plate. Even in this relatively simple geometry,

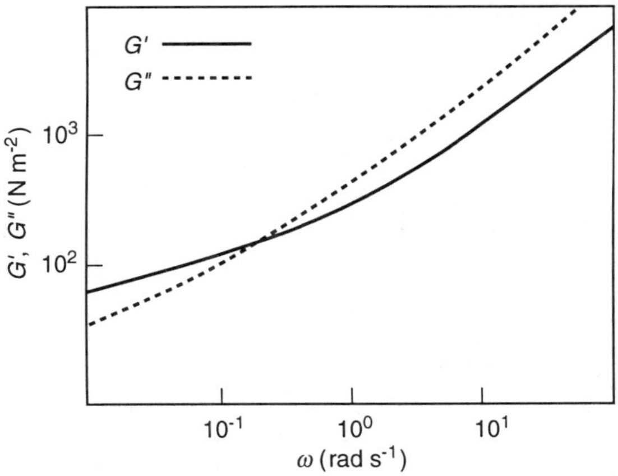

Figure 4.6 Storage (G') and loss (G'') moduli as a function of angular velocity (ω) for 73/27 HBA/HNA copolymer at 290°C in the linear region (from the results of Guskey and Winter [52]).

the behaviour of LCPs is much more complex and diverse than that of homogeneous isotropic polymers. It is useful, therefore, to discuss first the transient behaviour in these experiments before considering the steady-state data.

Representative start-up curves of the torque and normal force in such an experiment, for one thermotropic polymer [29], are shown in Fig. 4.7.

The torque (or shear stress) exhibits a sharp maximum at a small strain, of about two or three units of shear. After an undershoot, it then climbs to a broad second maximum at a large strain. The strain at which this second maximum is reached appears to be independent of the shear rate imposed. Upon stopping and reversing the flow, no small-strain maximum is observed.

The first maximum appears similar to that observed with isotropic polymers when the shear rate is higher than that for the onset of shear thinning [53]. It occurs at a comparable strain [54]. Furthermore, the magnitude of the overshoot increases with increasing shear rate, as shown in a detailed study by Viola and Baird [55] on another thermotrope. Also, the magnitude, or indeed presence, of the overshoot depends on the rest time in an interrupted shearing experiment [30, 54, 55].

In some respects, such as the influence of rest time during interrupted shearing, this first maximum is similar to the response of isotropic polymers [56]. The outstanding difference between isotropic and thermotropic polymers in this respect is that the onset of nonlinear

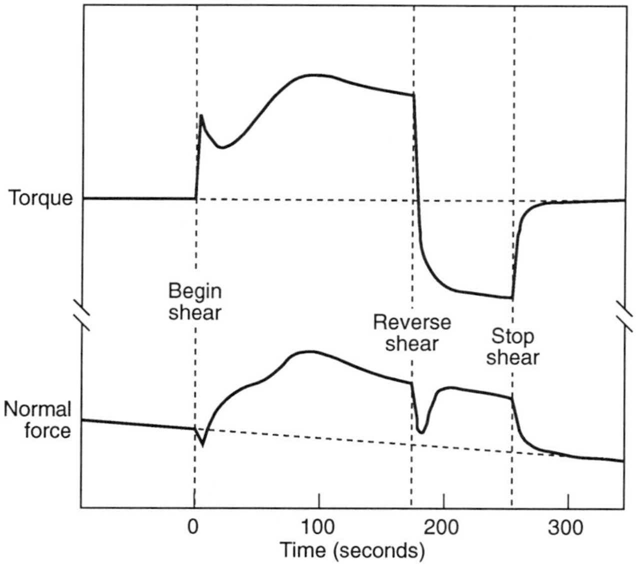

Figure 4.7 Typical transient behaviour of torque and normal force during start up, and reversal, of steady shear (based on the results of Wissbrun [29] using a polymer of molar composition 15% bis(4-carboxy-phenoxy) ethane, 15% terephthalic acid, 40% para-hydroxybenzoic acid, and 30% methyly hydroquinone).

viscoelastic behaviour, such as shear thinning, occurs at a much lower shear rate for thermotropic polymers than for isotropic polymers of comparable viscosity [46].

The second maximum in torque has no counterpart in the rheology of isotropic polymers. Although clearly visible in some cases [29, 55], it is not always observed unambiguously in others [52, 54, 57, 58], possibly because the experiments were not carried to sufficiently large strains; even in these cases, the shear stress was still climbing or perhaps going through a broad maximum when shearing was stopped. The possibility of thermal degradation reactions at the high measurement temperatures required for most thermotropic polymers is a barrier to shearing for long times. This difficulty was overcome by Kim and Han [59], who synthesized sufficient quantities of a flexible-spacer thermotropic polymer, with a low melting temperature, to make detailed studies of the flow behaviour. They observed the small strain maximum of shear stress, and multiple large strain oscillations, the amplitude and number of which varied with temperature and shear rate.

These maxima are usually seen in what is often referred to as the

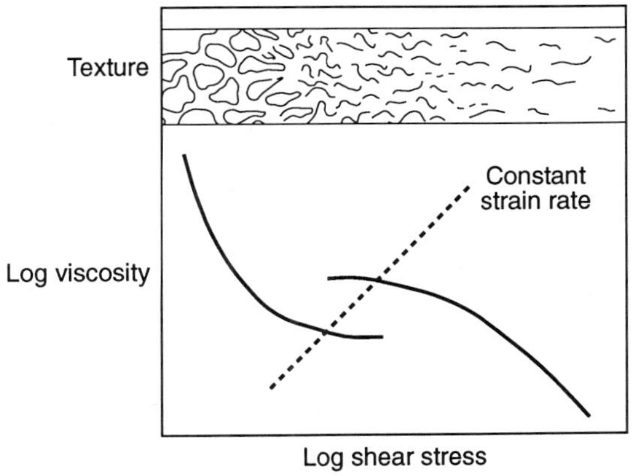

Figure 4.8 Qualitative relationship between morphology and rheology.

constant viscosity region of the flow curve. Here, the relationship between rheology and morphology is undergoing profound changes [30], shown qualitatively in Fig. 4.8.

The transient response of an experimental high-viscosity thermotropic polyester melt in this region is shown in Fig. 4.9 [30]. Figure 4.9(a) shows the response from rest, where the structure is being broken down; Fig. 4.9(b) illustrates the pseudo-equilibrium condition, when the flow is stopped and immediately restarted under the same conditions; and Fig. 4.9(c) demonstrates the effect of modest pre-shearing at a higher shear rate, when the structure is reforming during the experiment.

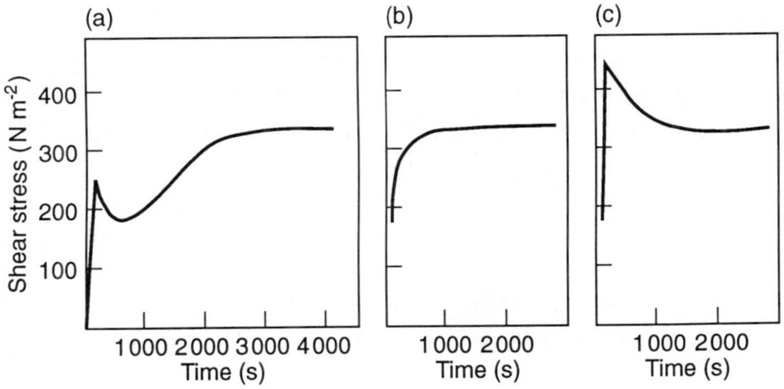

Figure 4.9 Transient response of an experimental, high viscosity, thermotropic polyester at a shear rate of $0.04\ s^{-1}$ at 250°C, equivalent to the strain rate indicated in Fig. 4.8: (a) from nominal rest state; (b) after 4200 s at $0.04\ s^{-1}$ followed by stress relaxation to $200\ N\ m^{-2}$; and (c) after 600 s at $1.14\ s^{-1}$ (stress level $5000\ N\ m^{-2}$) followed by stress relaxation to $200\ N\ m^{-2}$.

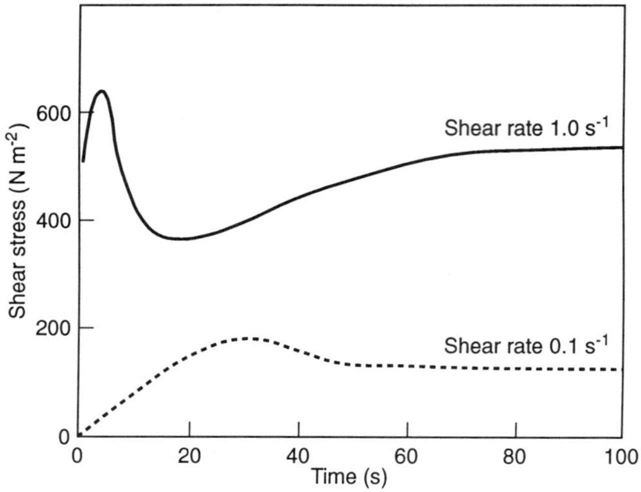

Figure 4.10 Transient behaviour during start up of steady flow in 73/27 HBA/HNA copolymer at 290°C and shear rates of $0.1\,s^{-1}$ and $1\,s^{-1}$, based on the results of Guskey and Winter [52].

The transient response for 73/27 HBA/HNA has been reported by Guskey and Winter [52] – Fig. 4.10.

Very similar maxima have also been studied by Moldenaers and co-workers with lyotropic systems; see [60] and Chapter 8. These have the advantage of being more amenable to rheo-optical investigations of structure changes during flow [60]. Burghardt and Fuller [61] showed that many of the transient shear stress observations were explicable by the classic Leslie–Ericksen theory of the rheology of liquid crystals which tumble, rather than align, in a shear flow.

Although it is complex, and although the details of the response may vary from one polymer to another, there is general agreement about the nature of the transient shear stress. The same cannot be said about the first normal stress difference (for the sake of brevity, from now on called 'normal stress' or 'N_1'). Most observers have noted a transient negative normal stress, which often, but not always, then becomes positive. Negative steady normal stresses are well known, and believed to be understood in lyotropic systems, as discussed by Marrucci in Chapter 2. They have also been reported for a number of thermotropic polymers [33, 62, 63].

For 'Vectra' A900, the one polymer which has been studied in a number of independent investigations, there is sharp disagreement as to whether the steady state value that is ultimately reached at large strains is positive or negative. Guskey and Winter [52] report that, at 290°C, N_1 is negative at shear rates below $0.5\,s^{-1}$, remaining virtually constant at

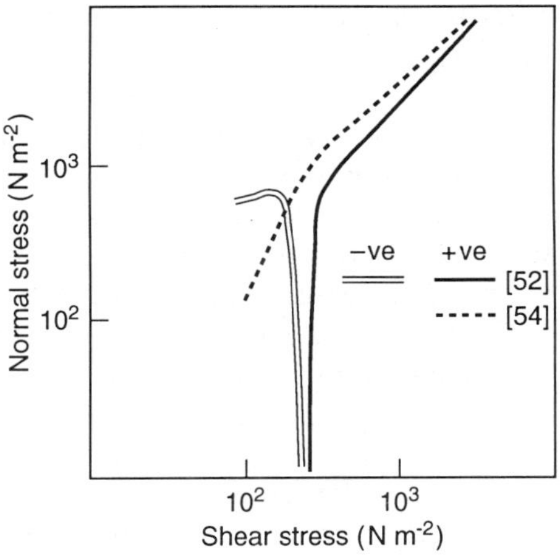

Figure 4.11 Normal stress as a function of shear stress for 73/27 HBA/HNA at 290°C [52] and at 310°C [54].

about 600 Pa below 0.1 s^{-1}, going to zero at 0.5 s^{-1}, becoming positive and large at higher shear rates. In the same issue of the *Journal of Rheology*, Cocchini and his colleagues [54], working at 310°C, find only positive values for N_1 over a comparable shear rate range (Fig. 4.11). Meissner [57] also finds positive normal stresses, at 305°C, at 0.01 s^{-1} and 1 s^{-1}.

All of these investigators agree that a critical source of error in the determination of N_1 is residual normal force at the beginning of an experiment. The principal source of this force is stress imposed on the sample during loading of the rheometer, in which generally a disc is squeezed to the shape of the gap. This force is relaxed by radial flow of the material. The stress relaxation is particularly slow for 'Vectra' A900, whose viscosity increases in power-law fashion at very low shear rates [27, 52, 54]. Different strategies were used to minimize this source of error. Guskey and Winter [52] attempted to achieve this in the following way:

1. vacuum moulding discs,
2. preheating the rheometer to the measurement temperature of 290°C,
3. loading the disc and heating it to 320°C for 60 s to melt out residual crystallites,
4. cooling to the test temperature,
5. moulding to the cone shape when the temperature reached 300°C,
6. and then equilibrating at 290°C for 300 s before beginning the test.

Cocchini and his colleagues [64] argue that this procedure does not relax normal force sufficiently, and that as little as 10 g force, corresponding to 400 Pa normal stress, would account for the negative values of normal stress recorded by Guskey and Winter. In reply to this argument, Winter and Wedler [65] performed measurements of the static normal force relaxation after various moulding procedures, and suggest that the normal force is sufficiently steady during a 500 second measurement time so as not to affect the measurement. Nevertheless, it does appear from those data [65] that variations on the order of 10 g or more can occur during the course of a measurement. Furthermore, as discussed further below, it is not clear that measurements of the static force are necessarily conclusive as to what occurs during a shearing experiment.

The procedure of Cocchini and his colleagues [54], for minimizing the residual normal stress, is based on their observation that, when an unrelaxed sample is sheared, the unrelaxed normal force is partially relaxed although, for unexplained reasons, some recovery does occur after cessation of shearing. They adopted the following protocol:

1. load and equilibrate the sample thermally for 1500 s;
2. pre-shear for 1200 s at 0.02 s^{-1}, followed by further shearing for 400 s at 0.5 s^{-1}; and
3. finish with a relaxation time of 400 s.

The total pre-conditioning and test time is, then, less than about 4500 s total. According to Cocchini, this time is sufficiently short to avoid significant increase of viscosity, presumably from polymerization at the exposed edges of the sample. That effect is decreased by 'flooding' the edges with excess polymer, probably because the polymerizing material is not sheared and therefore its increased viscosity does not contribute to the measured stress. It is not absolutely clear that they have prevented polymerization adequately, particularly since they monitored the complex viscosity at 10 radians s^{-1}, which represents a relatively high frequency compared with their measurement shear rates. Winter and Wedler [65] argue unequivocally that the measurement time required by Cocchini and his colleagues is too protracted for the chemical stability of the material.

Meissner [57] used the most direct procedure to minimize unrelaxed moulding stress. He:

1. cold-pressed the finely powdered polymer to the shape of the cone-plate gap;
2. sintered the preform at 320°C;
3. cooled it under pressure;
4. loaded into a preheated rheometer;
5. and equilibrated for 15 minutes under a vacuum to remove residual gases before beginning measurements.

Also, most noteworthy, Meissner used a fixture specially designed to hold fluctuations of temperature to 0.01°C, compared with the nominal 0.5°C variation of the commercial instrument [66]. Meissner was also able to photograph the edge of the sample during the measurement. He observed shear lines and irregularities developing even at the low shear rate of $0.01\,s^{-1}$. Nevertheless, as already mentioned, he sees an initial negative transient of N_1, followed by a positive, more-or-less steady value.

De'Nève, Narvard and Kléman [38] have measured the normal stress of 'Vectra' B950. They report that the normal force from the sample moulding relaxes in about 60 s, compared with the hours required for 'Vectra' A900. This observation is consistent with the difference in the flow curve shapes; the viscosity of 'Vectra' B950 levels off and becomes nearly constant at low shear rates, in contrast to the power-law behaviour of 'Vectra' A900. They report only positive values of N_1.

A number of other sources of error that may be contributing to the present state of uncertainty may also be mentioned, and should perhaps be investigated further:

1. A common test for proper instrument alignment is to observe the effect of reversal of rotation, which has apparently not been done by the three investigators of the normal stress of 'Vectra' A900. In the case of lubricating greases, this test revealed a directional effect, presumably related to a flow-induced structure [67].
2. Although bubbles, perhaps caused by degradation or condensation reactions, were looked for [54, 57] and discounted as sources of error, Meissner [57] also reports that an amorphous isotropic polyamide showed a transient negative normal stress, which was eliminated by more complete drying. The action of shearing would be expected to elongate any bubbles. Such elongated bubbles could divide into small bubbles, in which the effect of surface tension would increase the pressure in, and reduce the volume of, the trapped gas. Such a reduction in gas volume would be reflected in a shrinkage of the whole system that could be reflected in negative values of normal stress.
3. The importance of temperature variations, causing normal forces due to thermal expansion/contraction, is already well known [66]. Such temperature variation may occur as a result of the rotation of the instrument fixtures at the onset of a test, and result in erroneous normal forces. The pre-shearing procedure of Cocchini and his colleagues [54] may be effective in part because it helps to attain temperature equilibrium.
4. The effect of pre-shearing to accelerate relaxation of the moulding normal stresses is not understood. Could it be due to the small changes of the gap caused by the normal stresses generated in a shear flow, 'amplified' by the constrained cylinder effect analysed by Zapas

and his co-workers [68]? The gap dependence of the shear rate range of negative normal stress found by Gotsis and Baird [63] is consistent with this source of error.

In respect of measurements of normal stress, although the situation has not been resolved conclusively at this time, we incline to the view that the negative steady-state values are probably artefacts of the interaction of the unusual rheology of 'Vectra' A900, specifically its slow stress relaxation, with the mechanical, thermal and electronic characteristics of the measuring instrument. As has been pointed out before [41], a number of other systems that exhibit apparent yield stresses, and therefore will not completely relax normal forces by radial flow, exhibit apparent negative normal stresses.

While the normal stress response remains in need of clarification, there is a general consensus with respect to the behaviour of the torque response during transient flow. This is broadly similar to the response recorded, and extensively analysed for lyotropic systems, which observations are described by Moldenaers in Chapter 8.

4.3.3 Steady state shear viscosity

Qualitatively, the response of the viscosity to stress, and the associated changes of texture, has been shown in Fig. 4.8 [30]. At low shear stress, there is a region of progressive shear thinning, generally followed by a plateau, sometimes referred to as the constant viscosity region. Beyond that plateau pronounced shear thinning occurs. What is referred to as the constant viscosity region may actually hide quite complex response. This can include a minimum, followed by secondary maximum of the apparent viscosity [29, 30, 69]. It should be noted that not all of these features are necessarily observable with all materials and under all conditions. To obtain a full picture of the steady shear flow, it is necessary to integrate precise rheometric measurements with engineering approximations, such as capillary rheometry at high shear rate. Capillary rheometry is discussed in more detail in section 4.3.5.

Figure 4.12 illustrates the range of variation of flow curve shapes observable on a single material as the temperature is changed [29]. At 280°C, the viscosity has almost power-law behaviour whereas, at 340°C, the viscosity is almost constant. At intermediate temperatures, the 'three-region' flow curve first described by Onogi and Asada [40] are seen. There are also curves with apparent shear thickening viscosity, corresponding to the discontinuous behaviour illustrated in Fig. 4.8, which suggested two pseudo-equilibrium shear stress responses at a given shear rate, dependent on the structure present and whether that condition was reached by raising the imposed strain rate, or reducing it

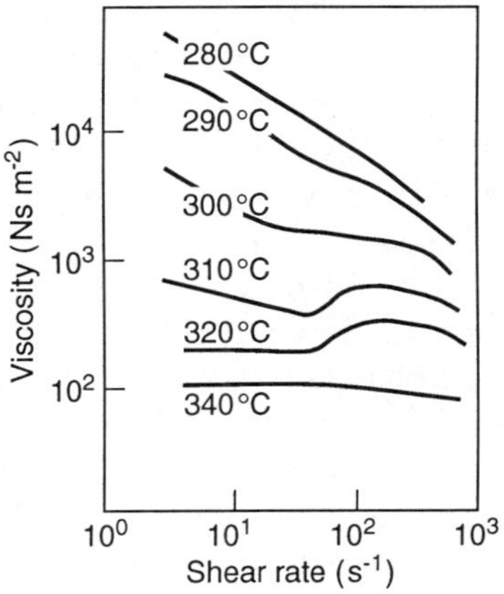

Figure 4.12 The range of variation of flow curve shapes for a single polymer as a function of temperature, based on the results of Wissbrun [29] using a copolymer based on para-hydroxybenzoic acid, 2,6-dihydroxynaphthalene, and terephthalic acid.

from a higher level. It is probable that the actual shape of the measured flow curves depends strongly on how it is measured with respect to the transient behaviour noted in the previous section.

The flow curve of 'Vectra' A900 is consistent with the 'three-region' behaviour. Figure 4.13 shows data from five groups of investigators [27, 52, 54, 70, 71] obtained by a variety of instruments, over a temperature range of 290–310°C and on different lots of material. Nevertheless, there is general agreement of power-law behaviour, with a slope of about

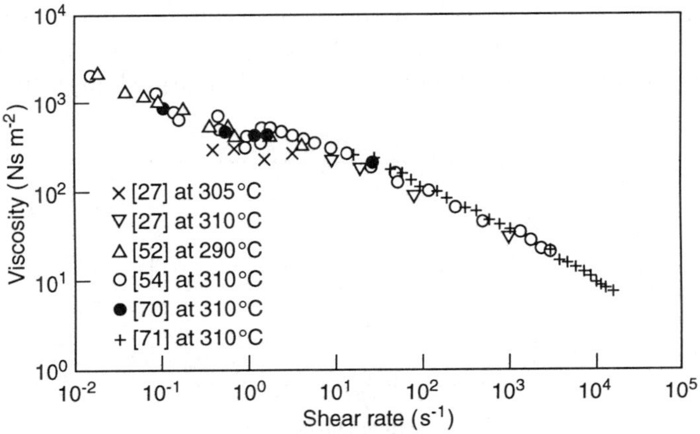

Figure 4.13 Flow curve of 'Vectra' A900, based on the observations of five teams of investigators [27, 52, 54, 70, 71].

−0.5, at shear rates above $10\,s^{-1}$, of a plateau region from about $0.1\,s^{-1}$ to $10\,s^{-1}$, and more limited data of another shear thinning region below $0.1\,s^{-1}$. The occurrence of a second plateau region at very high shear rates, above $10^4\,s^{-1}$, has been identified [27], but this must be considered as tentative because of uncertainties due to shear heating, end effects and other possible sources of error.

'Vectra' B950 has also been studied by a number of groups [31, 38, 72]. At temperatures above 290°C, above the melting temperature, the flow curve has a plateau region down to a shear rate of $0.1\,s^{-1}$, which goes into power-law flow at higher shear rates. Data to test whether the viscosity becomes shear-thinning again at still lower flow rates have not been published. At lower temperatures, only power-law flow, with a slope of about −0.5, is observed [31].

Flow curves of various grades of 'Xydar' have also been published [34], but only with limited number of data points over a narrow range of shear rates. The behaviour ranges from nearly constant viscosity, to power-law flow with a slope of −0.5.

4.3.4 Melt elastic response

Measurements of the elastic response, as indicated by strain recovery after deformation at constant stress [27], suggest a high elastic response as the material starts to flow, followed by a sudden collapse in that response (Fig. 4.14).

The observations of significant elastic response at low stress levels is consistent with considering deformation of the domains from an approximately spherical to ellipsoidal form. In that respect, the domain structure could be considered similar to an emulsion, which can have high elastic response due to surface tension effects between the phases, even when neither phase is itself significantly elastic. At higher stress level, as the domain structure breaks down, there is a dramatic change in the melt

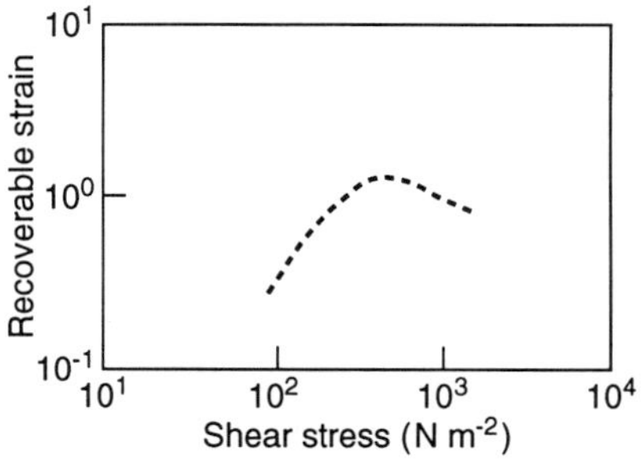

Figure 4.14 Recoverable shear as a function of shear stress in 73/27 HBA/HNA at 305°C.

elastic response of these materials. Larson and Mead [73] have postulated a limiting, or yield value, for the elastic strain to model their extensive observations of recoverable strain of lyotropic liquid crystals. The concept of a saturation value of the Ericksen number for the flow of tumbling nematics is also consistent with this observation [61]. Physically, it may be considered that this results from the length of a rod-like molecule as the lower limit of distance over which curvature of the nematic director can occur; there is a lower limit to the size of the domains or the distance between defects [74, 75].

Recoil after cessation of flow is a direct measure of elasticity in the sense of return to the state before a deformation was imposed. For viscoelastic materials, there exists a well-understood connection between elastic recovery and the normal stress in a steady-shear flow [76]. For viscoelastic materials, the normal stress increases quadratically with shear rate in the limit of low shear rates. Thermotropic polymers, on the other hand, have at most a linear dependence of (positive) normal stress on shear rate [29, 38, 54, 63], as do lyotropic polymers; see Chapter 8. This dependence is predicted theoretically by dimensional analysis, along with other scaling relationships, for materials without an intrinsic time constant [77]. For viscoelastic materials, the ratio of viscosity to elastic modulus is such a time constant; the liquid crystal theories do not provide equivalent quantities.

LCPs have been characterized as 'highly elastic' because they have measurable normal stress at very low shear rates, where, for isotropic polymers of comparable viscosity, the normal stress would be immeasurably low [58]. However, this is the result of the linear decrease of normal stress with shear rate, compared to the much steeper quadratic decrease for isotropic polymers. At very low shear rates, therefore, the normal stress is higher for thermotropic than for isotropic melts; at higher shear rates, in the range of interest for processing, the normal stress is lower for thermotropic polymers.

Another phenomenon, often used as a measure of melt elasticity, is swelling of an extrudate to a diameter larger than that of the die. Thermotropic polymers show very little, or even negative, swelling [29, 31, 33, 37, 69, 78, 79]. It has been proposed that these observations result from a stratified flow of concentric laminae of melt with different textures, which consequently have different viscosity characteristics [80]. This model, which has not been subjected to experimental tests, is discussed in section 4.3.5 as a possible explanation for anomalous behaviour of LCPs in capillary rheometry.

The magnitude of the storage modulus, G', from oscillatory flow experiments is another direct measure of melt elasticity. For viscoelastic melts, it is also connected theoretically to the normal stress. In the limit of low frequency, G' is equal to one half of N_1 measured at a shear rate numerically equal to the frequency of measurement of G'. For both lyotropic and thermotropic polymers, it has been found that G' is much smaller than 0.5 N_1, often by as much as a factor of ten or more [27]. Such observations also suggest that LCPs are only slightly elastic, and that the relatively high normal stress and the slow stress relaxation of LCPs originates from mechanisms other than the Brownian motion rubbery elasticity of isotropic polymer melts.

4.3.5 Capillary viscometry

Capillary studies are unanimous in finding strong sensitivity to shear rate, high values of entrance pressure drop (Bagley correction) and virtually no tendency to post-extrusion swelling. In all of these three major effects, the rheology of thermotropic LCP melts runs parallel to that of fibre reinforced materials.

There are some contentious areas. Most significant of these is the dependence of apparent viscosity on the dimensions of the capillary. In parallel studies in two laboratories [27], a considerable discrepancy was observed in the magnitude of the viscosity. By a process of elimination, this was traced to differences in the diameters of the dies in use in those laboratories. A detailed set of experiments then confirmed that smaller dies gave rise to significantly higher apparent viscosities (Fig. 4.15), the effect being most noticeable at low shear rates.

A similar dependence on die diameter has been noted in the flow of long-fibre reinforced thermoplastics [22]. In this case, the dependence on die diameter is not at all surprising, since the reinforcing fibres are long in comparison with the diameter of the die; the surprise is that the long fibre

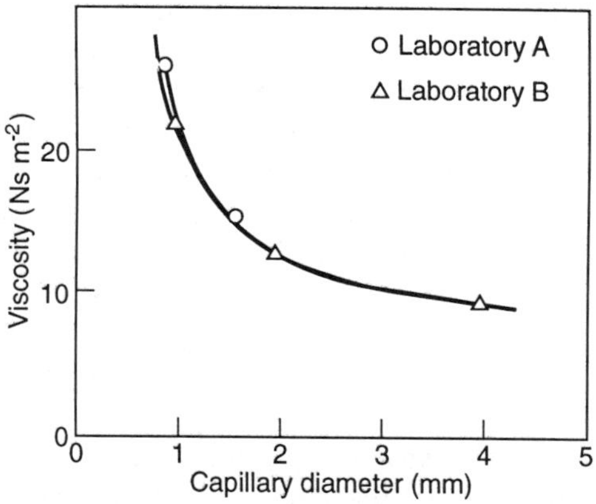

Figure 4.15 Apparent viscosity versus die diameter at a nominal shear rate of 15 s^{-1} and 310°C.

reinforced materials flow at all! This appears to be another area where the detail of the rheology of LCPs and of long fibre reinforced materials follow a similar, anomalous, pattern.

A third group of workers [31], studying thermotropic polymers independently, observed a dependence of apparent viscosity on die diameter but a fourth group [81], who set out to study the effect, did not detect it. In all cases, no systematic error can be found in the experimental techniques used that could account for the discrepancy, so that the results seem equally valid. The only obvious difference is that, in the work where no dependency on die diameter was found, the length to diameter ratio of the dies was rather longer, and the die diameters were smaller. Turek and Simon [32, 82] noted a significant effect of die length to diameter ratio on the strength of extrudates. Extrudate strength was reduced after extrusion from long dies. Turek and Simon argued that this reflected a reduction in orientation of the extrudate. The reduction of orientation was confirmed by X-ray diffraction measurements. They hypothesized that, because of the convergent flow in the entry region, the melt entered the die in a highly ordered low viscosity state and subsequently became disordered, and more viscous, during its passage through the die. Previously, La Mantia and Valenza [31] had suggested that such a change could account for anomalous Bagley plots, which showed excessive increase in pressure for longer dies. Optical examination of extrudates revealed the existence of transitions of texture [38]. These observations contributed substantially to a rationalization of these inconsistencies, both in the Bagley plot and in the effect of die diameter, in terms of the stratified laminar flow of two fluids [83].

The non-linearity of the Bagley plots of 'Vectra' B950 was attributed to the pressure dependence of viscosity by Zaldua and his colleagues [79]. Lefeuvre and co-workers [84] have proposed a refinement of this idea, assuming that the variously textured layers of the melt [38] have different pressure dependencies of viscosity. Neither of the two-fluid models [83, 84] has yet been subjected to independent tests of their validity.

There are different opinions as to detail, but there is general agreement about the level of viscosity and shear thinning behaviour at modest shear rates. Only two sets of experiments have been carried out at very high shear rates [27, 33]; these suggest a tendency to a constant limiting viscosity at shear rates in excess of 10^4 s^{-1}. It is possible to speculate that, under such conditions, no further increase in orientation can be achieved and that the material is flowing as a single domain. Typical capillary viscometry flow curves, and die entry pressure drops, as a function of shear stress are shown in Fig. 4.16; these have been extended at low shear stress by the data from cone and plate rheometry.

The main differences between thermotropic liquid crystals and

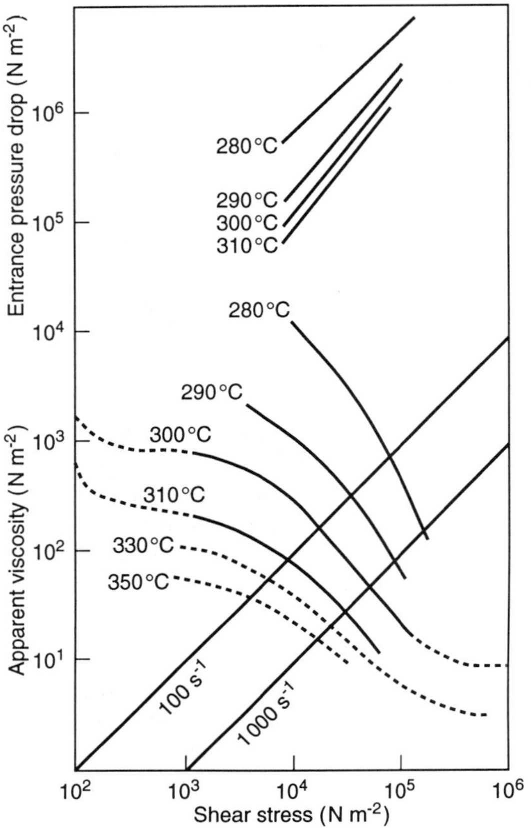

Figure 4.16 Apparent viscosity, corrected for entrance pressure drop, as a function of shear stress, and entrance pressure drop as a function of shear stress for 73/27 HBA/HNA at a range of temperatures based on data from Wissbrun, Kiss and Cogswell [27]. The high temperature and high shear rate data are estimated from uncorrected values reported in that paper.

conventional polymer melts, as experienced in capillary rheometry, are summarized in Fig. 4.17.

4.3.6 Elongational flows

Precise measurement of elongational viscosity is one of the most difficult experiments in rheometry, most particularly so where the material has a low viscosity. The first such measurements on thermotropic polymers were transient isothermal experiments using a rotary clamp extensional rheometer made by Wilson and Baird [85]. At low strains, the elongational viscosity of 'Vectra' A900 was close to three times the transient shear viscosity. Moderate strain hardening occurred at

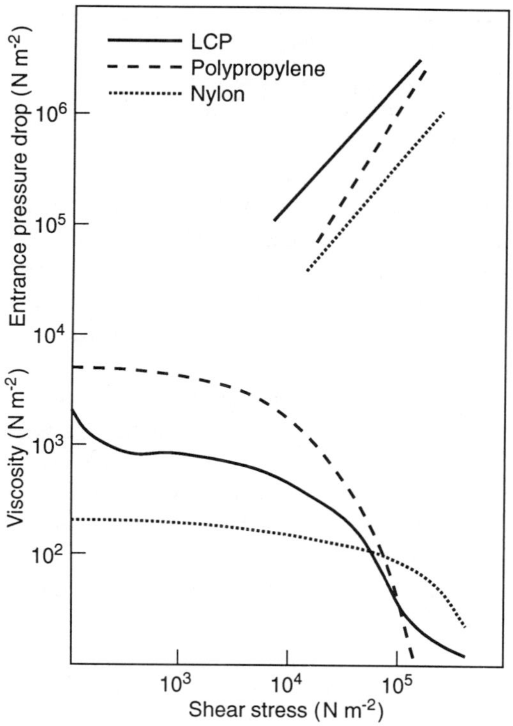

Figure 4.17 Comparison of rheometric studies on typical injection moulding grades of LCP, polypropylene and nylon. Of these polymers, LCP and nylon show very little tendency to post-extrusion swelling, while swell ratios for polypropylene are typically in the range 1.5–2.5, dependent on shear rate and die length/diameter ratio.

strains of 0.5–1, the elongational viscosity rising by no more than a factor of 2. Fracture occurred at strains of 1.2–2, so the question of whether a steady state elongational viscosity exists remains open. Wilson and Baird also observed interesting differences between the elongational viscosities of hydroxypropyl cellulose in the isotropic and liquid crystalline states. However, the differences could have been caused by differences in residual crystallinity at the two measurement temperatures.

Turek and his colleagues [86] have used two complementary, transient techniques to evaluate the extensional response of thermotropic melts: instrumented isothermal spinning and converging flow. For 'Vectra' A900 (73/27 HBA/HNA copolymer) the results of these techniques are in good agreement. They suggest that, at 300°C, the elongational viscosity is approximately independent of deformation rate, with a magnitude of 10^4 Ns m^{-2}. This is about 20 times higher than the apparent shear viscosity of the melt in the pseudo-Newtonian region (Fig. 4.18). It should be noted that their measurements were made at strain rates in the range of about 15–500 s^{-1}, and are thus not directly comparable to those of Wilson and Baird [85], which covered the range 0.02–0.5 s^{-1}.

Practical experience of melt spinning suggests that LCPs can readily be drawn into fine filaments. Although such practical deformations can be strongly stabilized by cooling of the thread line, melts with tension thinning viscosity behaviour tend to give unstable thread lines, while strongly tension stiffening response can lead to thread line rupture. Thus the conclusion of approximately constant or tension hardening elongational viscosity appears to be supported by practical experience.

During such extensional flows, the molecules are highly oriented. The question of how oriented rod-like molecules flow in extension is not a trivial one. At the microdomain, or individual molecule, scale, the flow must be simple shear parallel to the drawing direction. It is certainly not obvious that this should be a stable deformation in such a shear sensitive material as liquid crystal melts, much less one of constant viscosity.

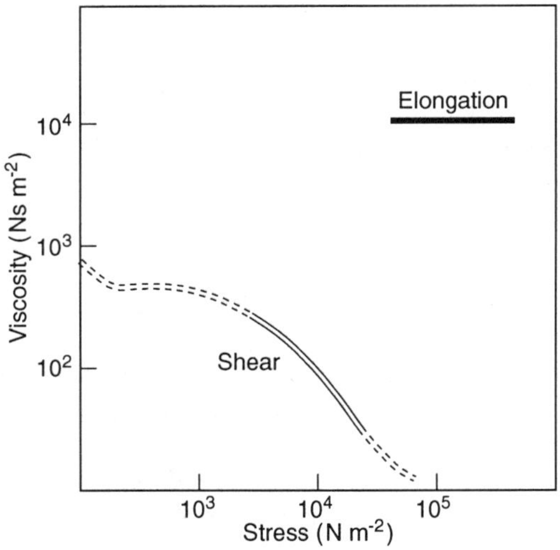

Figure 4.18 Comparison of apparent shear and elongational viscosities for a sample of 73/27 HBA/HNA copolymer at 300°C. The solid line shows elongational viscosity based on the data of Turek using an instrumented spin line and die entry flows. The open line represents the shear viscosity data from the same source, which has been extended on the basis of other data shown in Figs. 4.13 and 4.15.

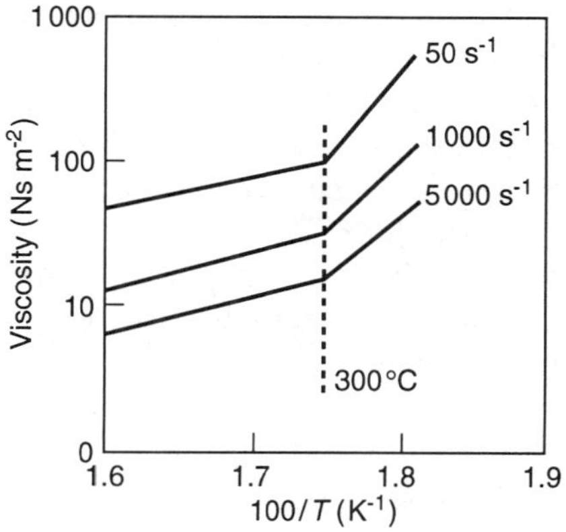

Figure 4.19 Arrhenius plots of viscosity of 73/27 HBA/HNA at various shear rates.

4.3.7 The influence of temperature and pressure

Arrhenius plots of the dependence of viscosity of 'Vectra' A900 on temperature for a range of shear rates, measured by capillary viscometry (Fig. 4.19), show an obvious transition in temperature sensitivity at about 300°C, 20°C above the peak melting endotherm at 280°C [27].

Similar plots using small amplitude dynamic data show similar temperature sensitivity above 300°C, but more complex response below that temperature. The mean value of activation energy, at constant shear rate, above 300°C is 11 kcal mol^{-1} (standard deviation 1.5 kcal mol^{-1}), not very different from the 13.5 kcal mol^{-1} of polyethylene terephthalate [87], an analogous isotropic polymer.

La Mantia and Valenza [31] have measured the temperature dependence of the melt viscosity of 'Vectra' B950 by capillary rheometry. This polymer also shows two distinctly different regions of behaviour. At high temperatures, above 290°C, near the maximum DSC melting peak, the activation energy is about 25 kcal mol^{-1}, independent of shear stress. Below this temperature, the activation energy jumps to the high value of about 90 kcal mol^{-1}, but depends somewhat on the shear stress of the measurement. For both of these thermotropic polymers, the high activation energy at, and below, the nominal melting temperature demonstrates that the small fraction of residual crystallinity has a powerful effect on the melt viscosity.

There has been no systematic study of the influence of pressure on the viscosity of LCPs. The melt has a similar compressibility to that of other polymers (see section 4.2.5) and, in the absence of any evidence to the contrary, it is reasonable to assume that the effect of pressure on viscosity will be similar to that of other polymer melts. Such data [88] suggest that a hydrostatic pressure of 1000 atmospheres will have the same effect on viscosity as reducing the temperature by 50°C. It has been suggested, however, that the flow of 'Vectra' B950 involves a collective motion of molecules and, as a result, requires a higher activation volume than do isotropic polymers [84]. This would result in a higher pressure dependence of viscosity, which is also dependent on the texture, or defect density, of the melt. Experimental data to test this interesting suggestion are obviously needed.

4.3.8 The effects of thermo-mechanical history

Any annealing of the melt close to the crystalline melting point can lead to large increases in resistance to flow [29, 44, 89, 90, 91, 92, 93]. A typical result [89] using the **Melt Flow Rate test**, a capillary measurement at constant stress that is widely used in quality control measurements, showed nearly three orders of magnitude change in fluidity for simple thermal treatments (Table 4.1).

Qualitatively similar results are found with 73/27 HBA/HNA [27]. Holding the 73/27 HBA/HNA melt at 300°C for 18 hours caused the viscosity to increase by an order of magnitude. In contrast, at 310°C, the melt viscosity appeared to be stable. Careful study of the intrinsic viscosity of those samples ruled out any possibility of molecular weight change; rather, the change in viscosity is associated with some annealing of the material associated with residual crystallite structure.

It should be noted that in studies conducted in closed systems, such as capillary rheometers, there is no opportunity for volatile condensation products to escape. A viscosity increase in this case is unlikely to be caused by a molecular weight increase. In an open system such as a cone-and-plate rheometer, on the other hand, the nitrogen sweep in the

Table 4.1 Sensitivity of fluidity to thermal history for an experimental thermotropic polyester

Sample history	*Melt flow rate 223°C/2.16 kg (grams/10 mins)*	*Extrudate quality*
As prepared	6.4	Fibrous skin with paste-like interior
18 hours at 223°C	0.3	Paste-like, brittle
Melted at 240°C, then cooled to 223°C	260	Fibrous throughout

oven can remove condensation products and shift equilibrium molecular weight. An increase of viscosity with time under these conditions, as observed by Cocchini and his colleagues [54] for example, may well be the result of polymerization at the exposed edge of the sample. A simple way of distinguishing between the crystallization and polymerization mechanisms for a viscosity increase is that they have opposite temperature dependencies; polymerization rate increases and crystallization rate decreases with increasing temperature. However, definitive discrimination requires independent estimation of molecular weight and/or crystallinity changes.

At higher temperatures, there is a tendency for degradation, but 73/27 HBA/HNA can be processed at temperatures up to 350°C without significant problems [27].

The influence of mechanical history on rheology falls into two groups:

1. the effect of modest mechanical history on the properties of the melt under low-stress conditions; and
2. the effect of intense working on the processability of the material in conventional shaping flows.

The stress transients observed during the start up of steady shear (section 4.3.2 and Chapter 9) show some of the effects of mechanical history on the viscosity. Modest mechanical history can lead to a small increase in viscosity which decays with time to its equilibrium value [30, 94], as is shown in Fig. 4.9(c), but it is more common to find significant reduction in viscosity as a result of preshearing. Modest preshearing, at shear rates of the order $1\ s^{-1}$, was found to be particularly effective in reducing the viscosity of an experimental polyester at a temperature near to its nematic/isotropic transition [57]. One group of experimenters [54] includes a standard modest preshear and relaxation regime before any rheometric studies are carried out. They particularly concluded that such history was essential to avoid what they considered to be anomalous determinations of negative normal stress [64]. It is impossible to ignore the fact that the squeezing of the sample as it is loaded into a cone-and-plate rheometer constitutes a mechanical history, and so may have some effect on the results that are measured. Allowing the sample a very long dwell time at temperature to forget such history may lead to other unwelcome effects [54], and the preshear conditioning of the material does appear to improve the reproducibility of results, but it is not clear that such conditioning necessarily gives rise to a well-defined ground state for the material.

A second area of adventitious mechanical history, though in this case somewhat better defined, is the deformation which the material experiences in the die entry region of a capillary extruder. This deformation includes a strong extensional component which induces high orientation. Turek and Simon [32, 82] demonstrated conclusively that the high orientation induced in that region subsequently decreases substantially during the passage through the die. As a result, the shear viscosity of the material towards the exit of the capillary is likely to be higher than at the entry. Relaxation of orientation, induced in the converging flow region, during passage through the die is well known in conventional polymers; it is clearly seen in the reduction of birefringence and the change in post-extrusion swelling from large, on extrusion through an orifice die, to modest, on extrusion from a long die. That significant relaxation of orientation should occur in LCPs during passage through a die is surprising. The transit time through the die is usually of the order of tenths of a second compared with texture relaxation times of hundreds, or indeed thousands, of seconds; there is virtually no mechanical relaxation on extrusion from orifice or long dies, as evidenced by post-extrusion swelling, which is very low for both short and long dies. A more likely explanation for the decreased orientation of LCPs is that tumbling creates defects and tends to randomize the overall orientation.

The picture of the influence of mechanical history becomes more complex when one considers intense preshearing of the material. It has been observed that preshearing, preferably at shear rates of the order 100 to 1000 s^{-1}, can greatly reduce the viscosity of the material in subsequent measurements [89]. Indeed, it was found that it was possible, by such preshearing, to induce thermotropic materials to form mobile liquid crystalline melts at much lower temperatures than they would naturally do so, and also to reverse the adverse effects of long thermal conditioning. Some lyotropic systems, which did not naturally form a liquid crystal state, could also be driven into a nematic phase by preshearing. However, to achieve this, the shear had to be applied by relatively moving surfaces. The application of much higher shear rates from pressure driven flows was found to be ineffective. Simple shear processes between relatively moving surfaces are homogeneous in that all the material undergoes a similar history, whereas the shear is inhomogeneous in pressure driven flows, so that most of the material is not subjected to any appreciable shear working. With thermotropic melts, such preshear is most effective at enhancing the fluidity of the melt at low temperatures. For 73/27 HBA/HNA copolymers at 260°C, a fourfold increase in mobility was achieved at a rate of 250 s^{-1} [27].

Frayer and Huspeni [34] testify to the importance of the thermomechanical history that LCPs experience in a screw extruder in determining the rheology of the material as it enters a die or injection mould. Because of such history, some commercial LCPs can be processed at temperatures where laboratory studies on materials at rest suggest that they are unprocessable. It has been noted [83] that the viscosity of LCP melts is lower when measured by an extruder-fed rheometer, which imposes a preshear history, than when measured conventionally by a pellet-fed rheometer [95, 96]. The die diameter dependence of the melt viscosity, when presheared by an extruder, is also in the opposite direction to that found by Wissbrun, Kiss and Cogswell [27, 96, 97].

4.3.9 Solid phase properties relevant to processing

In practical processing, some of the solid state rheological properties, such as friction, are important in determining the way in which material is conveyed through the feed zone of an extruder. Friction measurements on this class of polymer [27] suggest unusually high values at about 200°C associated with the softening of the polymer.

4.4 PROCESSING WITH THERMOTROPIC MELTS

This section will emphasize the specific interactions between melt rheology and plastics processing. That emphasis will focus on injection moulding, since it is as an engineering polymer for shaped articles that thermotropic aromatic polyesters are most extensively used at present. A more detailed discussion of the processing of LCPs and blends is given by Hull in Chapter 7.

4.4.1 Extrusion

The workhorse of most plastics processing operations is the single screw extruder. For ideal pumping in such extruders, the polymer should slip on the screw and adhere to the barrel wall. The veracity of this statement is more clearly seen if one considers the opposite case of adhesion to the screw and slip at the barrel wall, when the polymer will simply rotate as a plug with no forward movement. Because LCPs melt to a very low viscosity, and because the barrel wall is usually at a higher temperature than the screw, at least in the feed zone of the extruder, there is the danger that a low viscosity melt will be formed at the barrel wall which lubricates the flow and prevents satisfactory pumping. To avoid this, it is desirable to maintain the barrel temperatures in the feed zone below the melting point of the polymer. In section 4.3.9, we noted that there is a maximum coefficient of friction for 73/27 HBA/HNA at about 200°C, just as the material starts to soften. Uniform pumping is the first requirement of any screw extruder and, in low viscosity melts, this aspect of performance is dominated by the solids feeding zone.

As noted by Frayer and Huspeni [34], the screw extruder provides a mechanical history. Further, it is a mechanical history dominated by shear

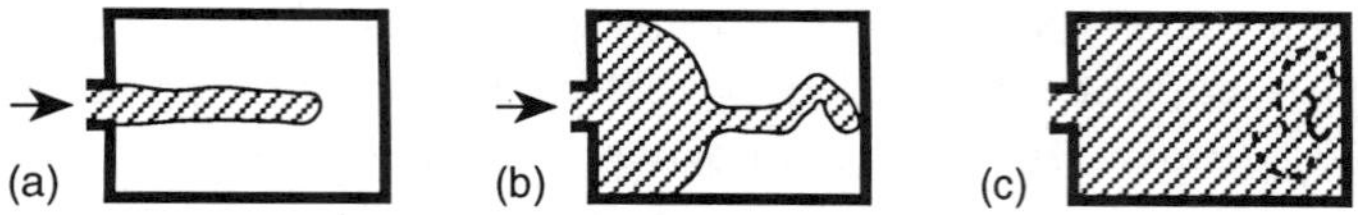

Figure 4.20 'Jetting' in injection moulding due to lack of post-extrusion swelling: (a) initially, material jets into the mould; (b) it eventually adheres to wall and starts to fill normally, meanwhile initial material starts to solidify; and (c) marking, and wield line weakness as a result of jetting.

between relatively moving surfaces – the very circumstances which favour modification of morphology of the polymer [89]. The shear rates in conventional screw extruders are usually calculated to be about 50 s^{-1}, although very much higher shear rates, of the order 1000 s^{-1} may occur in regions near the flight tip of the screw. This range of shear rates is that to which the structure of thermotropic materials seems to be particularly sensitive. For processes like injection moulding, where it is desirable to maximize the level of shear modification, and also the homogeneity of that step, it is desirable to use high screw-back speeds.

The absence of post-extrusion swelling is desirable in a simple extruded product, since shape can more easily be defined. It is less advantageous in injection moulding where, instead of smoothly expanding from the gate as it enters the mould, the melt may tend to jet forwards and then partially freeze before the mould is completely filled (Fig. 4.20). To avoid this defect, larger gate sizes than normal are frequently preferred, or gates are located so that the flow impinges on the cavity wall. Extrusion is dealt with more fully in Chapter 7.

4.4.2 Free surface flows

Much of the early commercial interest in thermotropic LCPs was as fibres. Aramid fibres such as 'Kevlar' are spun from lyotropic liquid crystals, and it was thought that, by removing the solution stage, a more cost effective and environmentally friendly manufacturing process would be achieved. For a variety of reasons, that early promise has not yet been completely fulfilled, but aromatic LCP melts can readily be spun into fine fibres of high stiffness and strength, such as 'Vectra', and future developments may well see increased interest in this field (see Chapter 7). Another family of thermotropic melts is being used to spin precursor material for high performance carbon fibres; this is 'mesophase' pitch. The wholly aromatic molecules of mesophase pitch are not rod-like, but plate-like in form. Although, because of thermal stability problems, those materials have not been characterized in such detail, their melts show essentially the same rheological behaviour as the polyester materials described here, including sensitivity to mechanical history.

Thermotropic LCPs are less suitable for films applications. Rod-like systems, which are readily oriented in one direction, are not so convenient for biaxial orientation. There are two types of film process: melt and solid phase drawing, which is usually carried out just below the melting point in crystalline materials. Melt processes need a moderately high viscosity and also an elastic melt in order to support the bubble and give it stability: here, the low viscosity of LCPs is an embarrassment. Bubble processes can operate by carrying the LCP 'piggy-back' on a sacrificial co-extruded film of high viscosity melt, but this is an expensive option. Stenter processes, where a uniaxially oriented sheet is gripped at the edges and drawn sideways, must be carried out where the material is sufficiently solid to support the hold of the stenter clips. For conventional semi-crystalline polymers, there is a 'rubbery' region below the final melting point where this can be carried out. No such 'rubbery' region exists in LCPs. Worse, once the material has been uniaxially oriented, it is relatively weak to transverse draw and the material is prone to splitting. To make biaxially oriented tubes, it is necessary to have recourse to innovative processing technology, such as a counter-rotating annular die and core [98], but such processes have not yet proved commercially attractive. Uniaxially oriented tape or sheet is, of course, a more simple matter, being an extension of simple fibre drawing, but from a slot die.

The very low viscosity of LCPs might, superficially, make them attractive for sintering flows, such as rotational casting from powder feedstock. Such flows do not induce high orientation and so would not be able to exploit the very high stiffness potential of this class of materials. Even so, the modulus of unoriented aromatic polyesters is high for conventional engineering plastics. What effectively limits the use of these materials in such low driving-force flow processes is the high viscosity at low shear stress of the material, which slows down or hinders completely the fusion of the particles. A possible variant on such technology could be envisaged as an extension of 'splat' forming. Here small, molten droplets would be fired at a surface. On impact they would flow out and orient. A succession of such 'splatted' droplets could be envisaged to build up, giving an in-plane oriented structure.

The high stiffness, good dimensional stability and tolerance of hostile environments, also makes LCPs attractive as a matrix for continuous fibre composite materials. During the manufacture of such systems, a large free surface area of fibres must be wetted by the matrix [99]. For thermotropic polyesters, the high viscosity at low shear stress is a hindrance in the impregnation of reinforcing fibre bundles by the melt. Only by appropriate design of the polymer [100] and process [101] can high quality composites be prepared.

Free surface flows also occur in injection moulding. At the advancing front of the flow, there is a fountain of melt, the front of which is strongly

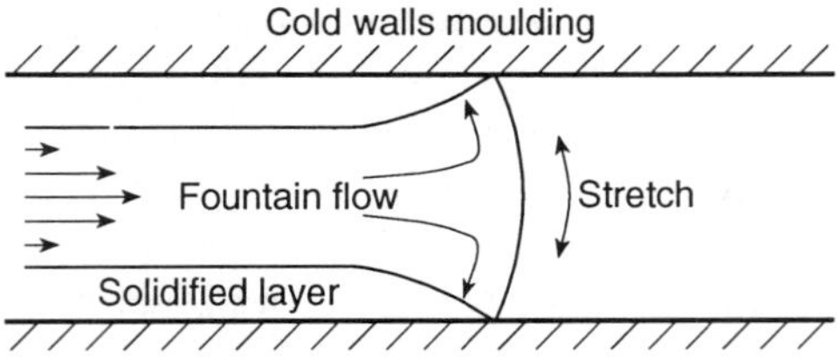

Figure 4.21 Fountain flow in injection moulding.

oriented in tension (Fig. 4.21). This highly oriented material is then immediately deposited and frozen on the mould wall.

4.4.3 Injection moulding

The combination of low viscosity and high stiffness are extremely attractive from the point of view of injection moulding of engineering thermoplastics. These are 'big' properties. Liquid crystal polyesters have other 'little' properties that ensure their position in that market place. Low thermal expansion coefficient is a tremendous advantage: it gives low warpage and shrinkage in the moulding, although moulds may have to be carefully tailored to that property if good mould release is to be achieved. Low shrinkage, in combination with the low viscosity of the melt, enables precision moulding of highly complex shapes that cannot be obtained with other materials. Other key 'little' properties include the insensitivity of such materials to conventional solvents, and their good resistance to burning. While the 'big' properties enable a material to be considered for an application, it is the 'little' properties that define its uniqueness in that field.

Control of orientation in a moulding depends on the flow path. Radial flow in an injection moulding will produce radial orientation in the surface layer because of the strong shear flows near the mould wall, but the same flow will induce hoop orientation in the centre, because of the stretching flows along the central plane (Fig. 4.22).

Such orientation patterns give the moulding a laminated structure that can be controlled and exaggerated by the use of technologies like multi live-feed injection moulding [102, 103]. Similar laminated structures are also seen in mouldings of fibre reinforced isotropic polymer melts [104], reinforcing the analogy discussed previously. More detail on orientation is given in Chapter 7.

Low viscosity during injection moulding is not always a boon. If it is easy to fill a mould, then it would also be expected that it would be easy for the material to flash over into the gaps at the mould parting surfaces (Fig. 4.23).

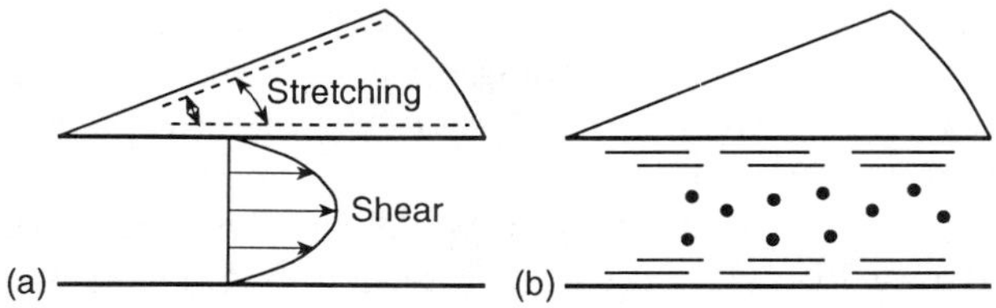

Figure 4.22 Radial and hoop orientation in centre grated disc mould: (a) combination of shear and stretching; (b) orientation pattern radially oriented on surface, transversely oriented along the centre.

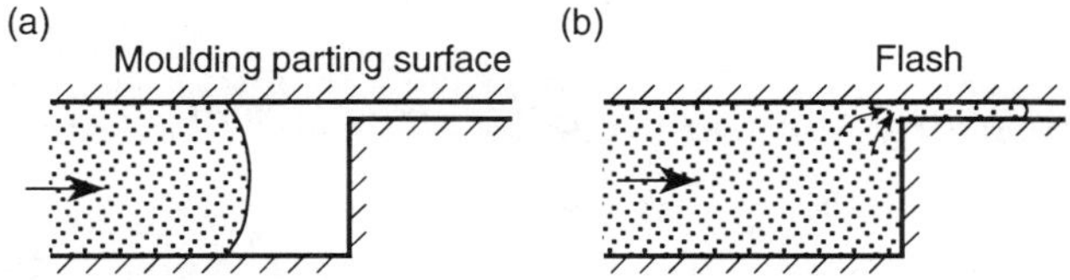

Figure 4.23 Flash in injection moulding: (a) material filling; and (b) when mould is full, high pressure tends to open the mould parting surface allowing the material to form flash.

Although this can be a problem if excess pressure is applied during mould filling, under optimum conditions it is much less than would be expected for a conventional isotropic melt of the same viscosity [105]. The rheological reason is that the flow into the flash region is a strong converging flow, similar to that at the entrance to a capillary which gives rise to the Bagley correction. Because the entrance pressure drop for liquid crystal melts is much higher than for isotropic melts, flow into the flash region is effectively constrained.

By far the most perplexing problem with injection moulding of LCPs is the control of weld lines in the material [106–108]. The problem is most clearly seen in the case of two advancing fronts meeting (Fig. 4.24).

Because of the fountain flow at the front, the polymer in the knit line ends up being oriented normal to the plane of the moulding – the weakest possible result. Other weld lines (e.g. when flows come together

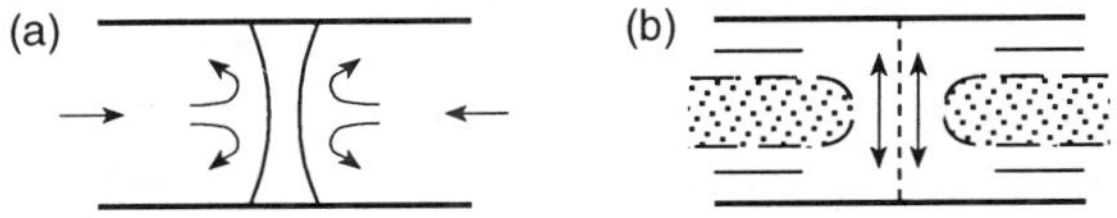

Figure 4.24 (a) Two advancing fronts about to meet in an injection moulding; (b) the likely resulting orientation pattern.

after passing an obstruction) are less severe, but even in those cases there is no driving mechanism to encourage the molecules to entangle across such a join. Because of the low viscosity of the melt, weld lines due to the need for multi-gating are rare. If multi-gating is necessary in order to fill a mould completely, then it is possible to eliminate weld lines by sequential gating in which the new gate is only opened once the flow path has passed that point. The avoidance of weld lines in complex moulds is less easy. Weld lines are always going to be points of weakness, and their positioning must be carefully considered when the mould is under design. Computer modelling of the mould-filling process can be useful for predicting weld line locations and designing moulds so as to place them in non-critical areas of the part [109].

4.4.4 Processing comparisons for LCPs and fibre reinforced plastics

Many of the issues encountered with processing of thermotropic melts are common to those of fibre reinforced plastics [110]. This is especially true of the control of anisotropy [111] and the problems associated with weld lines. Similar design strategies can be used with the two systems. Less dramatic, but no less important, are the positive aspects of good dimensional tolerance and low warpage: in fibre reinforced plastics, it is the presence of the fibre which suppresses the elasticity of the melt and keeps thermal expansion low.

4.5 CONCLUSIONS

4.5.1 Key characteristics of LCP melts

The rheological properties that are most significant for the processing of thermotropic polymers are:

1. low shear viscosity at high shear rates and high viscosity at low shear rates;
2. high extensional viscosity (especially its manifestation as a high Bagley end correction);
3. low die swell;
4. rapid solidification;
5. and, good dimensional stability.

The low viscosity at high shear rates is largely responsible for the ability to fill large moulds [30] and also moulds with long thin-walled flow paths [112]. Of course, the viscosity of any polymer can be reduced by decreasing its molecular weight. However, reduction of the molecular weight of isotropic polymers to a viscosity comparable to that of the thermotropic polymers also decreases their strength and toughness. The rod-like extension and rigidity of the thermotropic polymers allows them to achieve a low viscosity while retaining the necessary service performance.

A relatively high viscosity at low shear rates is also a desirable property. It is required for the melting mechanism in extruder screws, which relies on viscous dissipation to generate heat. High viscosity is also required for free surface processing operations (e.g. profile extrusion or melt spinning) in which it imparts resistance against sagging under the force of gravity but, unfortunately, it also impedes fusion across a weld line.

The high extensional viscosity, as manifested by the Bagley end correction in capillary flow, is useful in injection moulding of thin-walled parts (e.g. electronic connectors) because it retards the formation of undesirable flash [105, 113].

Low die swell is an advantage in extrusion of profiles with well-defined shapes. Because it favours jetting, resulting in undesirable and uncontrollable multiple weld-lines, it necessitates careful design and placement of gates in injection moulds.

The degree of crystallinity is low, requiring little removal of heat for crystallization to occur; this contributes to a rapid moulding cycle. On the other hand, that rapid solidification is also partly responsible for the difficulty of knitting strong weld lines. Another property, which also favours rapid processing, is the high rigidity below the crystallization temperature.

The small volume change on solidification, the high rigidity, and the low level of stored elastic energy all contribute to the excellent dimensional stability of moulded parts, particularly their small tendency to warp.

4.5.2 Similarities of LCPs and fibre reinforced melts

A number of similarities of the rheology and processing behaviour of thermotropic LCP melts and fibre reinforced isotropic melts have been alluded to, notably in sections 4.1.2, 4.3.5 and 4.4.4. To recapitulate:

- both have high shear sensitivity of viscosity, and large capillary flow end correction (elongational viscosity);
- both exhibit low die swell;
- both give highly anisotropic products from which the orientation does not relax;
- and, both have excellent dimensional stability.

The similarities are not constrained to the practical field; in respect of theory, the two-fluid model [83], which explains details of the flow for

LCPs, may also be applicable to the flow of fibre suspensions, which have the tendency to form a low concentration low viscosity layer near the wall.

The similarities of rheology, and the existence of strong anisotropy, also lead to similarities in processing. As a result of the interaction of the rheology of the material with the complex flow field during mould filling, both systems may produce mouldings with multiple laminae having different orientation directions. Both also have a tendency to weld line difficulties. A principal difference, in comparison to unfilled isotropic melts, is the low viscosity of LCP melts and the high viscosity of fibre filled melts. Nevertheless, even here there is an intellectually satisfying analogy in the surprising ease of processing of melts filled with long fibres, in which the high aspect ratio of the filler forces a parallel alignment, similar to the nematic ordering of rod-like molecules.

In a more general sense, both systems are examples of melts with 'structure'. Other members of that family include:

- immiscible polymer blends;
- phase-separated block copolymers;
- and, melts with residual microcrystallinity, such as polyvinyl chloride.

They have in common that their structure affects their rheological behaviour. In turn, the structure is affected by thermal and mechanical history. Sample preparation itself imposes a thermo-mechanical history which must be taken into account. In addition, measurements may require study of the details of the flow field, rather than making the assumptions that are used to derive the classical equations. The interaction between measurement and history complicates analysis of flow behaviour; it requires simultaneous measurement of both structure and rheology in order to understand the behaviour and properties of such systems.

Because of these interrelationships, the scientific benefit of studying LCPs, as a model for the study of other structured melts, may transcend the direct commercial utility of those investigations. At the same time, the study of what is apparent in the organization of reinforcing fibres in a filled polymer melt may give insights into the molecular organization and domain structure of the liquid crystal state.

REFERENCES

1. Cogswell, F.N. (1981) *Polymer Melt Rheology*, George Godwin, Ch 4.
2. Dealy, J.M. and Wissbrun K.(1990) *Melt Rheology and its Role in Plastics Processing*, van Nostrand Reinhold.
3. Hanson, D.E. (1969) Shear modification of polythene. *Polymer Engineering and Science*, **9**(6), 405–14.
4. Utracki, L.A., Bakerdjian, Z. and Kamal, M.R. (1975) The flow behavior

of semirigid and rigid poly(vinyl chloride) formulations. *Transactions of the Society of Rheology*, **19**, 173–99.

5. Busse, W.F. (1967) Mechanical structures in polymer melts. *Journal of Polymer Science A2, Polymer Physics*, **5**, 1249–81.
6. International Union of Pure and Applied Chemistry, Working Party on Structure and Properties of Commercial Polymers. (1980) *Sub-primary particles in PVC: identification and elucidation of their role during flow*. Proceedings of the 3rd International Conference on PVC, Cleveland.
7. Biais, R., Geny, C., Mordini, C. *et al.* (1980) X-ray and infrared investigation of the supramolecular structure of PVC. *British Polymer Journal*, **12** (4), 179–91.
8. Kraus, G. and Gruver, J.T. (1967) Properties of random and block copolymers of butadiene and styrene: melt flow. *Journal of Applied Polymer Science*, **11**, 2121–9.
9. Arnold, K.R. and Meier, D.J. (1970) A rheological characterisation of SBS block copolymers. *Journal of Applied Polymer Science*, **14**, 427–40.
10. Gihjsels, A. and Raadsen, J. (1979) *The melt rheology of Cariflex TR1102*. International Union of Pure and Applied Chemistry, Working Party on Structure and Properties of Commercial Polymers.
11. Folkes, M.J. (1971) The birefringence and mechanical properties of a'single crystal' from a three block copolymer. *Polymer*, **12**, 222–33; and (1976) **17**, 495–500.
12. Jackson, W.J. and Kuhfuss, H.F.(1976) Liquid crystal polymers – 1: preparation and properties of p-hydroxybenzoic acid copolyesters. *Journal of Polymer Science, Polymer Chemistry*, **14**, 2043–58.
13. McFarlane, F.E., Nicely, V.A. and Davis, T.G. (1979) Liquid crystal polymers –2. *Contemporary Topics in Polymer Science*, **3**, 109–41.
14. Cox, M.K. (1987) Liquid crystal polymers. *RAPRA Review*, **1** (2), Report 4, including 454 references.
15. Jackson, W.J. (1980) Liquid crystal polymers – 4: liquid crystal aromatic polyesters. *British Polymer Journal*, **12** (4), 154–62.
16. Griffin, B.P. and Cox, M.K. (1980) Thermotropic polyesters with non-linear links. *British Polymer Journal*, **12** (4), 147–53.
17. Jin, J.-I., Antoin, S., Ober, C. *et al.* (1980) Thermotropic liquid crystalline polyesters with rigid or flexible spacer groups. *British Polymer Journal*, **12** (4), 132–46.
18. McIntyre, J.E. and Milburn, A.H. (1981) Thermotropic liquid-crystalline polyesters and polyesteramides. *British Polymer Journal*, **13** (1), 5–10.
19. Chung, T.S. (1986) Recent developments of thermotropic liquid crystalline polymers. *Polymer Engineering and Science*, **26** (13), 901–19.
20. Chen, J., White, J.L. and Kyu, T. (1986) Rheology, extrusion and melt-spinning characteristics of thermotropic liquid crystalline cellulose ethers. *SPE Antec*, 723–5.
21. Flory, P.J., (1956) Phase equilibria in solutions of rod-like particles. *Proceedings of the Royal Society*, **A234**, 73–88.
22. Cogswell, F.N. (1987) The next generation of injection moulding materials. *Plastics and Rubber International*, **12** (3), 36–9.
23. Ophir, Z. and Ide, Y. (1983) Injection molding of thermotropic liquid crystal polymers. *Polymer Engineering and Science*, **23**, 792–801.
24. Garg, S.K. and Kenig, S. (1988) Development of orientation during processing of liquid crystalline polymers, in *High Modulus Polymers*, A.E. Zachariades and R.S. Porter (eds), Marcel Dekker, New York, 71–103.

25. Hsiung, C.M., Tian, J. and Cakmak, M. (1993) Detailed investigation of structural layering phenomena in injection molded thermotropic LCP. *International Polymer Processing*, **8**, 164–77.
26. Beery, D. and Denn, M.M. (1993) Weldline strength in a liquid crystalline polymer. Paper presented at 65th Annual Meeting of The Society of Rheology, Boston, October 17–21.
27. Wissbrun, K.F., Kiss, G. and Cogswell, F.N. (1987) Flow behaviour of a thermotropic liquid crystal aromatic polyester. *Chemical Engineering Communications*, **53**, 149–73.
28. Cocchini, F., Nobile, M.R. and Acierno, D. (1991) Transient and steady rheological behavior of the thermotropic liquid crystal copolymer 73/27 HBA/HNA. *Journal of Rheology*, **35**, 1171–89.
29. Wissbrun, K.F. (1980) Observations on the melt rheology of thermotropic aromatic polyesters. *British Polymer Journal*, **12** (4), 163–9.
30. Cogswell, F.N. (1983) Observations on the rheology of thermotropic polymer liquid crystals, in *Recent Advances in Liquid Crystals*, L. Chapoy (ed.), European Science Foundation Sixth Polymer Workshop, Copenhagen.
31. La Mantia, F.P. and Valenza, A. (1989) Shear and elongational characterization of a liquid crystal polymer. *Polymer Engineering and Science*, **29**, 625–32.
32. Turek, D.E., Simon, G.P. and Tiu, C. (1992) The effect of die aspect ratio on the rheological properties of a thermotropic copolyester. *Journal of Rheology*, **36** (6), 1057–78.
33. Gotsis, A.D. and Baird, D.G. (1985) Rheological properties of copolyester liquid crystalline melts – 2: Comparison of capillary and rotary rheometer results. *Journal of Rheology*, **29**, 539–56.
34. Frayer, P.D. and Huspeni, P.J. (1990) Processing preshear and residence time effects on the rheology of an LCP melt. *Journal of Rheology*, **34** (8), 1199–215.
35. Calundann, G. and Jaffe, M. (1982) *Anisotropic polymers, their synthesis and properties*. Proceedings of the Robert A. Welch Conference on Chemical Research, XXVL, Synthetic Polymers, Houston.
36. Cogswell, F.N. and Smith, C.P. (1980) Anisotropic melt-forming polymer, European Patent Office publication number 0 033 423.
37. Graziano, D.J. and Mackley, M.R. (1984) Shear induced optical textures and their relaxation behavior in thermotropic liquid crystalline polymers. *Molecular Crystals and Liquid Crystals*, **106**, 73–85.
38. De'Nève, T., Navard, P. and Kléman, M. (1993) Shear rheology and shear-induced textures of a thermotropic copolyesteramide. *Journal of Rheology*, **37**, 515–29.
39. Sawyer, L.C. and Jaffe, M. (1986) Structure of thermotropic polyesters. *Journal of Materials Science*, **21** (6), 1897–913.
40. Onogi, S. and Asada, T. (1980) *Rheology and rheo-optics of polymer liquid crystals*. Proceedings of the VIII International Congress on Rheology, Naples, 127–47.
41. Wissbrun, K.F. (1981) Rheology of rod-like polymers in the liquid crystalline state. *Journal of Rheology*, **25** (6), 619–62.
42. Turek, D.E. and Simon, G.P. (1992) Properties of a semi-crystalline and an amorphous thermotropic liquid crystalline polymer. *Polymer International*, **27**, 165–75.

43. Wissbrun, K.F. and Ide, Y. (1982) Processing of melt processible liquid crystal polymer by control of thermal history, US Patent 4,325,903.
44. Karis, T.E., Park, O.O. and Yoon, D.Y. (1992) Rheology of thermotropic liquid crystalline polyester Vectra E. *Journal of Rheology*, **36** (8), 1587–603.
45. Lin, Y.G. and Winter, H.H. (1991) High-temperature recrystallization and rheology of a thermotropic liquid crystalline polymer. *Macromolecules*, **24**, 2877-82.
46. Wissbrun, K.F. and Griffin, A.C. (1982) Rheology of a thermotropic polyester in the nematic and isotropic states. *Journal of Polymer Science: Polymer Physics Edition*, **20**, 1835–45.
47. Kim, S.S. and Han, C.D. (1993) Effect of thermal history on the rheological behavior of a thermotropic liquid crystalline polymer. *Macromolecules*, **26**, 3176–86.
48. Kim, S.S. and Han, C.D. (1993) Oscillatory shear behavior of a thermotropic liquid-crystalline polymer, private communication.
49. Cogswell, F.N. and Groves, D.J. (1988) *The melt rheology of continuous fibre reinforced structural composite materials*. Proceedings Xth International Congress on Rheology, Sydney.
50. Bailey, R.S., Bellamy, A.M., Groves, D.J. *et al.* (1992) *The characterization of shear flow in discontinuous fibre filled thermoplastic compounds*. Proceedings XIth International Congress on Rheology, Brussels, 838–40.
51. Benbow, J.J., Cogswell, F.N. and Cross, M.M. (1976) On the dynamic response of viscoelastic fluids. *Rheologica Acta*, **15** (5), 231–7.
52. Guskey, S.M. and Winter, H.H. (1991) Transient shear behaviour of a thermotropic liquid crystalline polymer in the nematic state. *Journal of Rheology*, **35** (6), 1191–207.
53. Graessley, W.W. (1974) The entanglement concept in polymer rheology. *Advances in Polymer Science*, **16**, 154.
54. Cocchini, F., Nobile, M.R. and Acierno, D. (1991) Transient and steady rheological properties of the thermotropic liquid crystal polyester 73/27 HBA/HNA. *Journal of Rheology*, **35** (6), 1171–89.
55. Viola, G.G. and Baird, D.G. (1986) Studies on the transient shear flow behavior of liquid crystalline polymers. *Journal of Rheology*, **30**, 601–28.
56. Dealy, J.M. and Tsang, W. K.-W. (1981) Structural time dependency in the rheological behavior of molten polymers (and references therein). *Journal of Applied Polymer Science*, **26**, 1149–58.
57. Meissner, J. (1991) Scherverhalten von Polymerschmelzen, Proceedings of Tagung Werkstoffe für die Bedürfnisse von morgen, April 17/18, reprinted in *Technische Rundschau*, Hallweg AG, Bern.
58. Meissner, J. (1992) Experimental problems and recent results in polymer melt rheometry. *Makromolecular Chemistry, Macromolecular Symposia*, **56**, 25–42.
59. Kim, S.S. and Han, C.D. (1993) Transient rheological behavior of a thermotropic liquid-crystalline polymer in the nematic state – 1: The start-up of shear flow. *Journal of Rheology*, **37**, 847–66.
60. Moldenaers, P., Fuller, G.G. and Mewis, J. (1989) Mechanical and optical rheometry of polymer liquid-crystal domain structure. *Macromolecules*, **22**, 960–5.
61. Burghardt, W.R. and Fuller, G.G. (1990) Transient shear flow of nematic liquid crystals: manifestations of director tumbling. *Journal of Rheology*, **34**, 959–92.

62. Prasadarao, M., Pearce, E.M. and Han, C.D. (1982) Rheological properties of liquid crystalline polyesters. *Journal of Applied Polymer Science*, **27**, 1343–54.
63. Gotsis, A.D. and Baird, D.G. (1986) Primary normal-stress difference for two liquid crystalline copolyesters. *Rheologica Acta*, **25**, 275–86.
64. Cocchini, F., Nobile, M.R. and Acierno, D. (1992) Letter: About negative first normal stress differences in a thermotropic liquid crystalline polymer. *Journal of Rheology*, **36** (7), 1307–11.
65. Winter, H.H. and Wedler, W. (1993) Note: About measuring the first normal stress difference in shear flow of a thermotropic copolyester. *Journal of Rheology*, **37** (2), 409–12.
66. Meissner, J., Garbella, R.W. and Hostettler, J. (1989) Measuring normal stress differences in polymer melt shear flow. *Journal of Rheology*, **33**, 843–64.
67. Hutton, J.F. (1975) On using the Weissenberg rheogoniometer to measure normal stresses in lubricating greases as examples of materials which have a yield stress. *Rheologica Acta*, **14**, 979–92.
68. Zapas, L.J., McKenna, G.B. and Brenna, A. (1989) An analysis of the corrections to the normal force response for the cone and plate geometry in single-step stress relaxation experiments. *Journal of Rheology*, **33**, 69–91.
69. Yang, D.-K. and Krigbaum, W.R. (1989) Rheological studies of thermotropic liquid crystal copolyesters: p-hydroxybenzoic acid/2-hydroxy-6-naphthoic acid. *Journal of Polymer Science Part B: Polymer Physics*, **27**, 819–35.
70. Nuel, L. and Denn, M.M. (1991) Effects of processing and particulate fillers on the rheology of a nematic polymer melt. *Rheologica Acta*, **30**, 65–70.
71. Giles, D.W. and Denn, M.M. (1994) An experimental study of the rheology of thermotropic liquid crystalline copolyesters using a novel pressurized capillary rheometer. *Journal of Rheology*, **38**, 617–37.
72. Izu, P., Muñoz, M.E., Peña, J.J. *et al.* (1993) Capillary and slit-die flow of two liquid-crystalline polymers. The effect of pressure on viscosity. *Journal of Polymer Science, Part B: Polymer Physics*, **31**, 347–56.
73. Larson, R.G. and Mead, D.W. (1989) Time and shear-rate scaling laws for liquid crystal polymers. *Journal of Rheology*, **33**, 1251–81.
74. Wissbrun, K.F. (1985) A model for domain flow of liquid-crystal polymers. *Faraday Discussions of the Chemical Society*, **79**, 161–73.
75. Marrucci, G. (1991) Tumbling regime of liquid-crystalline polymers. *Macromolecules*, **24**, 4176–82.
76. Laun, H.M. (1986) Predictions of elastic strains of polymer melts in shear and elongation. *Journal of Rheology*, **30**, 459–501.
77. Larson, R.G. and Doi, M. (1991) Mesoscopic domain theory for textured liquid crystalline polymers. *Journal of Rheology*, **35**, 539–63.
78. Jarman, R.E. and Baird, D.G. (1981) Rheological properties of copolyester liquid crystalline melts – I. Capillary rheometry. *Journal of Rheology*, **25**, 275–92.
79. Zaldua, A., Muñoz, S., Peña, J.J. *et al.* (1992) Slit die flow measurements of a liquid crystalline polyesteramide and its blends with polyarylate. *Polymer Engineering and Science*, **32**, 43–8.
80. Wissbrun, K.F. (1994) Note: Negative extrudate swell of liquid crystal polymers. *Journal of Rheology*, **38**, 247–52.

81. Kalika, D.S., Nuel, L. and Denn, M.M. (1989) Gap dependence of the viscosity of a thermotropic LCP. *Journal of Rheology*, **33**, 1059–70.
82. Turek, D.E. and Simon, G.P. (1992) Properties of a semi-crystalline and amorphous thermotropic liquid crystalline polymer. *Polymer International*, **27**, 165–75.
83. Wissbrun, K. (1993) A two fluid model for inconsistency of observations in capillary rheometry of thermotropic copolyesters. *Journal of Rheology*, **37**, 777–97.
84. Lefeuvre, P., De'Nève, T. and Navard, P. (1994) Bagley plot of the 'Vectra' B950 thermotropic polymer. *Journal of Rheology*, **38**, 169–75.
85. Wilson, T.S. and Baird, D.G. (1992) Transient elongational flow behaviour of thermotropic liquid crystalline polymers. *Journal of Non-Newtonian Fluid Mechanics*, **44**, 85–112.
86. Turek, D.E., Smejkal, F., Acierno, D. *et al.* (1993) Transient isothermal elongational flow of thermotropic liquid crystalline polymers. *Polymer Communications*, **34**, 204–206.
87. Gregory, D.R. (1972) Rheological properties of molten poly(ethylene terephthalate). *Journal of Applied Polymer Science*, **16**, 1479–87.
88. Cogswell, F.N. (1973) The influence of pressure on the viscosity of polymer melts. *Plastics and Polymers*, **41**, 30–43.
89. Cogswell, F.N. (1980) On the formation of a low viscosity state in stiff chain polymers. *British Polymer Journal*, **12** (4), 170–3.
90. Drappel, S., Yeung, B.W.A. and Sundararajan, P.R. (1993) Rheological and morphological studies of a thermotropic liquid crystalline polymer with low temperature transitions. *Journal of Rheology*, **37** (1), 89–101.
91. Gochanour, C.R. and Weinberg, M. (1986) Studies of the time dependent melt rheology of a liquid crystalline polymer. *Journal of Rheology*, **30**, 101–24.
92. Done, D. and Baird, D.G. (1987) Effect of thermal history on the rheology and texture of a thermotropic liquid crystal polymer. *Polymer Engineering and Science*, **27**, 816–22.
93. Lin, Y.G. and Winter, H.H. (1991) High-temperature recrystallization and rheology of a thermotropic liquid crystalline polymer. *Macromolecules*, **24**, 2877–82.
94. Driscoll, P., Masuda, T. and Fujiwara, K. (1991) Rheological properties of a homogeneous thermotropic liquid crystalline polyester: dynamic viscoelastic and interrupted-flow measurements. *Macromolecules*, **24**, 1567–74.
95. Geiger, K. (1990) *Influence of the shear and thermally induced texture on the rheological properties of liquid crystalline polymer melts*. Paper 10–09, Abstracts of the 6th Annual Meeting, The Polymer Processing Society, Nice, April 17–20.
96. Rietveld, J.X. and Tangyuengyong, P. (1992) Bagley and Mooney corrections for a HBA/HNA thermotropic copolyester measured using an extruder-based capillary rheometer. Report from the Polymer Processing Research Group, Department of Mechanical Engineering, University of Wisconsin-Madison.
97. Nuell, L. and Denn, M.M. (1991) Effect of processing and particulate fillers on the rheology of a nematic polymer melt. *Rheologica Acta*, **30**, 65–70.
98. Farell, G.W. and Fellers, J.F. (1986) Rotating annular die to control biaxial

orientation in melt processed liquid crystalline cellulose derivatives. *Journal of Polymer Engineering*, **6** (1–4), 263–89.

99. Cogswell, F.N. (1992) *Thermoplastic Aromatic Polymer Composites*, Butterworth Heinemann, section 3.2.3.
100. Wissbrun, K.F. and Orwoll, R.D. (1985) Method for the formation of composite articles comprised of thermotropic liquid crystalline polymers and articles produced thereby, US Patent 4,540,737.
101. Cogswell, F.N., Hezzell, D.J. and Williams, P.J. (1981) Fibre-reinforced compositions and methods of producing such compositions, US Patent 4,549,920.
102. Allan, P.S. and Bevis, M.J. (1987) Multiple live-feed injection moulding. *Plastics and Rubber, Processing and Applications*, **7**, 3–10.
103. Kaito, A., Kyotani, M. and Nakayama, K. (1993) Orientation distribution and layerlike morphology in extrusion-moulded sheets of a liquid crystalline copolyester amide. *Journal of Applied Polymer Science*, **48**, 2147–59.
104. McNally, D. (1977) Short fiber orientation and its effects on the properties of thermoplastic composite materials. *Polymer-Plastics Technology Engineering*, **8**, 101–54.
105. Wissbrun, K.F. (1991) Injection mold flashing of liquid crystalline polymers. *Polymer Engineering and Science*, **31** (15), 1130–6.
106. Engberg, K., Knutsson, A., Werner, P.-E. *et al.* (1990) Knit line fractures in injection molded liquid crystalline polymers. *Polymer Engineering and Science*, **30** (24), 1620–7.
107. Beery, D. and Denn, M. M. (1993) *Weldline strength in a liquid crystalline polymer*. Paper presented at 65th Annual Meeting of The Society of Rheology, Boston, October 17–21.
108. Engberg, K., Knuttson, A., Werner, P.-E. *et al.* (1990) Knit line fractures in injection moulded liquid crystalline polymers. *Polymer Engineering and Science*, **30**, 1620–9.
109. McChesney, C.E. (1993) quoted in Advanced thermoplastics producers regroup. *Chemical and Engineering News*, August 30th, 24–37.
110. Fisa, B. (1990) Injection moulding of thermoplastic composites, in *Composite Materials Technology*, P.K. Mallick and S. Newman (eds), Hanover, 265–320.
111. Bailey, R. and Rzepka, B. (1991) Fibre orientation mechanisms for long fibre composites, *International Polymer Processing*, **6** (1), 35–41.
112. Williams, G.E. (1989) Thermotropic liquid crystal polymers, in *Special Polymers for Electronics and Opto-Electronics*, M. Goosey (ed.), Elsevier, New York.
113. Reisch, M.S. (1993) Advanced thermoplastics producers regroup. *Chemistry and Engineering News*, August 30th, 24–37.

5

Rheological and relaxation behaviour of filled LC-thermoplastics and their blends

V.G. Kulichikhin, V.F. Shumskii and A.V. Semakov

5.1 INTRODUCTION

The use of liquid crystal polymers (LCPs) in blends with commercial thermoplastics for the modification of rheological and mechanical properties of the latter is a promising way for the development of a new generation of composite polymeric materials. Since we are primarily interested in the rheology, the decisive role of liquid crystal (LC) thermoplastics in blends can be illustrated by the results of papers [1–4], which were selected from numerous other papers as ascertaining that an LCP content of 3–30% brings about a considerable decrease in the blend viscosity, especially in capillary flow.

There are two more or less reasonable explanations of this effect. The first one originates from special conditions of LC jet formation in the convergent flow zone, penetrating the cross-section of the stream. The same situation takes place for the blend melts of usual incompatible polymers, but for them the final result will depend on the viscosity ratio of the components. If the viscosity of the dispersed phase is lower than the viscosity of the matrix, the resultant hydrodynamic resistance is lowered. In the opposite case, the blend viscosity can increase in

Rheology and Processing of Liquid Crystal Polymers
Edited by D. Acierno and A.A. Collyer
Published in 1996 by Chapman & Hall
ISBN 0 412 59640 7

comparison with the lower viscosity polymer. For the LC component, we have to keep in mind the viscosity and relaxation times anisotropies [5]. The point is that the apparent or 'bulk' viscosity measured in traditional rheological experiments 'consists' of at least two different coefficients: along and across the director. Inside the thin LC layers or jets, the viscosity in the direction flow will be essentially lower than the 'bulk' viscosity and, as a rule, lower than the melt viscosity of commercial thermoplastics. In other words, we can practically always expect the lowering of dissipative losses for blends containing LCPs.

For such blends, another effect was observed, namely migration of part of the LC phase to the periphery of the stream and the formation of an LC skin acting as a lubricant for the above mentioned reasons. Such enrichment of the outer layers of a stream by the LC component leads to a destruction of the parabolic velocity profile in capillary flow, realizing some elements of plug flow with uncertain kinematics [6].

The above mentioned approach is attributed to the dispersed phase, but another approach exists because of the matrix. This approach is based on the statement that, in polymer emulsions, 'gaps' between droplets of the dispersed phase exist in which increased local shear rates occur. For a viscoelastic matrix, this means a greater viscosity anomaly, leading to a decrease of the apparent blend viscosity [7]. In this case, the viscosity of the droplets must be higher than that of the matrix. The limiting examples of such two-component systems are suspensions containing solid particles. We believe that this factor will be much more pronounced for LCPs and their blends with commercial thermoplastics containing particulate and fibrillar fillers.

In spite of the interest of some huge industrial companies in filled LC thermoplastics (for example, some grades of Vectra or Rodrun contain fillers [8]), probably connected with some isotropization of LCPs during processing, systematic studies of rheological properties of filled LCPs are practically absent. We can comment only on the paper of Nuel and Denn [9] devoted to the copolyester of 2,6-hydroxynaphthoic and 4-hydroxybenzoic acids filled with carbon black (CB) and our early paper for the same filler and another matrix (copolyester of polyethyleneterephthalate with 4-hydroxybenzoic acid). The main result of these works consists of a decreasing viscosity of LCPs at low content of CB in comparison with neat polymers. So, at first sight, this result is similar to that obtained for polymer blends.

At the same time, we could not find literature sources in which filled polymer–polymer blends were described. We shall refer only to a study of rheological properties of a thermodynamically unstable two-phase blend of polyvinylacetate and ethylene-vinylacetate copolymer filled with silica [10]. The presence of particulate filler in such a blend increases the compatibility of the components. Blends of LCPs with flexible-chain polymers and the influence of fillers on their rheological behaviour are of interest from this standpoint as well.

In this work, the effect of particulate fillers on the rheological and relaxation (mechanical and dielectric) behaviour of filled LC-copolyesters and their blends with usual thermoplastic polymers has been investigated. There will be a brief discussion of the rheological behaviour of blends of LCPs with isotropic polymers in order to illustrate the effects of the fillers, but a fuller discussion of LC/isotropic polymer blends is given in Chapter 6.

5.2 EXPERIMENTAL

The LCPs used in this study were copolyesters (CPE) of polyethylene-terephthalate (PETP) and p-hydroxybenzoic acid (HBA) with a content of HBA of 70 (CPE-1) and 60% mol (CPE-2):

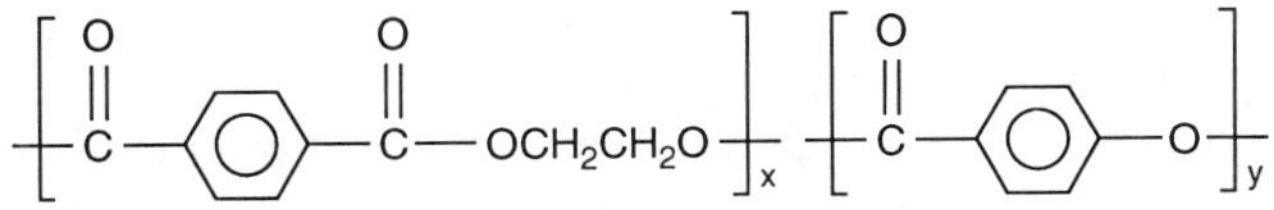

These were synthesized in the Research and Design Institute for Chemical Fibres and Composites (St Petersburg, Russia). The specific viscosity of these samples in a mixture of trifluoroacetic acid and chloroform was in the limits 0.5–0.8 dl g^{-1}. According to literature data [11–13], CPE has two main relaxations at $T_{g1} \simeq 70°C$ and $T_{g2} \simeq 160°C$, corresponding to the manifestation of large-scale mobility of chain segments enriched in PETP and HBA respectively. The CPE's softening (melting) points are between 190° and 220°C.

The thermally resistant polymer, aromatic polysulphone (PSF) was selected as the second component for the filled polymer-polymer blends. Its chemical formula is:

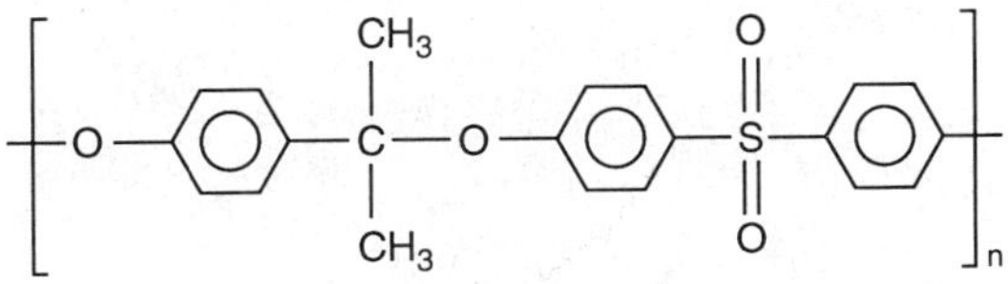

of molecular weight 32 000 and melt index 5.6 g/10 min. The glass transition point of PSF is approximately 220°C.

For the two-component mixtures, we used a set of fillers. As active structurizing fillers, CB and silica were chosen with specific surface areas of approximately 100 $m^2\,g^{-1}$ and 300 $m^2\,g^{-1}$ respectively. As an inactive filler, we used talc ($3MgO.4SiO_2.H_2O$) that had been put through a square mesh sieve (size of square 90 μm) and cut glass fibres. The talc particles are flaky hexagonal plates, with the length of the longest side greatly exceeding its thickness. The specific area of the talc, as calculated from the surface area of equivalent spheres [14], was approximately 10 $m^2\,g^{-1}$. Glass fibres (GF) were previously cut to sections of 2–2.5 mm in length. During a mixing procedure, they were crushed and the final dimensions of the glass particles in the mixture with the polymeric components were: d = 5.6 μm, l = 100–280 μm and l/d =18–50. As an example of GF distribution in the CPE-2, a micrograph is presented in Fig. 5.1. In this case, the GF content was 20%.

Prior to mixing, the fillers were dried for seven hours either in a high temperature furnace or at 120°C in a vacuum oven. The mixing was performed in a laboratory micromixer of a rotor-plunger type (the mixing volume is of 4 cm^3) at 230°C for approximately 30 s. The homogeneity of mixing was checked by optical and electron microscopy. Both the CPE-1 and the CPE-2 of various molecular weights were used as matrices for different fillers.

The same procedure was applied for the mixing of the three-component blends, but in this case the blends of CPE-2 and PSF were used as a matrix only for CB and GF. At the mixing temperature, both

Figure 5.1 Micrograph of CPE-1/glass fibres (80/20) mixture. Magnification × 145.

polymers have identical viscosities; this is essential for a uniform distribution of filler particles in the blend.

In preparing for rheological measurements, the compositions were pressed at 190°C into discs (32 mm in diameter and 0.7 mm thick) and into cylinders (9.4 mm in diameter and 15 mm high). These samples were used for testing the rheological properties in the regimes of rotational (Rheometer PIRSP-03, cone-and-plate operating unit, steady-state and low-amplitude oscillation modes of deformation [15]) and capillary flow (Microviscometer MV-2 of a melt indexer type, the capillary diameter is 1.26 mm and length 8.3 mm [16]) respectively.

For dynamic mechanical measurements of the solid samples (plates were prepared from extrudates by a thermoforming method at a pressure of 5 MPa and a temperature of 250°C), a spectrometer of an original design was used. A sketch of this apparatus is shown in Fig. 5.2. Steel bar 1 of cross-section 10.5 mm and length 40 cm was assembled on the soft suspension. The window was milled inside the middle of the bar where

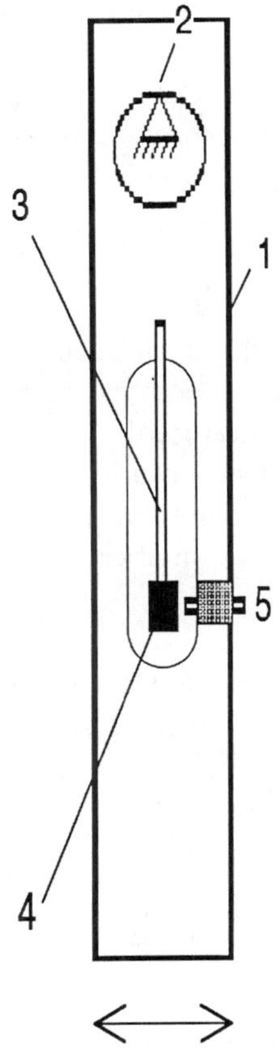

Figure 5.2 Sketch of apparatus for dynamic mechanical tests.

sample 3 is located. The cantilever sample is held by one end on the bar. Lead weight 4 is assembled by a forced fit on the other end. The mass of this load is chosen such that the natural frequency of oscillation of the loaded sample is within the working frequency range of the apparatus (10–240 Hz) in the temperature region −160°C to +240°C. Weight 4 acts as the electrode for the tuned-circuit instrument capacitor. Another insulated electrode is held stiffly on bar 1. The electromagnetic paraphase exciting system was applied to the bottom of the bar. The signal for excitation was from a precision low-frequency generator (GZ-110) with analogue and digital driving of frequency and amplitude. The oscillation period is measured by a digital frequency meter (ChZ-34). The amplitude of the forced oscillations of the sample is measured by a voltmeter (V2-34) after the signal from the variable capacitance transducer is detected and converted.

The traditional method of data processing for the 'tongue' mechanical spectrometer from the half-width of the resonance peak [17] has well-known disadvantages, especially at high levels of loss. For the version presented here, we can measure all values of loss tangent without artificial treatment.

The amplitude of induced oscillations for the linear mechanical system with one degree of freedom is expressed by the formula:

$$\alpha = \frac{F}{m\sqrt{(\omega^2 - \omega_0^2)^2 + \omega^2\omega_0^2 \tan^2\delta}}$$

where F is the amplitude of the impressed harmonic force acting on the system with frequency ω, m is the mass of the system, ω_0 is the natural frequency and $\tan\delta$ is the loss tangent.

If we measure at constant amplitude of external force, the amplitude of the induced oscillations for three various frequencies, it is possible to construct the system of three equations for the calculation of ω_0^2 and $\tan\delta$:

$$\omega_0^2 = \sqrt{\frac{(\alpha_1\alpha_2\omega_1\omega_2)^2(\omega_1^2 - \omega_2^2) + (\alpha_1\alpha_3\omega_1\omega_3)^2(\omega_3^2 - \omega_1^2) + (\alpha_2\alpha_3\omega_2\omega_3)^2(\omega_2^2 - \omega_3^2)}{\alpha_1^2\alpha_2^2(\omega_1^2 - \omega_2^2) + \alpha_1^2\alpha_3^2(\omega_3^2 - \omega_1^2) + \alpha_2^2\alpha_3^2(\omega_2^2 - \omega_3^2)}} \qquad (5.1)$$

$$\tan\delta = \frac{1}{\omega_0}\sqrt{\frac{\alpha_2^2(\omega_2^2 - \omega_0^2)^2 - \alpha_1^2(\omega_1^2 - \omega_0^2)^2}{\alpha_1^2\omega_1^2 - \alpha_2^2\omega_2^2}} \qquad (5.2)$$

where α_1, α_2 and α_3 are the amplitudes of the induced oscillations at frequencies ω_1, ω_2 and ω_3 respectively.

The elasticity modulus of the material is calculated using the formula:

$$E' = \frac{2m\omega_0^2 l^3}{\pi r^4} \tag{5.3}$$

where m is the mass of the load, l is the sample length and r is the radius of its cross-section.

The dc conductivity was measured with an electrical amplifer (U5-9); ac conductivity and electrical capacitance were measured with an automatic AC bridge (E7-9) at 1 kHz (all abbreviations refer to equipment manufactured in Russia).

The morphological and mechanical investigations were performed on extrudates prepared at fixed values of shear stress. The electron microscope observations were carried out on a scanning microscope (Tesla BS 301, Czechoslovakia). The tenacity characteristics of the extrudates at room temperature were obtained using an Instron 1122 testing machine (extension rate 10 mm min^{-1}).

5.3 RESULTS AND DISCUSSION

5.3.1 Rheology of the binary filled LCPs

The dependencies of the shear rate $\dot{\gamma}$ on the shear stress τ of CPE-1 and its mixtures with CB obtained on the PIRSP apparatus at 240°C and 260°C are presented in Fig. 5.3., which shows two families of flow curves. The main feature of the flow curves at 240°C is the presence of yield stresses as for CPE melts and mixtures. In our opinion, the viscoplastic behaviour of LCPs is as a consequence of their structure–orientation non-homogeneity, i.e. the presence of a disclination network. Its response to deformation action may be purely elastic [18]. After the destruction of this 'network' the value of yield stress decreases and for perfectly oriented LC samples it disappears as a whole [19]. The locations of flow curves for mixtures containing up to 10% of CB change monotonically in comparison with the curve for CPE-1. This is possibly due to the LC medium structure while the filler acts as a specific thickening agent. Unexpectedly, we obtained some indication of the stabilization of the flow for mixtures containing the low concentrations of CB (2–4%) as compared with CPE-1 flow, giving practically no oscillations of the torque during the measuring cycle and a smoother surface to the liquid jets in capillary flow (see below), and consequently, of the solid extrudates. The increase of viscosity in this range of CB content is insignificant. The increase of CB content up to 20% leads to a stiffening of the mixtures and a slipping of the material relative to the cone-and-plate at high shear stresses.

Another family of flow curves relates to 260°C and 280°C (the last

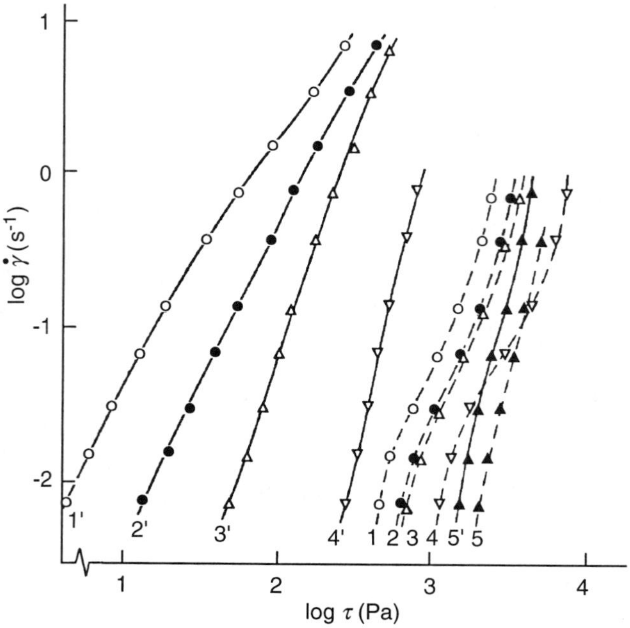

Figure 5.3 Flow curves of CPE-1/CB mixtures at temperatures 240°C (1–5) and 260°C (1′–5′) and CB content: 0(1,1′), 2(2,2′), 4(3,3′), 10(4,4′), and 20 wt% (5,5′).

case is not shown in the figure). For neat polymer, the yield stress is more pronounced, but the role of the filler in the manifestation of viscoplastic behaviour is more pronounced. With the increase of filler content, the slope of the log $\dot{\gamma}$/log τ graph increases as expected for filled flexible chain polymers [20, 21]. Such sharp differences in the rheological behaviour of CPE-1/CB mixtures at temperatures below and above 250°C indicates a change in the mixture structure at this temperature.

In Fig. 5.4, the dependencies of the relative viscosity η_r on filler content (C) are presented. At 240°C, the dependence $\eta_r(C)$ is practically linear and slightly growing with increasing CB concentration. Such a dependence is typical for a flexible chain polymeric matrix and roughly dispersed fillers, but in the present case we have a highly dispersed filler. On the contrary, at 260°C and 280°C the sharp increase in relative viscosity with filler content proceeds as for the usual polymers filled by structurizing agents. Moreover, the rate of viscosity growth increases at the transition from 260°C to 280°C.

In the same figure, the temperature dependencies of viscosity for different mixtures are presented as Arrhenius plots. They are non-linear and the apparent activation energy of CPE-1 is lowered with

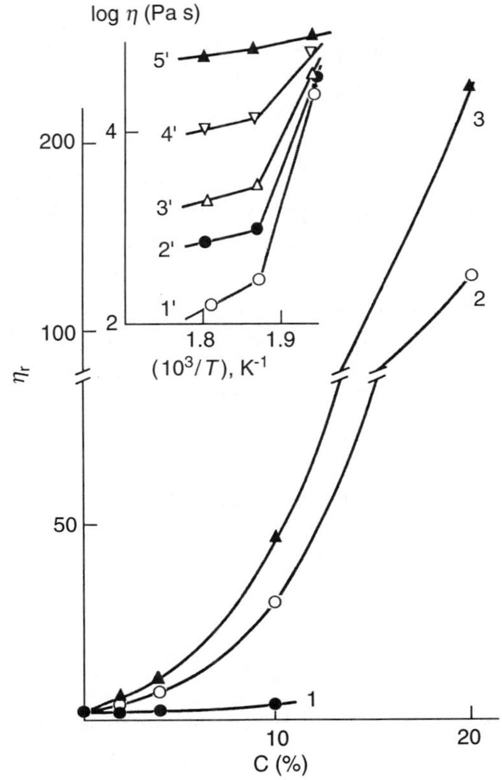

Figure 5.4 Concentration dependence of viscosity for CPE-1/CB mixtures at 240°C (1), 260°C (2) and 280°C (3) and viscosity versus temperature plot at different content of CB: 0(1′), 2(2′), 4(3′), 10(4′), and 20 wt% (5′).

increasing temperature from 510 kJ mol^{-1} at $T < 250$°C to 70 kJ mol^{-1} at $T > 250$°C.

All these data indicate a structural transition in CPE-1 in the temperature region 240–260°C. The main reason for such a transition may be the compositional heterogeneity of CPE-1, in particular the presence of the block sequences of HBA. We studied this process in detail [22] and showed that homopolycondensation of the hydroxy acid takes place during acetolysis and esterolysis of PETP. Special operations can help to decrease the fraction of the hydroxybenzoate blocks, but we have not rid ourselves of them completely. The existence of such blocks results in the preservation of the junctions of crystalline origin at the temperature region above the softening point of these copolyesters. This also gives two α-relaxation points for the species enriched by one of the components.

According to the rheological data, the melting of the residues of the crystalline junctions for CPE-1 takes place at approximately 250°C. Presence of the active filler prevents the blocks' moieties ordering; the difference between the activation energy values below and above 250°C decreases with the CB content and, at 20%, becomes negligible. This fact indicates the strong interaction of CPE-1 molecules with the surface of the filler, suppressing their capability of structural reforming. Besides, at higher CB contents and higher percolation concentration, the formation of a persistent cluster causes a sufficient contribution of such a skeleton to the elastic and dissipative properties of the mixtures.

The corresponding flow curves for CPE-2 and its mixtures with silica are shown in Fig. 5.5 (similar data are obtained for talc). The main distinction from the previous results consists in the less apparent viscoplastic behaviour of CPE-2 even at 240°C. It is likely that a decrease of HBA content in CPE decreases a fraction of the block structure, although two families of the flow curves remain. The $\log\dot{\gamma}/\log\tau$ dependencies at 240°C for mixtures containing up to 10% silica change in a symbiotic manner with respect to the same dependence for CPE-2. This apparently indicates that the structure of the LC medium predominates. The role of the silica particles or their aggregates causes an increase of the dissipative factor only.

At 260°C and 280°C, a tendency of the yield stress to appear is observed only at a silica concentration higher than 5%; for neat polymer, the yield stress is practically absent. In this case, as well as for CPE-1 in the high-temperature region, flow curves for the CPE-2 melt and its

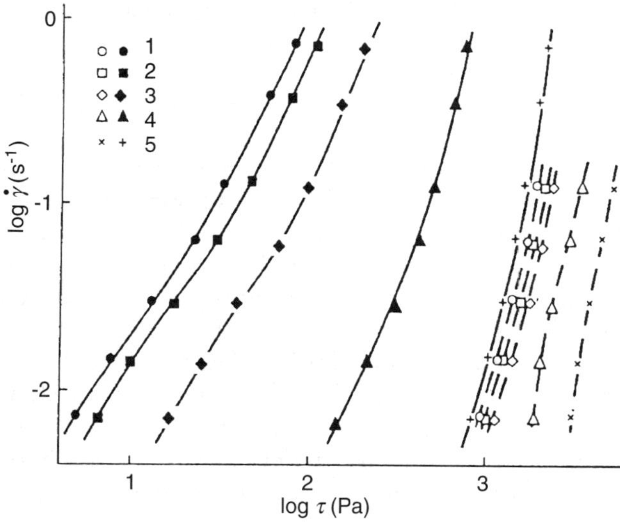

Figure 5.5 Flow curves of CPE-1/silica mixtures at 240°C (dashed lines) and 280°C (solid lines). Content of silica: 0(1), 1(2), 2(3), 5(4), and 10 wt% (5).

mixtures are not parallel, i.e. the dependence of viscosity on filler concentration is determined by shear rate. This indicates again a change of the CPE-2 structure at 240–260°C and a transition from two-phase structure to the homogeneous nematic structure. The role of the fillers in these temperature ranges is different. Useful information about this role can explain the data of dynamic mechanical spectroscopy in the melt state.

Let us consider now the results obtained under the conditions of harmonic oscillations at a low deformation amplitude. In Fig. 5.6, the values are shown of the storage and loss moduli as a function of frequency for mixtures of CPE-2 with talc at 240°C and 260°C. Similar results were obtained for other fillers and at 280°C. The shape and mutual locations of these dependencies $G'(\omega)$ and $G''(\omega)$ are very close to the corresponding ones presented in Fig. 5.6; that is why they are not shown here. From the data shown in Fig. 5.6, it follows that for dynamic characteristics, as well as for steady-state viscosity, two temperature ranges exist – below 250°C and above 250°C – where the different rheological properties are exhibited. At 240°C, the storage modulus and

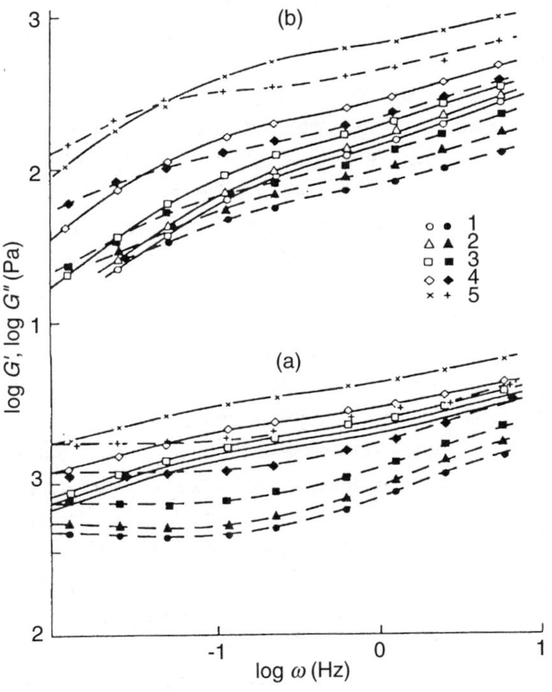

Figure 5.6 Storage modulus (solid lines) and loss modulus (dashed lines) as a function of frequency for CPE-1/talc mixtures at (a) 240°C and (b) 260°C. Content of talc: 0(1), 2(2), 4(3), 10(4), and 20 wt%.

the dynamic viscosity ($\eta' = G''/\omega$) are only slightly dependent on the filler concentration, and throughout the range of frequencies and concentrations the storage modulus exceeds the loss modulus. The magnitude of the slope of the storage modulus/frequency curve expressed in logarithmic form for the neat copolyester is 0.4 (in the range of low ω). However, this value decreases with an increase in the filler concentration.

It can be assumed that the corresponding plot for LC copolyester and its blends with fillers at 240°C is characterized by the specific relaxation plateau, which may differ in nature from the well-known rubber-like plateau typical of isotropic polymers with flexible chains. At 260°C and 280°C, the dynamic characteristics depend more strongly on frequency and on the concentration of fillers. In this case, the difference between the storage modulus and the loss modulus decreases and, at low frequencies, G' exceeds the storage modulus G''.

For this system, in the range of low ω and high temperatures, the power exponents of the functions, $G' \sim \omega^a$ and $G'' \sim \omega^b$, are smaller than for usual polymers. Thus, for the compositions of CPE-2 containing 10 wt% of talc at 260°C, $a = 0.86$ and $b = 0.45$. These values differ considerably from the theoretical values for flexible-chain polymers ($a = 2$ and $b = 1$ [23]) but are closer to values predicted for anisotropic systems, for which $a = b = 1$ [5, 24].

An attempt has been made to represent the dynamic characteristics as a function of frequency in generalized coordinates. With the aid of the classical WLF method for temperature–frequency reduction, it has been shown that the reduction holds true for the systems investigated only in the temperature region of 260–280°C. Reduction for the entire temperature range employed cannot be achieved because the CPE-2 itself and its blends below and above 250°C seem to be in different relaxation states. An example of the temperature–frequency reduction is given in Fig. 5.7 for a composition of CPE-2 with 10 wt% talc.

In a work of Lipatov *et al.* [25], an analogy was established for mechanical properties of various concentration–time (frequency) conditions for polymers containing fillers. This followed from the possibility of plotting a generalized curve describing the storage modulus as a function of frequency and filler concentration. The shift factor is the ratio of the relaxation times at different concentrations of the filler. We applied the same method [26] for the concentration–frequency reduction of the storage modulus as a function of frequency at various filler concentrations and temperatures (Fig. 5.8). The shift factor a_c is shown in Fig. 5.8 separately as a function of concentration. For the estimation of this factor, we used the ratio of frequencies ω_0 (CPE-2) and ω_c (filled melts) at $G' =$ constant. At 240°C, $\omega_0 = -0.64$, $\log G' = 3.2$; at 260°C, $\omega_0 = -0.64$, $\log G' = 1.9$; and at 280°C, $\omega_0 = -0.2$, $\log G' = 1.64$.

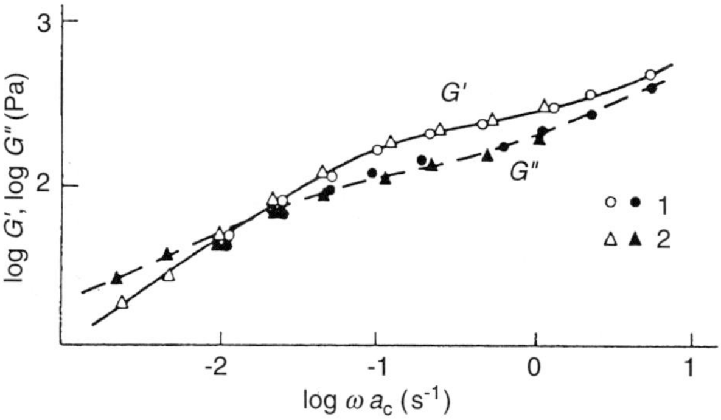

Figure 5.7 Temperature–frequency reduction of dynamic moduli for CPE-1 filled by 10% of talc at 260°C (1) and 280°C (2). Reduction temperature is 260°C, $\log a_T = -0.7$.

As can be seen, this approach for concentration–frequency reduction proves valid throughout the total concentration–frequency range at 240°C only. In other words, a unified concentration–frequency description of the dynamic storage modulus is possible at this temperature for fillers of different surface and structure-forming activities: structurizing CB and silica, and much less active talc. The polymer matrix is likely to be responsible for the relaxation properties of the filled compositions. It is important to note that the validity of such an approach renders it possible to predict the viscoelastic properties of CPE-2 containing the different fillers at any frequency. For other temperatures, this method works at low frequencies ($\log \omega < 0.1$), and then the divergence is evident, especially for highly concentrated dispersions.

Thus the temperature–frequency reduction is valid at $T > 250$°C where the 'usual' nematic melt is formed, while the concentration–frequency reduction works in the low-temperature region where the relaxation properties of the strong crystalline junctions of the polymeric matrix play an important part.

Under conditions of Poiseuille flow, we observed the unexpected effect of a decreasing CPE-2 viscosity upon the introduction of small amounts of the fillers. This effect is illustrated by the plot in Fig. 5.9 at 240°C, where the concentration dependencies of viscosity are presented for silica and talc (as we recall CB does not influence the CPE-1 viscosity). Similar effects have been described in papers [26–29] for LC and usual polymer matrices. These results were attributed to the formation of viscous surface layers around the dispersed particles, resulting in an increase of the free volume in the system [27, 28]. In the case of polymers that retain a globular (domain) structure in the molten state (e.g.

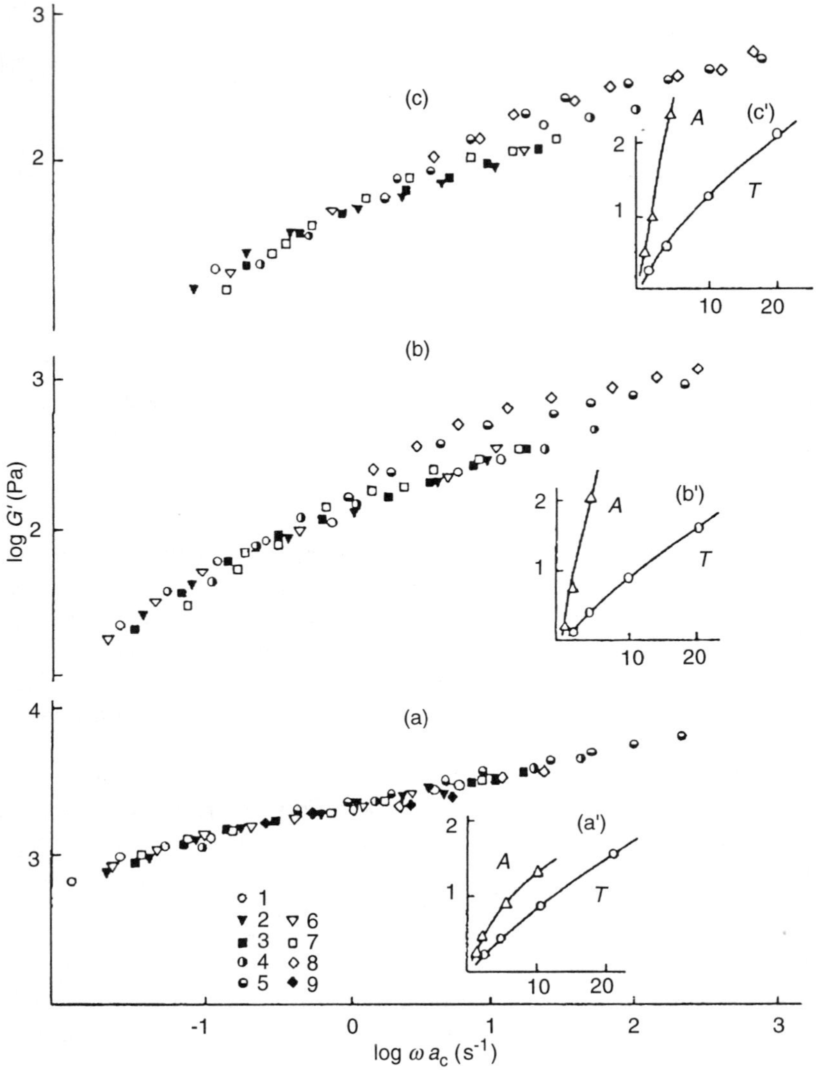

Figure 5.8 A generalized concentration–frequency description of the storage modulus for CPE-1 filled by talc (1–5) and aerosil (6–9) at (a) 240°C; (b) 260°C; and (c) 280°C. Content of talc: 0(1), 2(2), 4(3), 10(4), and 20 wt% (5); content of aerosil: 1(6), 2(7), 5(8), and 10 wt% (9). a′–c′ = dependencies of the factor of reduction log a_c on concentration of fillers.

polyvinylchloride), it was assumed that the viscosity decrease was caused by the possible accumulation of small amounts of filler in the inter-domain spaces destroying the links between domains [29].

In the case of LC matrices, we should not exclude another reason for

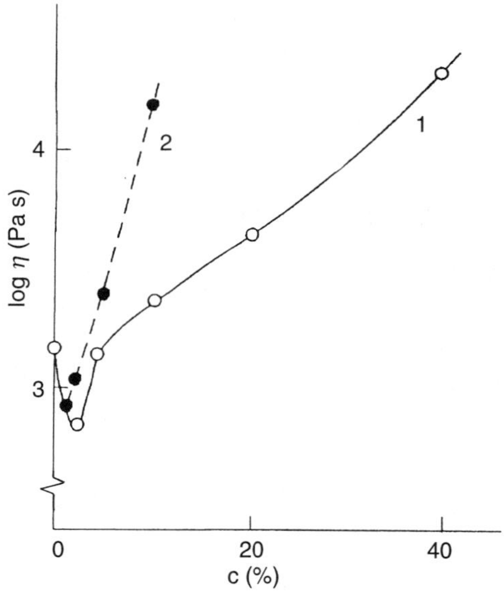

Figure 5.9 Viscosity in capillary flow versus content of talc (1) and silica (2) at $\log \tau = 3.87$ Pa and $T = 240°C$.

the viscosity decrease, namely the increase in molecular orientation, especially in the convergent flow in the presence of small amounts of fillers. Earlier, Nuel and Denn [9] observed such an effect for Vectra melts filled with CB in a Couette flow, but without a reasonable explanation. We shall return to a discussion of the orientation processes in flowing LC dispersions later, after a consideration of the mechanical characteristics of the extrudates obtained after capillary flow.

5.3.2 Comparison of mechanical and rheological properties of the binary filled systems

The dependencies of the relative values of tensile strength, σ, initial modulus of elasticity, E, and elongation at break, ε, of the products extruded at 240°C on filler concentration are shown in Fig 5.10. The magnitudes of σ, E and ε are normalized with respect to the corresponding characteristics of the unfilled matrix. The strength of the filled extrudates at room temperature slightly increases or remains unchanged up to approximately 1% silica, 2–4% CB and 6% talc, and upon further increases of filler concentration the strength gradually decreases. The

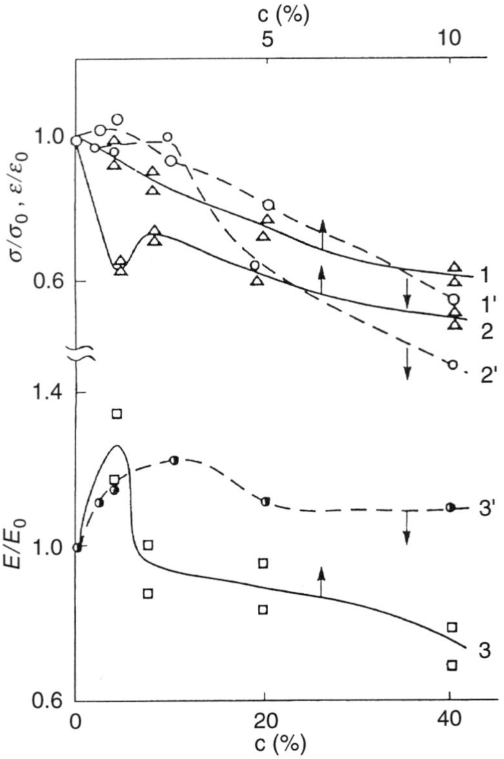

Figure 5.10 Mechanical properties of the extrudates filled by silica (1–3) and talc (1′–3′) $1, 1' = \sigma/\sigma_o$, $2, 2' = \varepsilon/\varepsilon_o$, $3, 3' = E/E_o$.

same also was obtained for commodity semi-crystalline thermoplastics reinforced by particulate fillers [30].

Regarding deformation at break, our results indicate that it either remains constant inside the narrow concentration interval then decreases, or it can start to decrease immediately at low filler loading. Electron micrographs of the fracture surfaces of extrudates containing different amounts of filler (in this case CB) can serve as a good illustration for understanding the reasons for the drastic decrease in ε (Fig. 5.11). The fibrillar morphology of the fracture surface is preserved up to 4% of CB. Inside this region, the tenacity and the elongation to break change negligibly. A transition to the brittle character of fracture is accompanied by a decrease of the ductile properties of the composites.

Of special interest is the dependence of the initial modulus of elasticity on the concentration. In the low concentration region of fillers, where the strength of the extrudates changes very little and the viscosity of the

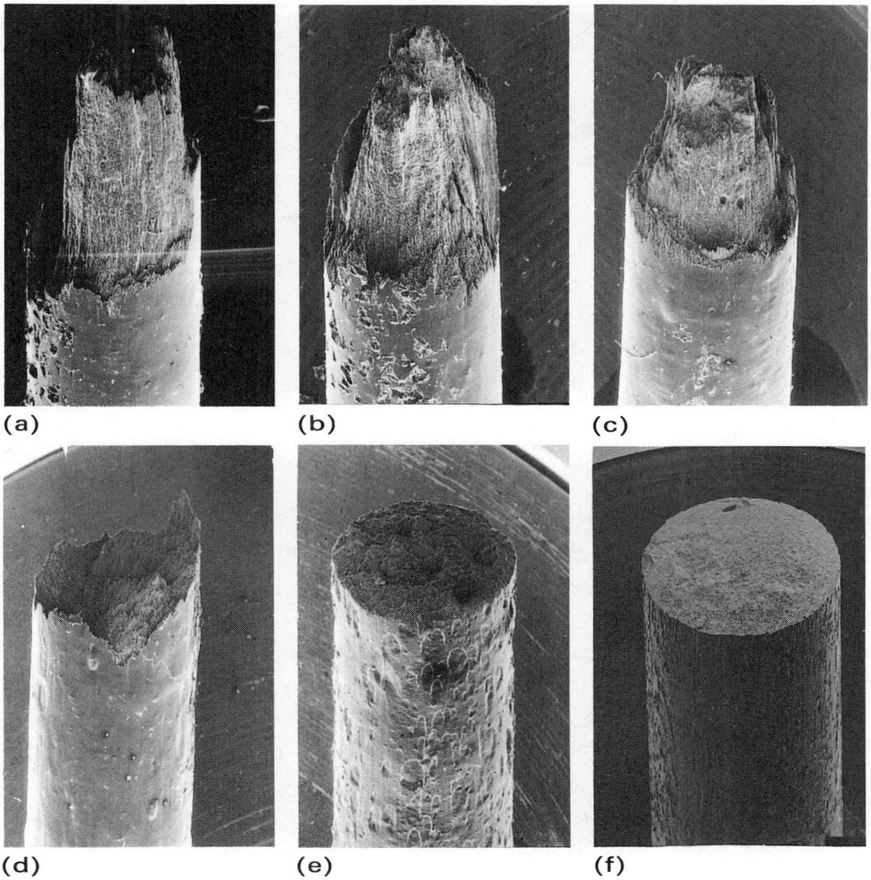

Figure 5.11 Fracture surfaces of the extrudates filled by (a) 0; (b) 2; (c) 4; (d) 10; (e) 20; and (f) 30 wt% of CB.

composites in the melt state decreases, the extension modulus of the solid composites increases, undergoing a turning point. The maxima of the elasticity modulus are slightly shifted in the direction of the higher filler concentrations as compared with the concentrations corresponding to the viscosity minima. Nevertheless, the order of such specific concentration shifts is the same as for the viscosity turning points: silica less than carbon black less than talc. It should be noted that the position of the maximum of elasticity modulus, as well as the viscosity minimum, along the concentration scale for LC polymers containing fillers is determined, first of all, by the specific surface of the filler particles (the degree of dispersity). It shifts into the regions of higher concentrations with a decrease in the specific surface.

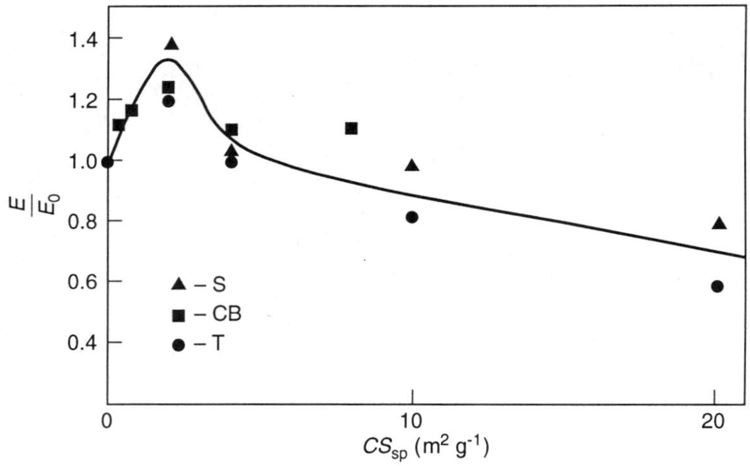

Figure 5.12 Reduced dependence of the extension elasticity modulus on the product of concentration of filler and its intrinsic surface for silica (S), carbon black (CB) and talc (T).

The confirmation of such a suggestion is the unified dependence of the relative modulus on the product of filler concentration and its specific surface (Fig. 5.12). The generalized curve shows the common tendency of an influence of the active filler on the mechanical properties of the extruded products.

A more complicated situation exists for the rheological properties. Depending on the chemical composition of the LC copolymers, and even on their molecular weight, two kinds of concentration dependence of viscosity are observed. The first one with a minimum only was shown in Fig. 5.9. Another is shown in Fig. 5.13 for the same matrix but with a lower molecular weight. In this case, at 240°C, we can observe two turning points, a maximum and a minimum. At 280°C, the viscosity of the CPE/CB mixtures remains practically unchanged for filler concentrations of 2–4%. Using the corresponding data for other fillers, we tried to construct the generalized concentration dependence of viscosity in the same coordinates as for the initial modulus (Fig. 5.14). There are two distinctions in the shape of the viscosity-concentration dependencies for the two polymers with various molecular weights but, as a rule, the viscosity minimum practically corresponds with the maximum in the elasticity modulus.

In spite of the fact that the surface nature, and hence the surface physical-chemical activity, is different for the particulate fillers used, it is enough for the absorption of CPE macromolecules on the particles' surfaces. Conditions for the formation of the pre-wall layer are also

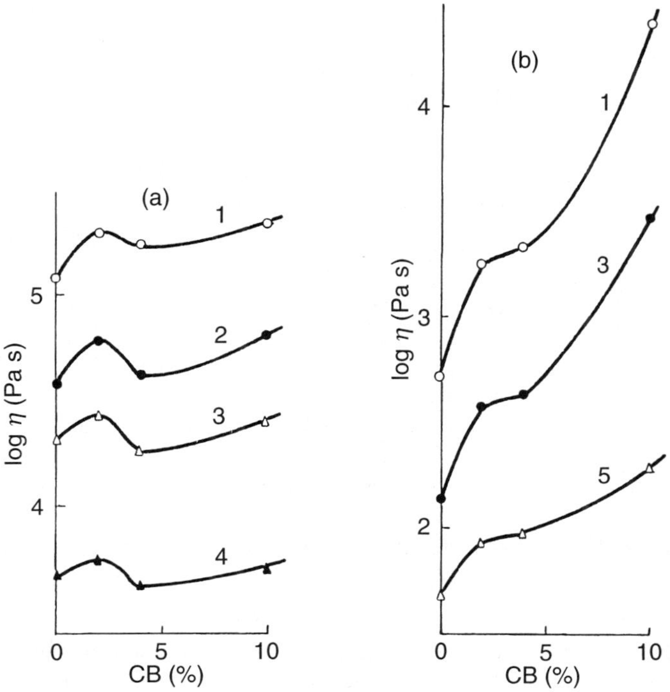

Figure 5.13 CPE-2 viscosity dependence on CB content at (a) 240°C; and (b) 280°C at various shear rates: 7.32×10^{-3} (1), 2.93×10^{-2} (2), 1.32×10^{-1} (3), 1.46(4), and 6.58 s^{-1} (5).

changing due to the variation of the melt viscosity (in the last example). So, the main reason for the common changes of rheological and mechanical properties may be the presence of the particulate fillers only. The role of the solid particles may be considered as additional sources for increasing the molecular orientation in the anisotropic melt. From this point of view, we can explain both effects: the decrease (or increase and decrease) in viscosity and the increase of the initial modulus.

Since the effects of the changes of these characteristics are most prominent in capillary flow, we can suggest a very simple approximate scheme taking into account the existence of the convergent flow zone. This scheme is presented in Fig. 5.15. Two situations are probable at low particulate filler concentration. The first describes the case when absorbed layers lead to increasing the effective hydrodynamic volumes of the particles without their touching (interpenetrating). This case is accompanied by a viscosity increase. The other concerns the partial or total overlapping of the absorbed layers of two neighbouring particles. Under

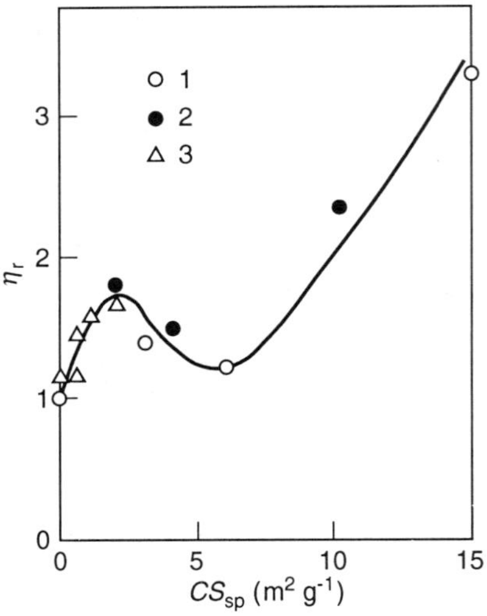

Figure 5.14 Generalized dependence of viscosity for CPE-2 filled by silica(1), CB(2), and talc(3) at 240°C.

the conditions of convergent (accelerating) flow, where the distance between particles increases during their displacement towards the entrance of the capillary, it is possible to expect additional molecular orientation. In this region, the micro-volume between any pair of particles is increased due to the stretching flows.

Owing to the viscosity anisotropy effect, there is a decrease in the effective viscosity in the melt state and, owing to the anisotropy in the

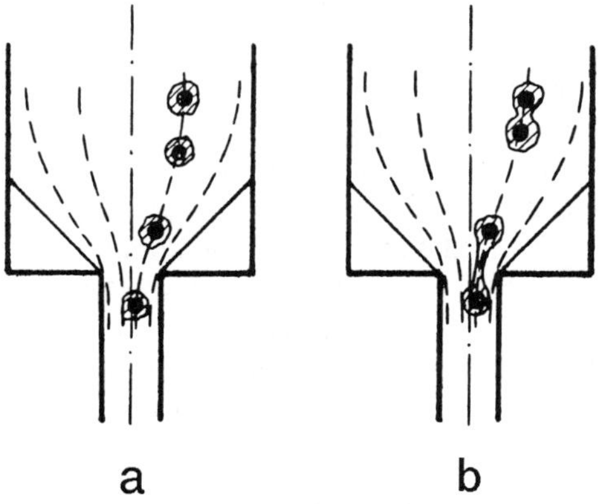

Figure 5.15 Approximate scheme of the filled polymer melt flow in the entrance zone to a capillary at (a) low; and (b) moderate concentration of particles.

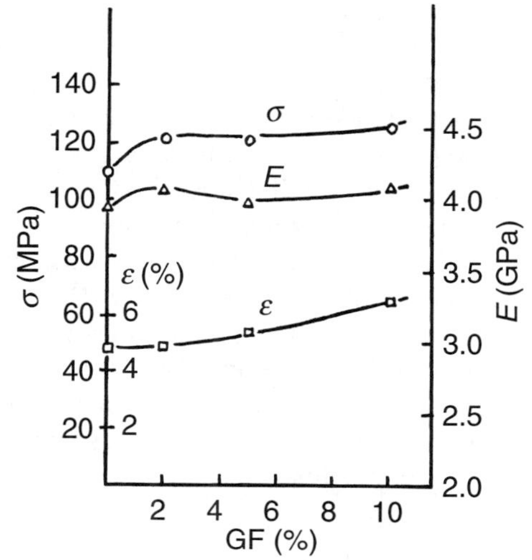

Figure 5.16 Mechanical properties of CPE-2 filled by GF at various concentration of filler.

mechanical properties, an increase of the extension elasticity modulus would occur. Our attempts to register the additional orientation effect for LCPs in the presence of small amounts of fillers were unsuccessful because the usual X-ray techniques do not pick up such small changes in molecular orientation.

Such insensitivity of the structural method to the probable increase of molecular orientation practically closes the possibility of observing the additional orientation of the LC matrix in the 'gaps' between the particles surrounded by 'non-mobile' absorbed layers, although, in the case of the matrix orientation, we could expect that the fraction of such oriented medium would be sufficient to be detected by an X-ray scattering method.

The statement about the decisive role of the absorbed layer formation in the appearance of both effects – the decrease in viscosity and the increase in elasticity modulus at low filler concentrations – can be confirmed by the absence of similar behaviour for CPE-2 filled with glass fibres. As an example, the variation of mechanical characteristics with GF content for CPE-2 with low intrinsic viscosity ($0.51\,\mathrm{dl\,g^{-1}}$) is presented in Fig. 5.16. Neither strength nor modulus have a clear tendency to show an extrema, although a slight rise in these characteristics, and especially elongation at break, may indicate an increase in orientation of the matrix located in the spaces between anisotropic particles that are inactive to CPE molecules.

The speculations presented above concerning the roles of the absorbed layers can, more or less, explain the role of the filler types on the variations of viscosity and tenacity with filler concentration. Meanwhile, the absorbed layer itself is capable of inducing the reinforcing effect. The point is that the field of the particle can promote the additional orientation inside the absorbed layers and lead to the reinforcing of the system as a whole. An improvement in mechanical properties at low content of particulate fillers was observed even for composites based on isotropic matrices [31]. But the joint change in the rheological and mechanical properties, in our opinion, needs an explanation from the standpoint of the unusual orientation effects realized in the stream. Nevertheless, we shall touch on the problem of filled polymer reinforcement again during the discussion of the relaxation properties of the solid composites.

5.3.3 Mechanical and dielectric relaxation in CB-reinforced CPE-1

Here, we focus on the effects of the active particulate, electroconductive filler on viscoelastic and electrophysical properties of the thermotropic LCP in the solid state. Figures 5.17 and 5.18 present the data on dynamic viscoelastic properties of the CB-reinforced composites as functions of temperature and concentration: $\log(E'(T, \varphi))$ and $\log(\tan\delta(T, \varphi))$. The spacial representation of the viscoelastic functions of the above composites allows an insight into the general features of mechanical relaxation. Among them are the following.

1. In the temperature range −150 to 200°C, three main regions of molecular relaxation are observed: low-temperature region I, the main relaxation process II and the high-temperature relaxation III. (Similar results were earlier reported by Benson *et al.* [32] for the neat copolyester of the same chemical composition, although the high-temperature relaxation was not mentioned.)
2. Reinforcement of the CPE-1 with CB is accompanied by cocomitant changes in intensity of the relaxation processes.
3. At a low content of the filler (2–4% of CB), the storage modulus exceeds the storage moduli of the CPE-1 and other composites over the entire temperature range studied.

The last result coincides with the above mentioned increase of the composites' tensile modulus in static tests.

To understand the effect of the CB filler on the relaxation properties of the LC copolyester, it is necessary to distinguish between the three superimposed processes, I, II and III. This problem was solved under the assumption that the well-known model of the standard linear viscoelastic

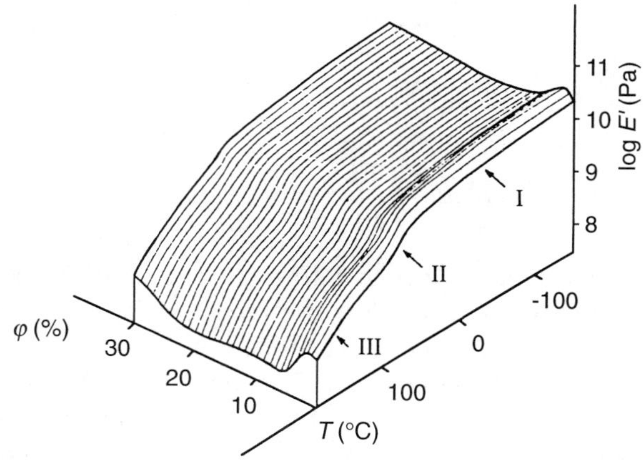

Figure 5.17 Dependence of dynamic modulus of elasticity on content of CB and temperature. The locations of relaxation transitions are indicated by arrows.

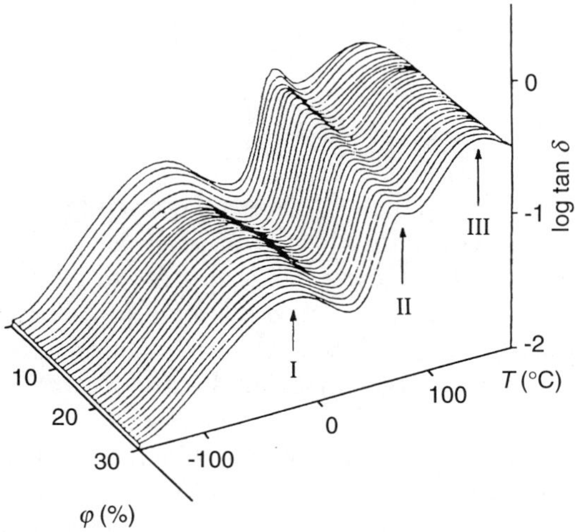

Figure 5.18 The loss properties of CPE-1 filled by CB as function of concentration of filler and temperature.

body with a discrete spectrum of relaxation times [33] could be invoked to describe the viscoelastic properties of the composites studied.

According to this model, the temperature dependencies, $E'(T)$, $E''(T)$ and $\tan\delta(T)$, for a set of n processes can be represented as follows:

$$E' = \sum_{i=1}^{n} E_{0i} + \frac{E_{1i}\omega^2\tau_i(T)}{1+\omega^2\tau_i^2(T)} \tag{5.4}$$

$$E'' = \sum_{i=1}^{n} \frac{E_{1i}\omega\tau_i(T)}{1+\omega^2\tau_i^2(T)} \tag{5.5}$$

$$\tan\delta = \frac{E''(T)}{E'(T)} \tag{5.6}$$

where ω is the frequency of cyclic deformation, E_{0i} and E_{1i} are the elastic constants of the model and $\tau_i(T)$ is the temperature dependence of the relaxation time for the corresponding process.

Computer simulation shows that the best fit between the model calculation of $E'(T)$ and $\tan\delta(T)$ and the experimental data is observed if the temperature function of the relaxation times is selected in such a manner that process II is described by the WLF equation (5.7) and processes I and III obey the Arrhenius type equations, (5.8) and (5.9) respectively:

$$\tau_2(T) = \tau_{02}\exp\left(\frac{B}{T_2 - T}\right) \tag{5.7}$$

$$\tau_1(T) = \tau_{01}\exp\left(\frac{U_1}{RT}\right) \tag{5.8}$$

$$\tau_3(T) = \tau_{03}\exp\left(\frac{U_3}{RT}\right) \tag{5.9}$$

where τ_{0i}, U_i, B and T_2 are the experimental constants.

Hence, one may conclude that process II is similar to the glass transition in amorphous polymers, i.e. it appears to be a main relaxation transition in the studied CPE-1. Process I, which is characterized by the Arrhenius temperature dependence of the relaxation times and by low activation energy, is likely to be associated with local molecular mobility. The possible origin of process III will be discussed below.

Relaxation parameters of each process, which are calculated from the corresponding constants of equations (5.1)–(5.5), are summarized in Table 5.1, where $\tan\delta_{\max i}$ is the maximal loss tangent of the process, $T_{\max}$ is the temperature corresponding to the maximal loss modulus, U is the activation energy of processes I and III and $U(T_g)$ is the apparent activation energy of the main relaxation process at $T_g = T_2 + 51.6$ (by analogy with the determination of the structural glass transition temperature for amorphous polymers).

Table 5.1 indicates that the introduction of CB into the copolyester neither conspicuously shifts the temperature regions of either of the three processes, nor does it change their activation energies, However, the

Table 5.1 Parameters of the relaxation processes I–III evaluated from the mechanical tests for the CB-reinforced CPE-1

CB content (*wt%*)	*tan δ max*	T max	*U* (*kJ mol*$^{-1}$)	*tan δ max*	*T max*	T_g	B, (K)	*U(Tg)* (*kJ mol*$^{-1}$)	*tan δ max*	*T max*	*U* (*kJmol*$^{-1}$)
		I				II				III	
0	0.06	−30	13	0.12	73	57	390	133	0.087	128	40
2	0.054	−30	12	0.109	71	57	380	129	0.117	130	40
4	0.051	−30	13	0.112	71	57	380	129	0.106	128	40
10	0.055	−30	13	0.101	71	57	380	129	0.125	128	40
20	0.062	−30	12	0.089	71	57	380	129	0.151	128	40
30	0.052	−30	13	0.07	72	58	350	120	0.167	128	40

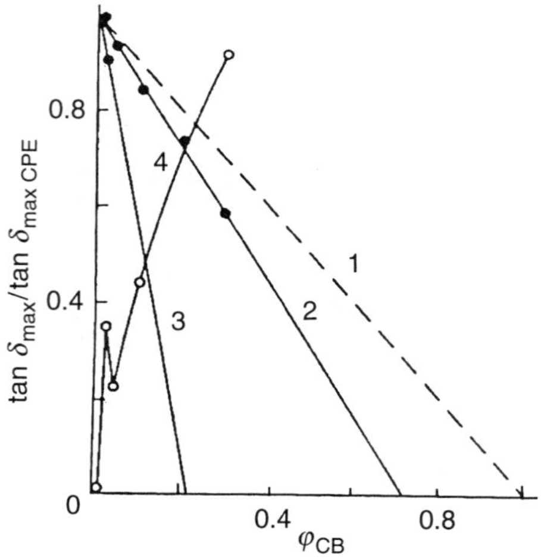

Figure 5.19 The effect of the filler content on excessive loss tangent for processes II(1–3) and III(4) in CB-reinforced CPE-1. Line 1 is the approximation for the inert filler, lines 2 and 3 correspond to mixtures containing more than 4 wt% and 2 wt% of CB respectively.

intensity of the main process II tends to decrease and the loss tangent of process III increases. Figure 5.19 illustrates the relative changes in tan δ with CB content.

If a filler exhibited no surface activity, the intensity of the main process would have decreased linearly with the filler content and levelled off to zero at 100% of CB content (Fig. 5.19, line 1). However, for all composites containing CB, the above dependence appears to be much lower (lines 2 and 3). It suggests that the particles of the filler assume an active role restricting or hindering the motion of the kinetic units responsible for the occurrence of the process.

This approach was developed by Slutsker *et al.* [34], who suggested a simple method for evaluating the transition layer thickness, or the effective radius of the field of the particle within which the motion of the kinetic units is hindered. To do this, it is necessary to determine the critical concentration of the filler φ_c at which the extrapolated dependence of tan δ_m becomes zero. The thickness of the transient layer, h, can be evaluated from:

$$h = \frac{d}{2}\left(\sqrt{\frac{1}{\varphi_c}} - 1\right) \tag{5.10}$$

where d is the particle diameter and φ_c corresponds to φ at $\tan\delta(\varphi) = 0$. Assuming that the average dimensions of the CB particles are

30 ± 2 nm and the thickness of the transition layer $h_2 = 1.7$ nm for dependence II, the sample containing 2 wt% of CB does not fit the above dependence. Similar estimates give the layer thickness $h_3 = 10$ nm.

Let us compare the distance of the effective action of the particle field, h, with the length of the mesogenic fragment, L. The copolyester studied is characterized by $L \simeq 2.5$ nm. For the composites containing more than 4 wt% of CB, the radius of action of the particle field is comparable with the length of the mesogenic fragment, whereas the main relaxation is hindered. Hence, one may suggest that the minimal kinetic unit responsible for the primary relaxation is the mesogenic fragment of the macromolecule.

At 2% CB content, the effective space of the particle field action $h \simeq 4L$. In this case, the main relaxation process is even more restricted and the concomitant increase in the storage modulus (by approximately 40%) of the composite occurs. These two phenomena seem to be closely related. At low contents of CB, when agglomeration of the particles is poor, some processes accompanied by an increase in the storage modulus may take place.

The first phenomenon, connected with the increase in molecular orientation during convergent flow, was discussed earlier. The second one is associated with adhesion forces, provided the scales of the field of a particle and of the structural elements of the material are comparable. Furthermore, the particles of a filler, being the attractive sites for the individual fragments of the macromolecules, may promote the formation of additional physical intermolecular contacts. This conclusion is supported by the above-mentioned phenomenon concerning the increased loss tangent in the high-temperature range, corresponding to molecular relaxation III, as the CB content increases (Fig. 5.19, curve 4).

Let us assume that, initially, this process III in the PET-HBA copolyester is associated with relaxation of the network formed by the junctions of crystalline nature arising from the presence of p-hydroxybenzoate sequences [11–13]. This assumption is corroborated by the characteristic features of the λ-relaxation, which involves the mobility of supersegmental structure [35]: the activation energy of process III achieves about 40 kJ mol^{-1}, and it appears to be practically independent of the filler content.

Because the intensity of dissipation of mechanical energy is decided by the number of kinetic units involved in the process, one may anticipate, at low filler content, that additional junctions are produced and, hence, the overall density of the network increases. As a result, a growth in loss tangent is observed. Such a physical network may give rise to an increased storage modulus in the composites only at low filler content, when no agglomeration of the particles takes place. As the filler content increases, the process of agglomeration becomes more pronounced, and the efficiency of the particle field tends to decrease and, hence, the rigidity of such a quasi-network decreases.

The indirect indication of the specific effect of the filler on the dissipative properties in the liquid composites is the variation of the temperature dependencies of the viscosity with increasing filler concentration; see Fig. 5.4. Already, at 2% CB, the high-temperature activation energy decreases and then the range of such decreased temperature coefficient of viscosity extends throughout the temperature region. This process is completed at 10–20% of filler content, i.e. corresponding to CB network formation. It means that the physical intersegmental network acts as a network of contacts between particles, probably through the absorbed layers.

To summarize for the CB-reinforced composites based on CPE-1, the role of the filler in the molecular relaxation involves both the formation of supermolecular structures of the polymer and the direct influence on its segmental mobility. The above phenomena were observed over the entire composition range: 0–30% CB. The self-organization of the filler appears to exhibit an indirect influence on the viscoelastic properties of the composite. For instance, a transition from the isolated particles to their aggregates manifests itself indirectly via a decrease in efficiency of the stress field induced by the CB particles, whereas the formation of the network consisting of the CB particles is not accompanied by any marked changes in the viscoelastic properties.

While the dynamic mechanical properties of the blended systems are not 'susceptible' to structure formation by CB in the CPE-matrix, electrophysical properties of the composites appear to be more sensitive. Figure 5.20 shows that in the composition range of CB 10–20 wt%, a jump in the dc conductivity (by approximately 10 orders of magnitude) and in the dielectric constant (by approximately 2.5 orders of magnitude) occur. Similar phenomena were observed earlier for composites containing a dispersed electroconductive filler, and the percolation theory of conductivity was invoked to give a plausible interpretation of the above data [36, 37]. The transition from ionic conductivity of the matrix inside the CB concentration range 0–10 wt% to the metallic one, in the range 20–30 wt%, may be attributed to the development of an infinite conducting cluster consisting of CB particles.

The formation of this infinite conducting cluster is accompanied by the appearance of a well-developed electroconductive surface on which the electrical charge may be distributed. One may imagine the situation when an infinite conducting cluster with a finite conductivity may be divided into two subsystems of carbon chains. Some of these chains have their ends at one external electrode, while the other end is at the other electrode. Hence, the achievement of the percolation threshold in the

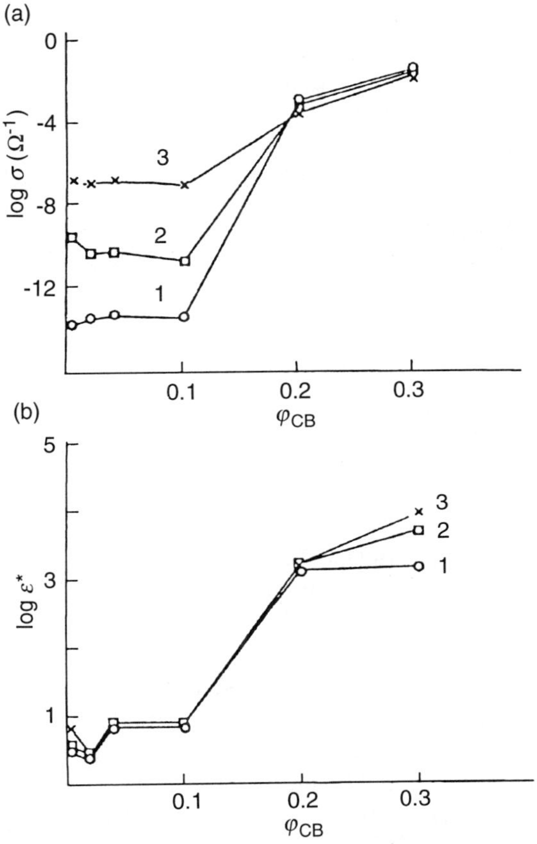

Figure 5.20 The effect of CB-content on (a) DC conductivity; and (b) effective dielectric permittivity of composites at 50°C (1), 120°C (2) and 240°C (3).

system with finite conductivity is accompanied by a jump not only in conductivity but also in the electric capacitance.

The temperature dependence of dielectric relaxation in reinforced composites in the composition range below the percolation threshold (Fig. 5.21) suggests that the subsystem of dielectric relaxers in the region of the main relaxation process behaves differently from the subsystem of mechanical relaxers. Indeed, as the content of the filler increases the intensity of this process at least does not decrease, and a marked shift of the intensity maximum of this process towards lower temperatures occurs. It means that CB is not inert with respect to the system of dielectric relaxers, as was observed for the mechanical relaxers, but assumes an active role: it shifts the spectrum of relaxation times to smaller values.

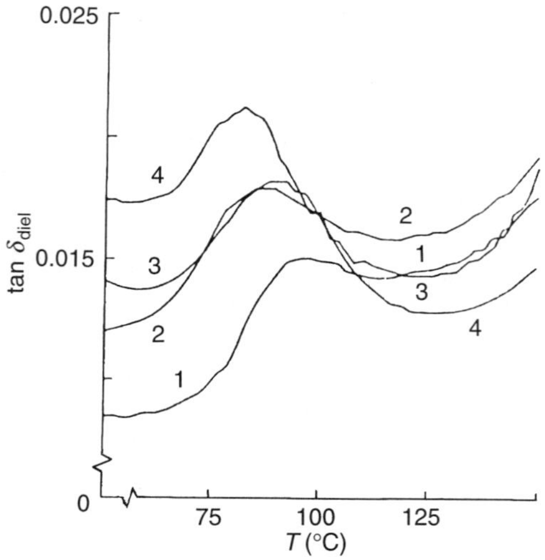

Figure 5.21 Dielectric loss tangent versus temperature plot for CPE-1 containing 0(1), 2(2), 4(3), and 10 wt% of CB.

This phenomenon seems to be most pronounced for the dielectric relaxation at the percolation threshold of the mixtures. While there is no noticeable difference between the dc and ac conductivities for the composites containing 20% CB (Fig. 5.22, curves 1 and 2), the composite containing 30% CB exhibits a well-pronounced frequency dependence of conductivity (curves 3 and 4). The presence of strictly linear sections of temperature dependence of conductivity allows us to use this feature in practical problems or, at least, to think about it. The role of the anisotropic matrix still remains unexplained, but we do know of similar effects in filled commercial thermoplastics.

To conclude, the introduction of an active dispersed filler into the LC copolyester may be considered as an effective means of modifying its solid-state properties. It may also be helpful in studying the processes of molecular relaxation.

5.3.4 Rheology and mechanics of polymer blends filled with carbon black

For a better understanding of the role of particulate and fibrillar fillers in the rheological properties of PSF/CPE blends, we must know the corresponding properties of the unfilled blends of the neat polymers. A fuller discussion of the rheological properties of LCP blends is given in Chapter 6.

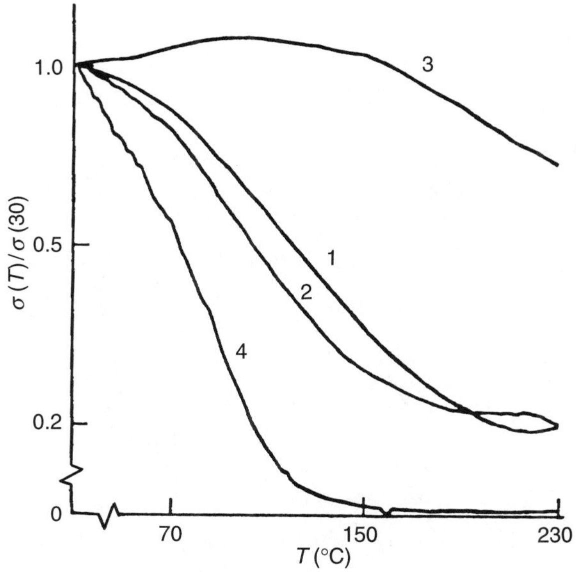

Figure 5.22 Temperature dependence of DC (1, 3) and AC (2, 4) conductivity for CPE-1 filled by 20 (1, 2) and 30 wt% of CB (3, 4).

From Fig. 5.23, which presents flow curves for CPE-2 and PSF and their blends at temperatures of 240°C and 280°C, it is clearly seen that the characters of the flow curves of the LC copolyester and the polysulphone are sharply different. The log $\dot{\gamma}/\log \tau$ dependence for CPE-2 is steeper than for the PSF. The shape of the LCP copolymer flow curve shows again its viscoplastic behaviour, which is due to the existence of the yield stress at both temperatures under study.

The temperature dependencies of viscosity for neat polymers used in this part of the work are presented in Fig. 5.24. As in the previous case, for CPE-2 the temperatures fall into two regions on the log $\eta/1T$ curve, where the flow activation energies are quite different. In the temperature range 220–250°C, $E_{CPE} = 700\,\text{kJ mol}^{-1}$, whereas in the range 260–280°C, $E_{CPE} = 70\,\text{kJ mol}^{-1}$. For PSF, in the temperature range 240–280°C the activation energy is approximately $140\,\text{kJ mol}^{-1}$. We can state that the features of the rheological behaviour of CPE-2 with temperature originate from structural changes in the melt. To a heterogeneous state of the material in a temperature range 220–260°C (mixture of nematic and crystalline phases) there corresponds a high value of the flow activation energy and a high value of the viscosity. The viscoplastic behaviour of the CPE melt under such conditions is augmented by the existence of a structural network, whose junctions are local crystallites. Low values of viscosity and activation energy at temperatures above

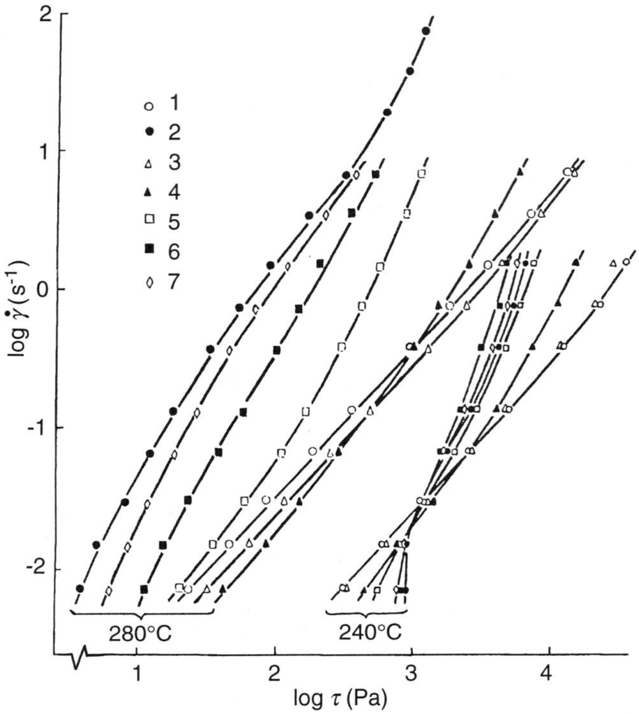

Figure 5.23 Flow curves of polysulfone (1), CPE-2 (2), and their blends: PSF/CPE = 9/1 (3), PSF/CPE = 7/3 (4), PSF/CPE = 5/5 (5), PSF/CPE = 3/7 (6), and PSF/CPE = 1/9 at (7) at 240°C and 280°C.

260°C correspond to a homogeneous nematic state. Polysulphone, over practically the whole investigated range of shear rates (stresses) at temperatures of 240°C and 280°C behaves as a newtonian fluid, i.e. its viscosity is independent of the flow conditions. A departure from the newtonian flow regime is observed only at the high shear rates.

On the addition of CPE-2 to PSF, the locations of the flow curves at 240°C change such that they rotate around a hinge 'secured' in the region of $\log \dot{\gamma} = 1.5$–1.6 and $\log \tau = 3.0$–3.1. For this reason the viscosity of the blends at high shear rates and stresses decreases and, at low ones, increases as compared with the viscosity of the initial PSF. At the same time, the viscosity anomaly increases, ending, in the low shear rate region, with the appearance of the yield behaviour.

A temperature increase of 40°C results in a decrease of the CPE-2 viscosity below that of PSF over the whole investigated shear rate range. The determining factor here is a sharp difference in the flow activation energies for the neat components. Also, at a temperature of 280°C, CPE-2 exhibits a viscoplastic behaviour and PSF is a newtonian fluid. The

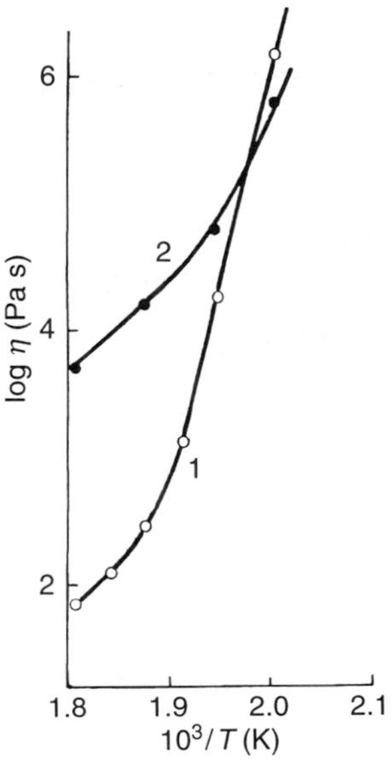

Figure 5.24 Temperature dependencies of viscosity for CPE-2 (1) and PSF (2).

shape of the flow curves for the blends at 280°C is determined by the component whose content in the blend exceeds 50–70%, although a tendency of the high shear rate sections of the flow curves to come closer to the corresponding region for CPE is observed beginning at a CPE concentration of over 30%.

The dependencies of the viscosity on the composition of the PSF/CPE-2 blends at temperatures of 240°C (curves 1 and 2) and 280°C (curves 3 and 4) and various shear rates are shown in Fig. 5.25(a). Owing to a strong viscosity anomaly of CPE-2, its viscosity exceeds that of PSF in the low shear rate region (at 240°C) and decreases below the PSF viscosity at the high shear rates. This fact is the main cause of the blends' viscosity exceeding that of PSF at $\dot{\gamma} = 7.32 \times 10^{-3}\ s^{-1}$ (Fig. 5.25(a), curve 1). On the whole, it follows the log additive rule. At $\dot{\gamma} = 7.32 \times 10^{-1}\ s^{-1}$ the CPE-2 addition, on the contrary, reduces the blends' viscosity with a negative departure from the log additive rule at a CPE-2 content over 10%.

At a temperature of 280°C and shear rate of $7.32 \times 10^{-3}\ s^{-1}$ (curve 3), there occurs a positive departure from the log additive rule over the whole range of compositions. As the shear rate increases, the $\eta(C)$ dependence transforms into an S-shaped curve (curves 4 and 5) which, as adopted, is due to a phase inversion.

Of fundamental importance to understanding such a complicated

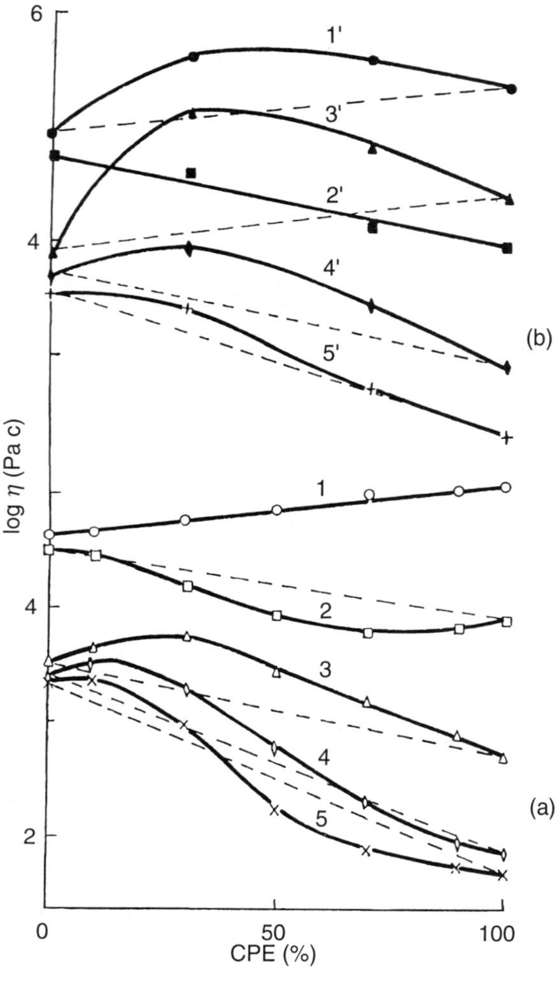

Figure 5.25 Variation of viscosity of (a) PSF/CPE-2 blend; and (b) of PSF/CPE-2/CB blend (CB content 10%) at 240°C (1,1′,2,2′) and 280°C (3–5,3′–5′) and various shear rates (s^{-1}): 7.32×10^{-3} (1,1′,3,3′); 7.32×10^{-1} (2,2′, 4,4′), and 6.58 (5,5′).

transformation of the concentration dependence of viscosity with varying temperature and shear rate is information on the thermodynamic state of the compositions in the stages of preparation of the blends and a measurement of the rheological characteristics within a certain temperature–concentration region. If we assume that the neat polymers are thermodynamically incompatible, i.e. their blends are two-phase systems, then the features of viscosity variation with composition, temperature and flow conditions may be ascribed to the different heterogeneity levels occuring not only at the blend preparation stage but also in the course of the shear action in the operating unit of the rheometer.

We shall accept that the positive departures of the blend viscosity from the log additive rule at 280°C reflects the state of the emulsion, characterized by a strong interaction at the CPE-2/PSF interface. At the same time, at 240°C, the departure from the log additive rule is either zero or negative, i.e. in accordance with the above-mentioned principle, a strong interphase interaction is absent. The shear rate here plays a substantial role. As reported in the literature [38], a decrease in the degree of miscibility of blend components and a microphase separation of the blend occur in a shear flow. The filler may influence this process.

We pass now to a discussion of the rheological properties of carbon black-filled CPE-2/PSF blends. Figures 5.26 and 5.27 show flow curves for the filled PSF/CPE = 3/7 and PSF/CPE = 7/3 blends at 240°C (broken lines) and at 280°C (solid lines). It follows from the figures that the shape of the $\log \dot{\gamma}/\log \tau$ dependence for such systems is determined first of all by the polymeric component, which is a matrix in the bicomponent system. For the former blend, this is the LC copolyester and, for the latter blend, polysulphone.

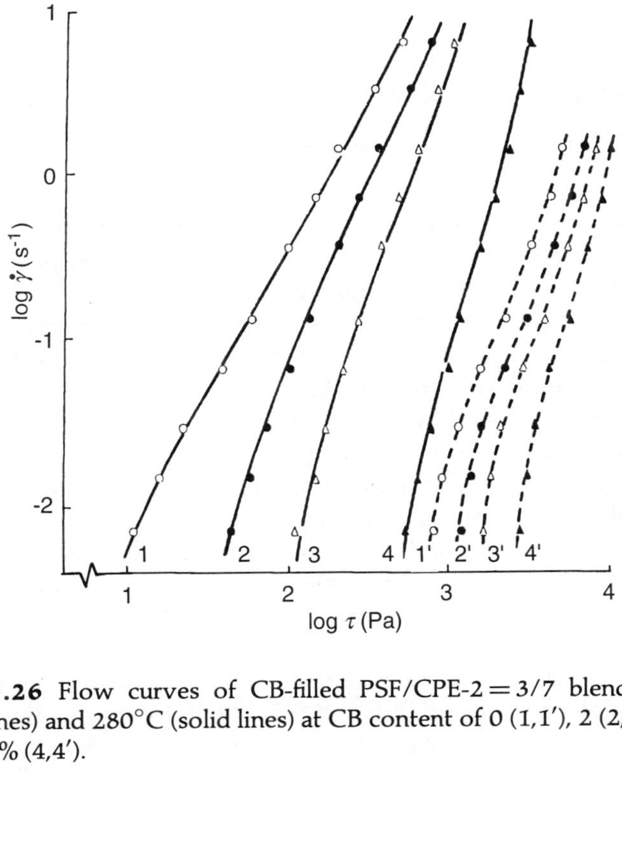

Figure 5.26 Flow curves of CB-filled PSF/CPE-2 = 3/7 blend at 240°C (broken lines) and 280°C (solid lines) at CB content of 0 (1,1′), 2 (2,2′), 4 (3,3′), and 10 wt% (4,4′).

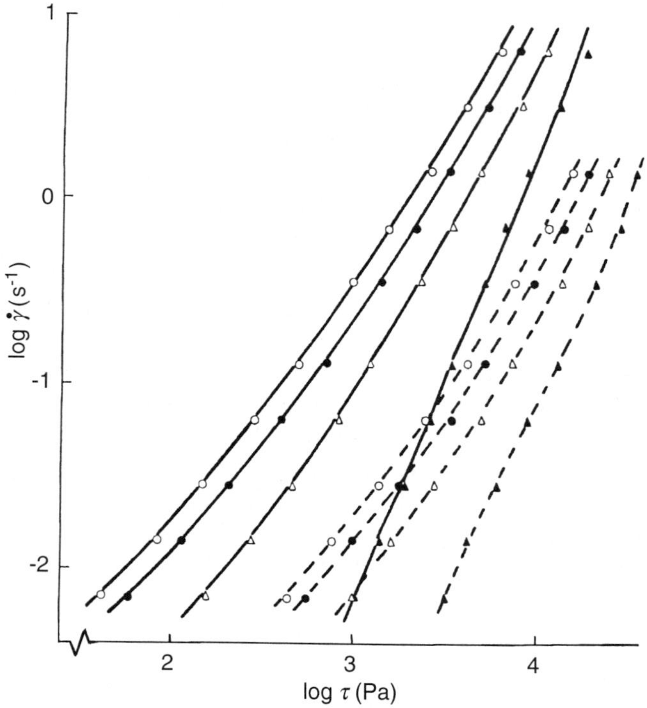

Figure 5.27 Flow curves of CB-filled PSF/CPE-2 = 7/3 blend at the same conditions as in Fig. 5.26.

The variation of viscosity of a filled (10% CB) PSF/CPE-2 blend with blend composition at various shear rates and temperatures is shown in Fig. 5.25(b) (curves 1′–5′). Comparison of these curves with curves 1–5 for unfilled blends indicates a considerable increase in the degree of positive departures of experimental values from log additivity for comparable shear rates and temperatures.

Choosing, as a criterion, the interphase interaction reinforcing the positive deviation of the viscosity-concentration dependencies and supposing the increase of CPE-2/PSF interaction in the presence of CB in the melt state, it is reasonable to expect the occurrence of these properties in the solid state also. To check this suggestion, we have investigated the dynamic mechanical characteristics of the same polymer blends (CPE-2/PSF = 3/7 and 7/3) containing 2%, 4% and 10% CB. Besides, such an approach allows us to compare the relaxation properties of the above-mentioned systems with data obtained previously for PSF/CPE blends [39] and mixtures of CPE and CB [40]. Returning to the previous data, we should like to recall the important results for further analysis:

1. There is a definite level of interphase interaction in blends of PSF/CPE, probably as a result of similar-to-similar attraction.
2. CB is a strong structurizing agent for CPE, blocking the main relaxation process and participating in the formation of the additional interchain contacts.

For both filled blends, the dependencies of tan δ on T and CB content are shown in Figs. 5.28 and 5.29 in a manner used earlier for the binary

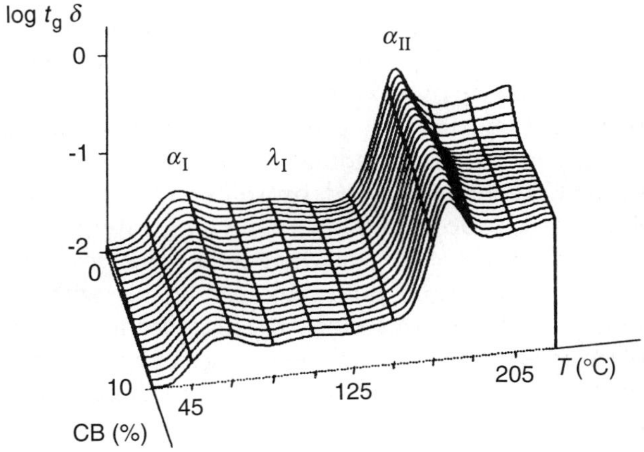

Figure 5.28 Loss tangent as a function of CB content and temperature for PSF/CPE-2 = 3/7 blend matrix.

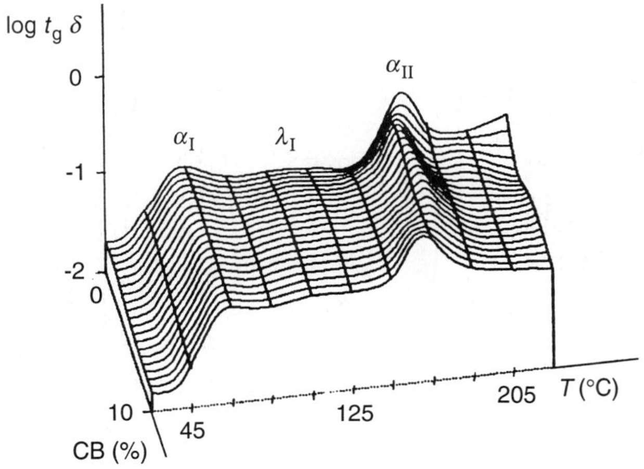

Figure 5.29 Loss tangent as a function of CB content and temperature for PSF/CPE-2 = 7/3 blend matrix.

systems. It is possible to identify reliably two processes of molecular relaxation, α_I and α_{II}, corresponding to the main relaxation transitions in CPE-2 and PSF respectively. Between these two distinctive tan δ peaks the third weak relaxation process λ_I may be observed. Presumably, it is attributed to an entanglement network in the blend matrix. Comparing these dependencies, we can postulate that the variation of polymer matrix composition essentially redistributes the damping levels of relaxation processes inside the range of positive temperatures.

Introducing CB to the blend matrix leads to the appearance of more delicate effects, which may be discovered by the analysis of so-called topographic maps of equal loss levels (Figs. 5.30 and 5.31). On the one hand, with increase in CB content, the loss level for the α_I process decreases but, for the λ_I process, it increases. It means the growth of intermolecular contact density, probably as a consequence of the absorption phenomena. But, depending on matrix composition, the change of the losses with CB content proceeds by a different route. Thus, a decrease in loss level for the α_I process takes place over a greater extent for blends with lower concentrations of CPE-2, whereas the enhancement in losses for the λ_I process is much more noticeable for blends with predominant concentrations of CPE. For this blend the significant decrease of T_g of PSF takes place as well (the maximal departure of approximately 8°C is observed for the composition with 2% CB). Since this effect is practically absent for the blends with the higher PSF contents, we may conclude that

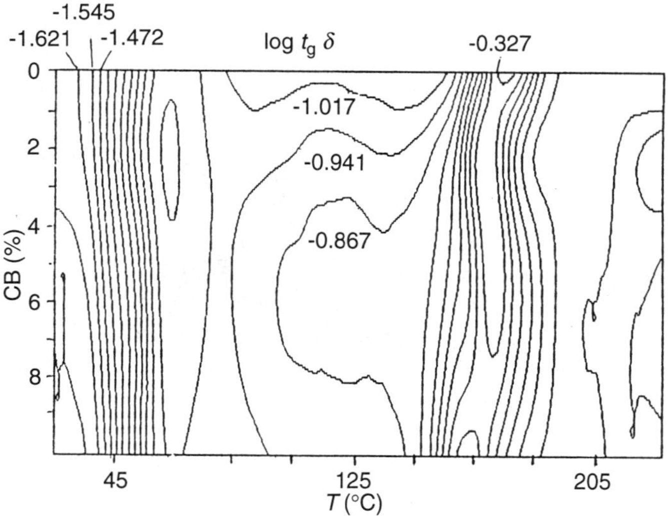

Figure 5.30 Loss tangent isolines map for PSF/CPE-2/CB = 3/7 composites. The pitch between isolines is equal to 0.076.

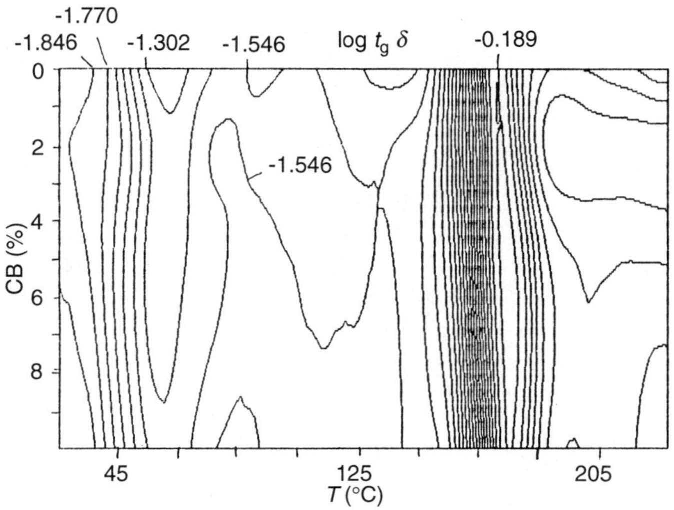

Figure 5.31 Loss tangent isolines map for PSF/CPE-2/CB = 7/3 composites with the same pitch as in Fig. 5.30.

the LCP phase is the more sensitive to the presence of the active filler and, due to a change of the CPE-2 relaxation properties, the interface thickness between CPE and PSF becomes larger. The latter event leads to a decrease in the glass transition temperature of the PSF. We observed the same decrease in the glass transition temperature of PSF in binary blends of PSF/CPE [39]. The above-mentioned results and discussion confirm the active role of the CB particles as a structurizing agent for CPE and as a compatibilizer for CPE/PSF blends.

Another situation occurs for the fibrillar filler (GF). Unfortunately, in this case, we used the sample of CPE-2 with the lower intrinsic viscosity ($0.51\,dl\,g^{-1}$ against $0.80\,dl\,g^{-1}$ in the former blends), and the common picture of the filler's influence on the rheological behaviour of the PSF/CPE blends are as shown in Figs. 5.32 and 5.33. In these figures the flow curves were obtained on a rotational rheometer for filled neat polymers and for filled PSF/CPE blends with various contents of CPE at 280°C. As seen from these plots, the characters of the dependencies of $\log \gamma/\log \tau$ inside the high shear rate region are more or less similar. In the low shear rate region, where the shape of the flow curve for CPE is determined by yield behaviour while the PSF remains a newtonian fluid, the dependence of viscosity on GF content is stronger for the filled CPE than for the filled PSF.

For the binary matrices, the shape of the flow curves is determined by which of the polymeric components is the more predominant: PSF for

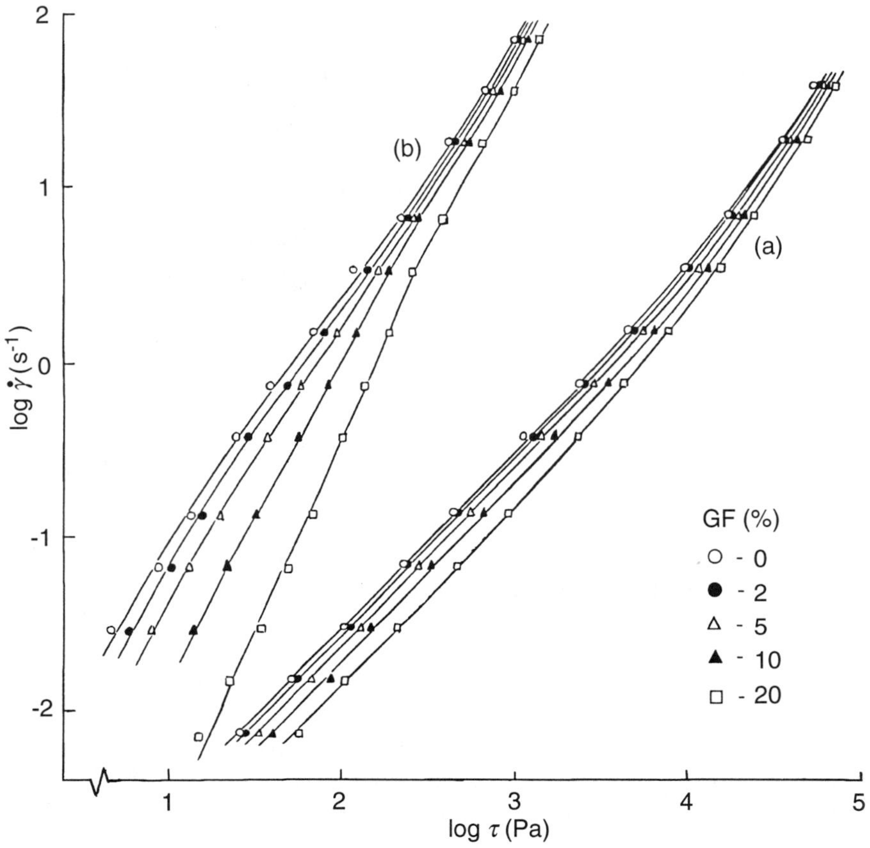

Figure 5.32 Flow curves of PSF/CPE-2 = 3/7 blend filled with GF at (a) 240°C and (b) 280°C.

the composition PSF/CPE = 7/3 and CPE for the binary matrix PSF/CPE = 3/7.

Concentration dependencies of viscosity for the initial blends and blends containing 20% of glass fibres at $T = 240$°C are shown in Fig. 5.34. Comparing the corresponding curves for the neat blends with those filled with GF, we note some diminishing scales of positive and negative departures from the log additive rule. Using the standard approach, we can declare that GF is practically inert from the viewpoint of an influence on the miscibility of the polymer components. The origin of such passivity lies in the very weak interaction of both components with the surface of the glass fibres. This result confirms the above conclusion concerning the necessity for a definite level of adhesion for an influence on the compatibility of the polymer components.

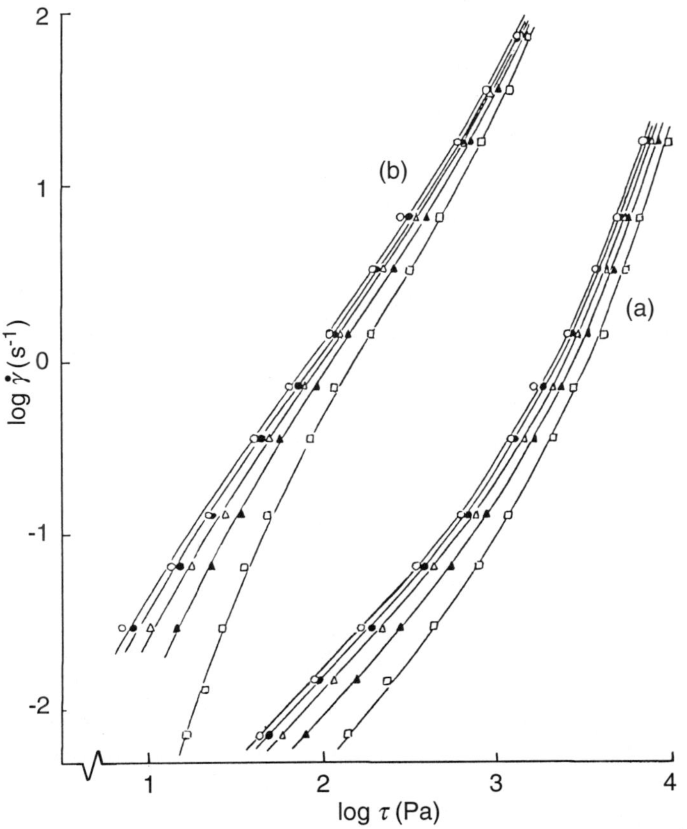

Figure 5.33 Flow curves of PSF/CPE-2 = 7/3 blend filled with GF at (a) 240°C and (b) 280°C. Designations are the same as in Fig. 5.32.

Figures 5.35 and 5.36 show dependencies of the relative viscosity of the studied systems, $\eta_r = \eta/\eta_0$ (η is the viscosity of the filled blend and η_0 is the viscosity of the matrix, i.e. CPE-2, PSF or PSF/CPE-2) on the filler (CB and GF) concentration. For CPE-based mixtures at 240°C and PSF-based mixtures at 240° and 280°C (Fig. 5.35, curve 1), this dependence is practically linear with a slightly increasing function of CB concentration, as is the case for ordinary thermoplastics and elastomers containing low disperse fillers. At 280°C, on the contrary, η_r rises steeply with increasing CB content for a filled CPE-2 which is typical for fillers having a high structurizing activity in thermoplastics and rubbers. So, the apparent activity of some fillers changes, depending on the CPE melt's structure. The η_r(C) dependencies for PSF/CPE mixtures occupy an intermediate position between the corresponding dependencies of CPE at the temperatures 240° and 280°C.

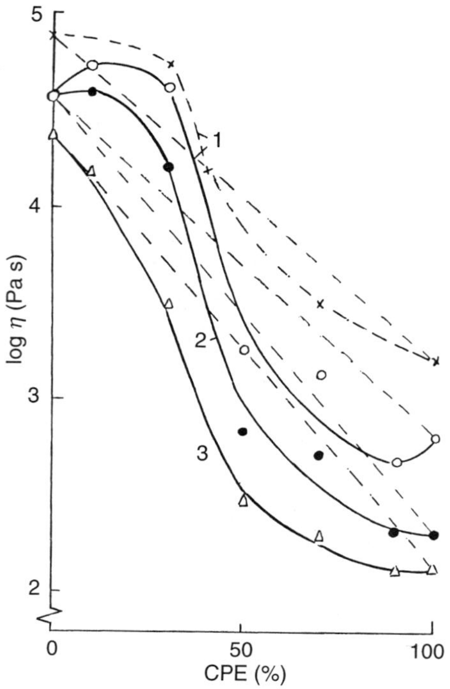

Figure 5.34 Viscosity versus composition plot for PSF/CPE-2 blend at shear rates ($\log \dot{\gamma}\ s^{-1}$); −2.14 (1), −0.88 (2), and 0.52 (3). Dotted line 1 relates to the PSF/CPE-2 composite filled with 10 wt% of GF.

In this sense, fibrillar and particulate fillers give very similar results. But, in the common case, the rate of viscosity increase for the fibrillar filler is lower than that for the particulate active filler. Comparison of curve 3 in Fig. 5.36 and curve 5 in Fig. 5.35 for the same matrix and 10% content at 280°C shows a much more intensive viscosity rise for CB (35-fold) than for GF (2.5-fold).

It has been noted above that the characteristic feature of the rheological behaviour of LC polymers is the existence of a yield stress. At the same time, such a feature is a characteristic of filled systems with polymeric and low molecular weight dispersion media, beginning at a certain content of filler. It is interesting to trace the variation of the yield stresses of the copolyester and its blends with PSF and CB. The yield stress τ_0 can be determined from Casson's equation [41]:

$$\tau^{\frac{1}{2}} = \tau_0^{\frac{1}{2}} + a\dot{\gamma}^{\frac{1}{2}} \tag{5.11}$$

where a is the constant of the system.

The dependencies of calculated τ_0 values on CB content for CPE/CB

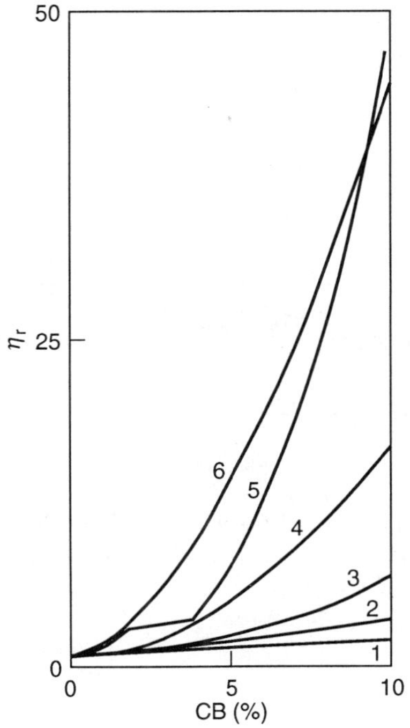

Figure 5.35 Dependence of relative viscosity on CB content for different matrices: CPE-2, 240°C and PSF, 240–280°C (1), PSF/CPE-2 = 3/7, 240°C (2), PSF/CPE-2 = 7/3, 240°C (3), PSF/CPE-2 = 7/3, 280°C (4), CPE-2, 280°C (5), and PSF/CPE-2 = 3/7, 280°C (6).

mixtures at various temperatures are presented in Fig. 5.37. Corresponding dependencies of yield stress on PSF content for PSF/CPE-2/CB blends are shown in Fig. 5.38. As seen, a temperature change of 40°C results in a change in the yield stress of neat LC copolyester of three orders of magnitude (see curves 1 and 1′). This change is attributed to the presence in the CPE melts at lower temperatures of an additional structural network formed by junctions of crystalline nature. These junctions are stronger than those formed by agglomerated CB particles. An indication of such a correspondence of network strength is the weak τ_0 dependence on CB concentration at 240°C (curve 1) and the strong dependence at 280°C (curve 1′), when structurization of the particulate filler itself in the nematic CPE phase occurs.

The concentration dependence of the yield stress for matrices based on PSF/CPE-2 blends is determined by the temperature and the content of the polymeric components in the binary mix. Such a dependence is more pronounced at a high temperature and a predominating content of

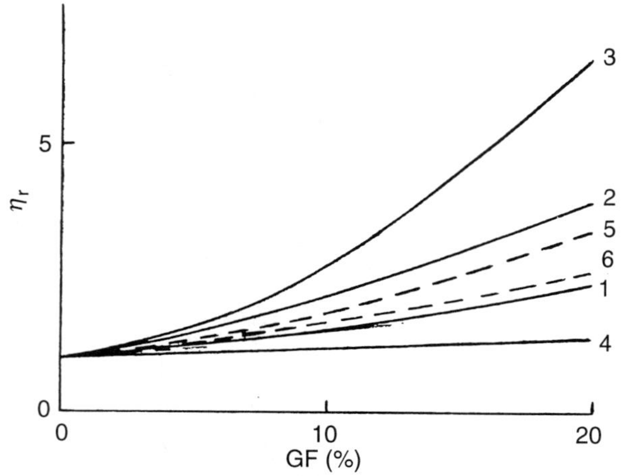

Figure 5.36 Dependence of relative viscosity on GF content for matrices: CPE-2 at 240°C (1), 260°C (2), 280°C (3); PSF at 240–280°C (4); PSF/CPE-2 = 3/7 at 280°C (5); PSF/CPE-2 = 7/3 at 280°C (6). Shear rate, $\log \dot{\gamma} = -1.53\ s^{-1}$.

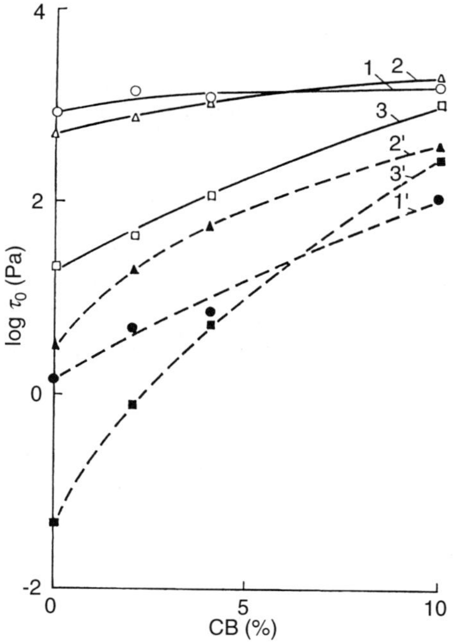

Figure 5.37 Yield stress variation with CB content for CPE-2 (1, 1′) and blends: PSF/CPE = 3/7 (2, 2′) and PSF/CPE = 7/3 (3, 3′). Temperature: 240°C (1–3) and 280°C (1′–3′).

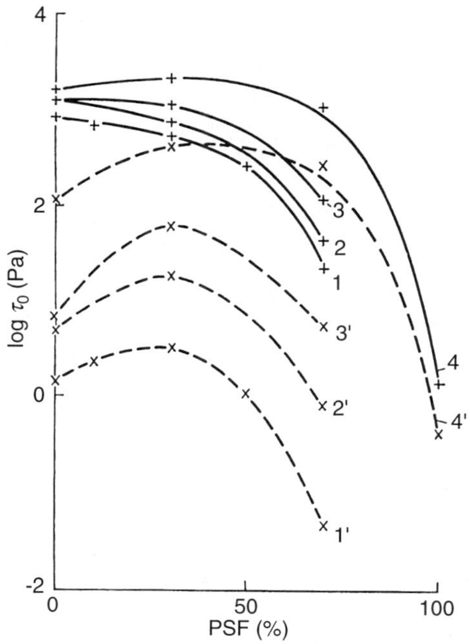

Figure 5.38 Yield stress variation with content of polymeric component in PSF/CPE/CB blends at various CB concentrations: 0 (1, 1′), 2 (2, 2′), 4 (3, 3′), and 10% (4, 4′). Temperature: 240°C (1–4) and 280°C (1′–4′).

the LC polymer; see Fig 5.37, curves 2, 2′, 3, 3′ and Fig. 5.38, curves 1–4 and 1′–4′. A τ_0 increase at 280°C and a PSF content of up to 30% in the dispersed medium (Fig. 5.38, curves 1′–4′) is attributed to a considerable interaction between PSF and CPE-2 at the interface in this range of compositions. This result correlates with the conclusion about the nonzero level of the phase interaction in the binary blends PSF/CPE observed earlier [39, 42].

Now we shall compare the effect of CB on the mechanical characteristics of the two matrices: PSF/CPE = 7/3 and PSF/CPE = 3/7 at 240° and 270°C (Fig. 5.39). In the first case, the strength and the elongation at break are decreasing, and the initial modulus is slightly increasing. This picture is similar to the usual stiffness enhancement that takes place in the usual isotropic polymeric media. For the matrix containing 70% CPE-2, there is a tendency for maximum turning point in σ and E while the elongation at break is decreasing. The maxima of strength and modulus correspond to approximately 4% CB. At first sight, these dependencies are similar to those observed for the pure LC matrices (Fig. 5.10). In our opinion, an increase in CPE-2 concentration in the binary matrix greatly influences the mechanical characteristics. The

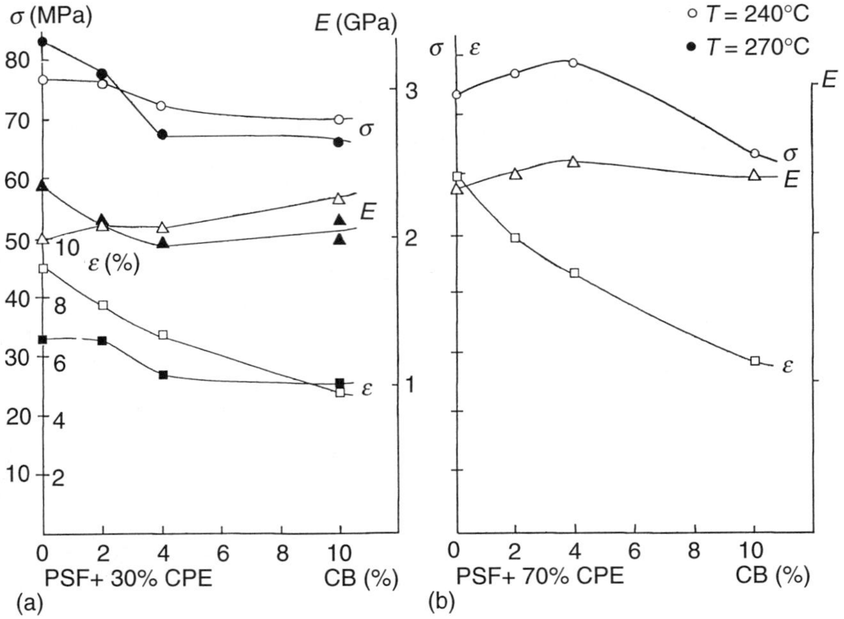

Figure 5.39 Mechanical characteristics of PSF/CPE-2/CB extruded composites for two matrices: (a) PSF/CPE-2 = 7/3; and (b) PSF/CPE-2 = 3/7, both prepared at 240°C and 270°C.

reason for such a prominent influence may be due to the predominant absorption of the LC component at the surface of the fillers and its partial ordering, as already described above.

5.4 CONCLUSION

The results obtained make it possible to draw some general conclusions concerning the difference in the rheological behaviour of the filled compositions on a base of CPE at low and at high temperatures. It is evident that the noted features are connected with non-continuous changes in the structure and in the degree of order of the CPE as a function of temperature. In the low temperature region, the local crystalline residues are distributed in the nematic melts. The source of such local crystallites is the block sequences formed by the HBA. The existence of such a strong structural network reinforces the strong viscoplastic behaviour, and the relaxation properties of the matrix controls the response of the filled melts. The presence of the fillers becomes prominent at high concentration (greater than 15%), where the formation of the non-relaxing skeleton takes place.

On the other hand, at 280°C, we have a homogeneous nematic melt and, in these conditions, filler begins to play an essential role much earlier. If we obtain the extrudates from the high-temperature region, it is impossible to ascertain reliably the reinforcing effects. Thus, the distribution of filler particles is more or less uniform in this case but, in the low-temperature region, we cannot expect such homogeneity. A possibility that cannot be excluded is that the superposition of the additional orientation effect inside the pure nematic medium in the presence of a filler and the existence of the strong (but rare) crystalline junctions lead to the summary reinforcement of the filled LC copolyesters.

We should like to note that the non-homogeneity of the LC melts for the PETP-HBA copolyesters is not the outstanding feature inherent in this polymer only. There are many papers dedicated to so-called rheological instabilities of thermotropic LC melts [43, 44]. One of the probable explanations of the viscosity increase with residence time is connected with non-equilibrium structure of such polymers at room temperature and an ordering or crystallization of them above the softening point. From this point of view, the results of this chapter are in common agreement with other workers.

Dynamic mechanical data show that particulate fillers, especially carbon black, are active from the point of view of polymer absorption at the surface. Increase of the loss tangent for high-temperature relaxation process III reflects the ordering of the mesogenic fragments of the CPE in binary filled mixtures. This process is, to some extent, similar to the formation of the crystalline junctions by the block moieties of the hydrobenzoic acid. The existence of such absorbed layers under conditions where they touch each other leads to the formation in convergent flow of a specific orientation network. In the melt state, this network promotes a decrease in viscosity and, in the solid state, it reinforces the elasticity modulus. The adhesion forces only (without convergent flow) may increase the stiffness of the particles but to a lesser degree.

The hypothesis concerning the role of the gaps between particles (or liquid droplets in the case of polymer–polymer blends) was not reliably confirmed, although for specific systems, e.g. blends of two LCPs, it may be valid [7].

Dielectric relaxation is more sensitive to the presence of conductive fillers, especially above a threshold concentration corresponding to infinite conducting cluster formation. The linear sections of the temperature dependencies of dc- and ac-conductivity may be used for practical applications.

As to the filled polymer blends, the analysis of obtained data should lead to the basic factors affecting the rheological behaviour of the binary matrix. The first of them is the compatibility of the components in the melt or the degree of interphase interaction between them in a micro-heterogeneous system. The second factor is the influence of the filler both on the character of the structurization in the system and on the compatibility of the components. As shown by experimental results, rheological properties of the blends differ substantially in the character of their departure from the log additive rule. At 240°C, the departures are either zero or negative whereas, at 280°C, they become positive. This shows different degrees of interaction between the components at temperatures below and above 250°C, if we accept *a priori* that the positive deviation reflects an increased interphase interaction.

It seems natural that, in the pure nematic phase, the system becomes more mobile, i.e. more inclined to the formation of interphase layers with the PSF-chains containing the related phenyl groups. At the same time, in the low-temperature region, the existence of a strong enough crystalline network in the CPE melts restricts the mass transfer processes in the blend melts. The shear rate promotes the limiting of the interaction between components, especially within the concentration region where the LCP becomes the matrix. A probable reason for this influence may be the creation on the interface of additional tangential stresses, hindering the mutual diffusion of the two kinds of macromolecules [38].

On the addition of a particulate filler to the blend melt, the situation remains practically the same, i.e. at 280°C the interphase interaction is higher than at 240°C. The absolute scale of the positive deviation from log additivity in the filled blends is higher than in neat blends. This tendency is reflected in the solid state, leading to a more preferable absorption of the LC component on the particles' surfaces and the mutual diffusion of polymers on to the interface. The last effect results in a decrease of the PSF glass transition temperature. However, we cannot so far be sure from this observation about the interaction between the polymers in the presence of CB without some additional and sophisticated experiments. We cannot know exactly the particles' distribution between polymer phases, although the simple test with selected dissolution of the PSF phase from the filled CB blend indicates an increased concentration of CB in the CPE phase (approximately 40% at a content of about 10% CB).

Following from the data of Figs. 5.35 and 5.36, it is reasonable also to assume that the nematic CPE melt is much more structured by a filler as compared with the PSF melt. So we can suggest that the surface layers around particles are enriched in CPE, but the mechanism of the interaction between the absorbed layers and the blend matrix remains unknown. The answers to these questions will be found only after additional investigations.

REFERENCES

1. Blizard, K.G. and Baird, D.G. (1987) *Polym. Eng. Sci*, **27**, 653.
2. Weiss, R.A., Huh, W. and Nicolais, L. (1987) *Polym. Eng. Sci.*, **27**, 684.
3. Acierno, D. and La Mantia, F.P. (eds) (1993) *Processing and Properties of Liquid Crystalline Polymers and LCP Based Blends*, ChemTec Publishing.
4. Kulichikhin, V.G. and Plate, N.A. (1991) *Vysokomol. Soed.*, **A33**, 3 (in Russian).
5. Kulichikhin, V.G., Volkov, V.S. and Plate, N.A. (1992) in *Comprehensive Polymer Science*, First Supplement, Pergamon Press, p. 385.
6. Kulichikhin, V.G., Vasil'eva, O.V., Litvinov, I.A. *et al.* (1991) *J. Appl. Polym. Sci.*, **42**, 1363.
7. Zabugina, M.P., Parsamyan, I.L., Litvinov, I.A. *et al.* (1990) *Vysokomol. Soed.*, **B32**, 765 (in Russian).
8. Suenaga, J. (1990) *Polym. News*, **15**, 201.
9. Nuel, L. and Denn, M.M. (1991) *Rheol. Acta*, **30**, 65.
10. Lipatov, Yu.S., Shumsky, V.F., Rosovitsky, I.P. *et al.* (1993) *J. Appl. Polym. Science*, **47**, 941.
11. Jackson, W.G. and Kuhfuss, H.F. (1976) *J. Polym. Sci., Polym. Chem. Ed.*, **14**, 2043.
12. Sugigama, H., Lewis, D.N., White, J.L. *et al.* (1985) *J. Appl. Polym. Sci.*, **30**, 2329.
13. Viney, C. and Windle, A.H. (1982) *J. Mater. Sci.*, **17**, 2661.
14. Nielsen, L. (1974) *Mechanical Properties of Polymers and Composites*, M. Dekker, Inc., New York.
15. Vinogradov, G.V., Malkin, A.Ya., Plotnikova, E.P. *et al.* (1978) *Vysokomol. Soed.*, **A20**, 226 (in Russian).
16. Konyukh, I.V., Vinogradov, G.V. and Konstantinov, A.A. (1963) *Plast. Massy*, **10**, 45 (in Russian).
17. Perepechko, I.I. (1973) *Acoustic Methods for Polymer Investigation*, Moscow, Khimiya, p. 60 (in Russian).
18. Kulichikhin, V.G. (1993) in *Liquid Crystal Polymers* (ed. N. Plate), Plenum Press.
19. Kulichikhin, V., Polushkin, E., Antipov, E. *et al.* (1992) *Makromol. Chem., Macromol. Symp.*, **56**, 79.
20. Malkin, A.Ya. (1990) *Adv. Polym. Sci.*, **96**, 69.
21. Leonov, A.I. (1990) *J. Rheol.*, **34**, 1039.
22. Yul'chibaev, B.A., Perfilov, A.N., Urinov, E.U. *et al.* (1992) *Vysokomol. Soed.*, **A34**, 69 (in Russian).
23. Vinogradov, G. and Malkin, A. (1980) *Rheology of Polymers*, Moscow, Mir Publ.
24. Volkov, V.S. and Kulichikhin, V.G. (1990) *J. Rheol.*, **34**, 281.
25. Lipatov, Yu.S., Babich, V.F. and Rosovizky, V.F. (1974) *J. Appl. Polym. Sci.*, **18**, 1213.
26. Shumskii, V.F., Getmanchuk, I.P., Parsamyan, I.L. *et al.* (1992) *Polymer Science USSR*, **34**, 56.
27. Prokopenko, V.V., Petkevich, O.K., Malinskii, Yu.M. *et al.* (1974) *Dokl. Acad. Nauk SSSR*, **214**, 389 (in Russian).
28. Prokopenko, V.V., Titova, O.K., Fesik, N.S. *et al.* (1977) *Vysokomol. Soed.*, **A19**, 95 (in Russian).
29. Guzeev, V.V., Rafikov, M.N. and Malinskii, Yu.M. (1975) *Vysokomol. Soed.*, **A17**, 804 (in Russian).

30. Dreval, V.E., Tsidvintseva, M.N., Kerber, M.L. *et al.* (1987) *Mekhanika Kompozit. Mater.*, **3**, 505.
31. Lipatov, Yu. (ed.) (1982) *Macromolecules onto Interface*, Moscow, Khimiya, p. 230 (in Russian).
32. Benson, R.S. and Lewis, D.N. (1987) *Polym. Commun.*, **28**, 289.
33. Ward, I.M. (1973) *Mechanical Properties of Solid Polymers*, Wiley-Interscience.
34. Slutsker, A.I., Polikarpov, Yu.I., Fyodorov, Yu.N. *et al.* (1990) *Vysokomol. Soed.*, **B32**, 177 (in Russian).
35. Akopyan, L.A., Gronskaya, E.V., Zobina, M.B. *et al.* (1985) *Vysokomol. Soed.*, **A27**, 400 (in Russian).
36. Shuvaev, E.P., Scherbak, V.V., Pugachev, A.K. *et al.* (1989) *Vysokomol. Soed.*, **A31**, 990 (in Russian).
37. Kotosonov, A.S., Kuvshinnikov, C.V., Chmutin, I.A. *et al.* (1991) *Vysokomol. Soed.*, **A33**, 1746 (in Russian).
38. Gilbert, E.F. and McHugh, A.J. (1984) *Macromolecules*, **17**, 2657.
39. Semakov, A.V., Kantor, G.Ja., Vasil'eva, O.V. *et al.* (1991) *Vysokomol. Soed.*, **A33**, 161 (in Russian).
40. Semakov, A.V., Kulichikhin, V.G., Kantor, G.Ja. *et al.* (1992) *Vysokomol. Soed.*, **A34**, 104.
41. Casson, N. (1959) in *Rheology of Disperse Systems*, (ed. C.C. Mill), Pergamon, Oxford.
42. Vasil'eva, O.V., Dobrosol, I.I., Parsamyan, I.L. *et al.* (1991) *Plast. Massy*, **10**, 13 (in Russian).
43. Polushkin, E.Yu., Antipov, E.M., Kulichikhin, V.G. *et al.* (1990) *Dokl. AN SSSR*, **315**, 1413 (in Russian).
44. Lin, Y.G. and Winter, H.H. (1988) *Macromolecules*, **21**, 2439.

6

The morphology and rheology of liquid crystal polymer blends

A.A. Collyer

6.1 INTRODUCTION

When two simple fluids are mixed together to form a blend the resulting viscosity sometimes obeys a mixture rule of a log additive nature:

$$\log \eta_{12} = \phi_1 \log \eta_1 + \phi_2 \log \eta_2 \tag{6.1}$$

where ϕ_1, ϕ_2 and η_{12}, η_1 and η_2 are the volume fractions of the two phases and the zero shear rate viscosities of the blend and the two phases. In many cases, this is not true, especially when the molecules involved are large, and the rheological behaviour depends on the morphology of the blend. This is particularly true of polymers, with a more complex situation occurring for partly crystalline polymer blends. The morphology of the blends is determined by:

(a) the concentration of the dispersed phase;
(b) the viscosity ratio of the two phases;
(c) their elasticity ratio;
(d) the size and form of the dispersed phase;
(e) the miscibility and presence of compatibilizers;
(f) the character and size of the interphase regions, and
(g) the method of processing, which has some influence on (d).

Rheology and Processing of Liquid Crystal Polymers
Edited by D. Acierno and A.A. Collyer
Published in 1996 by Chapman & Hall
ISBN 0 412 59640 7

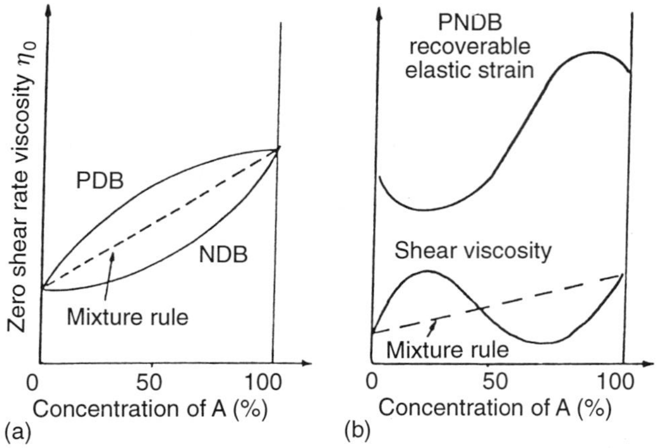

Figure 6.1 Deviations from the mixture or log additive rule (Equation (6.1)) as shown by blends of polymer A and B in a melt.

The types of behaviour that occur are shown in Fig. 6.1 and summarized as follows:

1. behaviour governed by the log additive rule, equation 6.1;
2. an increase in viscosity over that predicted by the log additive rule, as shown by miscible blends and immiscible blends in which there are strong interactions across the interface (referred to as **positive deviation blends (PDB)**);
3. a decrease in viscosity below the values given by the log additive rule, as shown by immiscible blends where adhesion at the interface is poor and the interaction weak, giving gross delamination between phases – **negative deviation blends (NDB)**); and
4. both positive and negative deviations, which occur usually when there is a concentration-dependent change in morphology such as a phase reversal or a change from a discrete dispersed phase to an interlocked structure (PNDB).

These deviations usually occur such that the blend viscosity is intermediate between those of the viscosities of the two components. This seems to be true in shear rheometer configurations such as the cone-and-plate and the Couette. However, in capillary rheometry, which involves an entrance flow comprising tensile stresses, the blend viscosity may be even lower than that of either of the blend components, and this often occurs when one component is in a very low concentration range. This gives a minimum in the graphs of viscosity versus concentration. These variations in viscosity with concentration are accompanied by variations in the elasticity of the blends such that when the viscosity is a maximum, the elasticity is a minimum and vice versa. In the case of LCP blends with isotropic polymers, there is little published work on this elasticity aspect as most of the measurements are made from the swelling ratios, which for LCPs are very small.

In the same way that the addition of a small amount of one isotropic polymer to another may lead to a blend viscosity lower than that of either of the constituents, an addition of a small amount of a liquid crystal polymer (LCP) to an isotropic polymer may give the same effect but, in many cases, this effect is enhanced. Thus it may be possible to use LCPs as processing aids in which the viscosity of a blend with a high viscosity polymer may be reduced to make processing easier or even to process an intractable polymer. Cogswell, Griffin and Rose [1] of ICI patented this idea in 1983, citing the use of LCPs with many high viscosity isotropic polymers. Siegmann, Dagan and Kenig [2] first noted that a minimum occurred in the viscosity–concentration graph in their blends of an amorphous polyamide traded under the name Trogamid T (Dynamit-Nobel, Germany) and Vectra A900 (Hoechst–Celanese) at a concentration of Vectra of about 5%.

It would be of great benefit in polymer processing to be able to understand the circumstances leading to this great reduction in viscosity so that the appropriate conditions can be generated when processing difficult melts. Unfortunately, research to this end is not yet fully developed. In this chapter, the observations described are used to formulate the prerequisites for minima to be found in viscosity–concentration curves. Often, blend viscosities are intermediate between those of the constituent polymers; sometimes, a large viscosity reduction may occur at low concentrations of the LCP; occasionally, yield stresses are observed. In all but rare cases, the blends show NDB. The review is not exhaustive but, hopefully, covers the interesting points.

A relatively small amount of research has been carried out on elongational flows and on oscillatory rheometry of LCP blends; these topics are covered at the end of the chapter.

6.2 RESULTS FOR CAPILLARY FLOWS

As a brief history of the beginning of this interest has already been given, the following discussion will be conducted from a material point of view. Various commercial LCPs and isotropic polymers have been used by the different research teams.

6.2.1 Polystyrene/LCP blends

Weiss *et al.* [5, 6] studied blends of polystyrene (PS) with an LCP synthesized by Processor Sirigu at the University of Naples. A nematic mesophase was revealed between 158°C and 251°C by X-ray diffraction and DSC. Couette viscometry and capillary rheometry results were obtained for blends at 220°C. The die used in the capillary rheometry had an entrance angle of 100° and a length-to-diameter ratio ($L:D$) of 57 : 1. The LCP domains were spherical at low shear rates and not deformed by the shear field, with blend viscosities higher than that of the PS. These workers attributed this to the dissipation of energy in the rotation and tumbling of the LCP domains.

At the higher shear rates in the capillary rheometer, the viscosities of the blends were lower than that of the PS, decreasing with increasing LCP concentration. For 10 wt% LCP, a reduction in viscosity of 40% was obtained. This reduction was attributed to the deformation and orientation of the LCP domains caused by the tensile stresses at the die entrance and the relatively long relaxation times of the LCP, which maintain this deformation and orientation, providing slip surfaces to lubricate the flow.

This work highlights the different flow regimes involved in Couette flow and in capillary flows, where tensile fields at the entrance elongate the LCP domains, thereby providing large interfacial areas over which slip can occur. This is a prerequisite for viscosity minima to occur. For this reason, mainly capillary flows will be discussed.

6.2.2 X7G/polycarbonate blends

Much of the early research was carried out with the LCP X7G (Tennessee Eastman-Kodak), which is a copolymer of 60 wt% hydroxybenzoic acid (HBA) and 40 wt% polyethyleneterephthalate (PET). The PET acts as a flexible spacer for the stiff HBA mesogens and it allows them to align in elongational and shear fields. The flexible spacer between the stiff mesogens reduces the nematic transition temperature to 200°C [7] and reduces the tensile modulus and tensile strength of the LCP below that of LCPs without flexible spacers in the main chain. This limits the effectiveness of X7G as an engineering LCP.

In general, blends involving X7G do not show a minimum in the viscosity–concentration curves, although Acierno *et al.* [8] indicated the presence of minima in the results from blends of X7G/PC examined by Blizard and Baird [9–12]. However, in the work of Acierno *et al.* [13, 14] on similar blends, no viscosity minima were observed, with blend viscosities intermediate between those of the constituents (NDB). Comments made by Acierno *et al.* [8] include the observations that, in

Couette viscometry, only a droplet morphology was apparent and that as the $L:D$ ratio of a die used in capillary extrusion was increased the LCP domain shape changed from fibres to droplets. This would be expected as the tensile stresses relax along the length of the die.

Glass transition measurements [13–15] revealed a small but significant decrease in the T_B values of the PC in the presence of the X7G. This was attributed to a plasticizing effect [13] or to a partial miscibility of the PET-rich phase of the LCP as it dissolves the PC matrix [15]. Neither physical nor chemical interaction between the two phases were observed by SEM or IR [13–15].

6.2.3 Vectra/Trogamid T blends

The Vectra range of LCPs (Hoechst–Celanese) are based on copolymers of hydroxybenzoic acid (HBA) and 6-hydroxy-2 naphthoic acid (HNA). The latter molecule is a crankshaft molecule which frustrates the chain packing of the LCP and reduces the nematic transition temperature to 275°C for Vectra A950 [16]. Vectra A900 used by Siegmann *et al.* [2] in the blends with Trogamid T (Dynamit Nobel) is similar to A950. These LCPs are straighter and stiffer than the X7G polymers as there is no flexible spacer unit in the main chain. The experiments were carried out at 260°C, below the nematic transition temperature, in a die with $L/D = 33:1$, and no entrance pressure corrections were made. Their results are shown in Figs 6.2 and 6.3. It was indicated by Hawksworth *et*

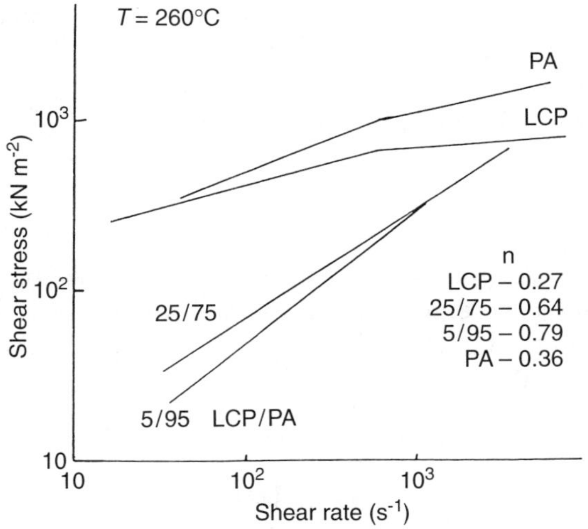

Figure 6.2 The variation of log shear stress with log shear rate for Vectra A900 and Trogamid T and two of their blends at 260°C using a capillary die of $L/D = 33$ [2]. The shear thinning index is given for each extrusion [2].

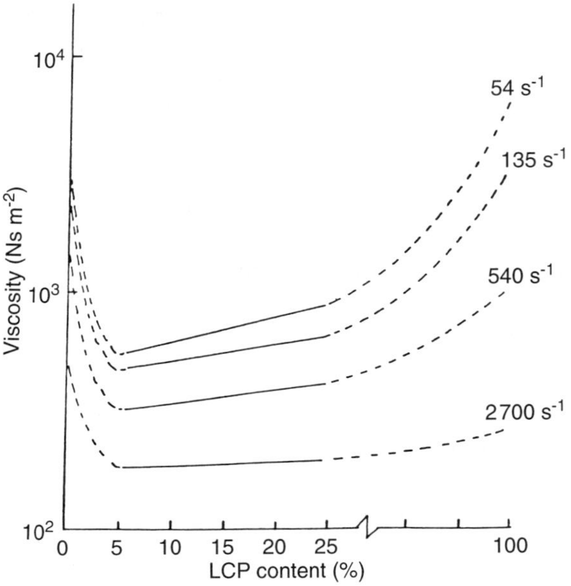

Figure 6.3 Concentration dependence of the viscosity of Vectra A900/ Trogamid T blends as a function of shear rate at 260°C [2].

al. [4] from Fig. 6.3 that the shear stresses involved were high, higher than those experienced by other workers, giving rise to flattened curves. This was attributed to the possibility of shear heating, which could raise the temperature of the sheath region to above the nematic transition temperature of the LCP. This is probably why the Vectra behaved as an LCP at the extrusion temperature. The estimate of the shear heating was made by Hawksworth *et al.* [4] as follows [17, 18]:

$$\Delta T = \frac{\Delta p}{\rho C_P}$$

where ΔT is the temperature increase due to shear heating at a pressure difference between the ends of the die of Δp (this may be an overestimate as no correction for entrance pressure was made [4]), a melt density of ρ and a specific heat capacity of C_P. The shear stress is given by

$$\sigma = \frac{\Delta p}{4\left(\frac{l}{d}\right)}$$

$$\Delta T = \frac{4\sigma\left(\frac{l}{d}\right)}{\rho C_P} \qquad (6.2)$$

It is assumed that C_P for Trogamid T is $2500\,\mathrm{J\,kg^{-1}\,K^{-1}}$ similar to that of PA11 and PA12, with a density of $1100\,\mathrm{kg\,m^{-3}}$ [17]. For values of σ of around $5 \times 10^5\,\mathrm{N\,m^{-2}}$, $\Delta T \simeq 24°\mathrm{C}$, quite sufficient to raise the temperature of the LCP above the nematic transition temperature.

Siegmann *et al.* [2] observed elongated LCP domains of a few microns in diameter and a sheath/core morphology, which implies that the LCP has softened. They noted a good interfacial adhesion, but this was during mechanical tests and not at the extrusion temperature.

6.2.4 Vectra A950/polycarbonate blends

Isayev and Modic [19], using L/D ratios of 30 and 40 and temperatures above the nematic transition temperature (280°C and 310°C), did not correct for entrance pressure. They noted minima in viscosity at low and high shear rates for 310°C. They remarked that, as the L/D ratio is decreased, the tensile strength of the extrudates increased and the uniformity and the number of fibrils increased. At concentrations of LCP of 5–10% the extrudates peeled, and they attributed this to the formation of nearly continuous fibres of diameter 2–5 μm distributed randomly across the extrudate.

Nishimura and Sakai [16] took measurements close to and below the transition temperature (250°C, 260°C, 270°C and 280°C) using a die with $L/D = 10$. There were no corrections for entrance pressure or to the apparent shear rate. Minima were seen at 50/50 mixtures. At 260°C, the LCP domains were lumpy and irregular whereas, when the LCP had softened at 280°C, the domains were rod-shaped and of diameter 1–8 μm.

Malik *et al.* [20] investigated similar blends, finding that the LCP additive reduced the viscosity of the PC matrix but, in all cases, the viscosities were intermediate between those of the components. The flow curves showed a greater degree of shear thinning behaviour at the higher LCP concentrations. Beery *et al.* [21] used the same sample of LCP and they obtained similar results to those of Acierno *et al.* [13] and Nobile *et al.* [14].

Kohli *et al.* [22] used a different form of Vectra in PC. Vectra RD500, synthesized from HBA/HNA/terephthalic acid(TA)/hydroquinone(HQ), which has a transition temperature at 236°C. They carried out experiments at 270°C in capillary and cone-and-plate flow. They obtained viscosity minima at high shear rates in dies for 5–10% LCP, where the morphology consisted of elongated particles, but at low shear rates in capillary flow and in the cone-and-plate rheometer, a droplet morphology occurred and no minima were found.

6.2.5 Vectra B950 in polyamide 6, polybutyleneterephthalate blends and polyamide 12

Vectra B950 is a polyesteramide comprising units of 60HNA/20TA/20 aminophenol (AP). The AP produces some hydrogen bonding but, apart from that, its behaviour is very similar (with a similar transition temperature) to copolyesters in which the HQ replaces the AP [23]. The transition temperature of B950 is 285°C [24]. The presence of the AP may make it possible to map the position, shape and size of the LCP domains in a non-amide matrix by looking at the NH stretching modes or the NH structure.

La Mantia *et al.* [25–27] extruded blends of Vectra B950/PA6 in a capillary viscometer (Rheoscope 1000) using a die with $L/D = 40$ at temperatures 260°C and 290°C. Figure 6.4 shows the variation in torque with LCP concentration. The maximum at about 1% LCP is an interesting feature of this graph and will be discussed later. At higher concentrations, the viscosity falls and remains low for 10–20% LCP. These investigators noted that, even when extruding the LCP below its transition temperature, there may be fibres formed on passing through a zero length die with a conical inlet at a shear rate of $1200\,s^{-1}$. They attributed the viscosity minimum to the fibre formation and interfacial slip, with a possibility of migration of the LCP phase to the regions of higher shear rate. They stated that the fibrils can be lost in long dies because the large stresses transmitted from the matrix phase to the dispersed phase break

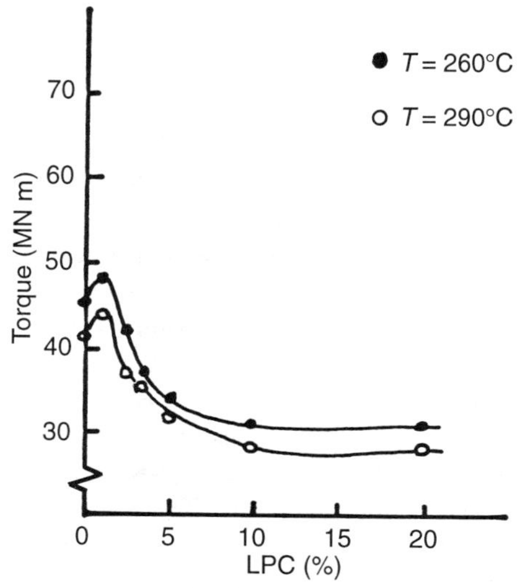

Figure 6.4 The variation in torque with LCP concentration in Vectra B950/PA6 blends at 260°C and 290°C [25, 26].

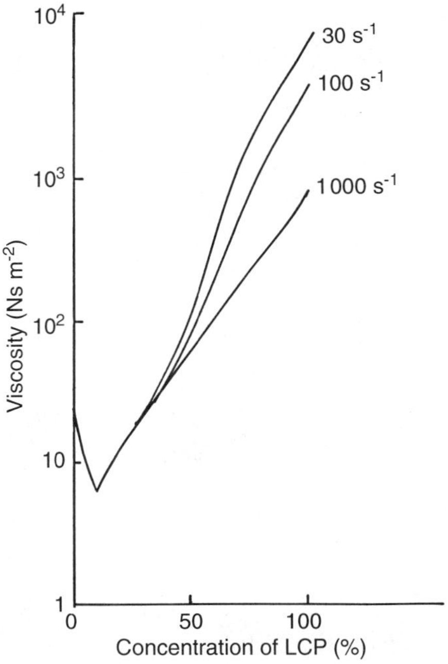

Figure 6.5 Viscosity–concentration curves for blends of Vectra B950 with polybutyleneterephthalate at 260°C [25, 26].

the fibrils, or when the orientational relaxation time of the LCP is less than the average time to flow through the die.

They noticed that, at low shear rates, the blend behaviour indicated the presence of a yield stress. The viscosity minimum and the yield stress were both apparent even at extrusion temperatures lower than the transition temperature of the LCP.

La Mantia *et al.* [25–27] also blended Vectra B950 with polybutylene-terephthalate (PBT) and obtained the viscosity minima shown in Fig. 6.5. The viscosity of the LCP is higher than that of the PBT as it was in the case of the PA6. There was no maximum produced as with the former blend but there was evidence of a yield value.

Using Vectra A, Chung [28] investigated blends with PA12. The viscosity–concentration curves obtained are shown in Fig. 6.6 and are very complex, with both maxima and minima present. Depending on the shear rate, minima are found both at low and high LCP concentration. Chung [28] suggested that, at low concentrations, the LCP acted as a lubricant of the PA matrix whereas, at high concentrations, an interlocked morphology was developed, which increased the viscosity.

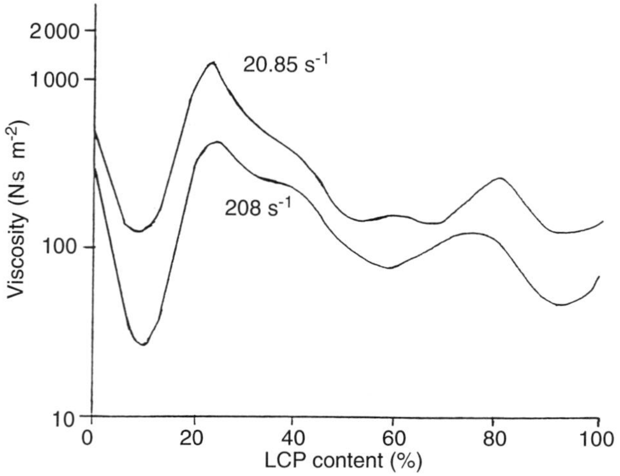

Figure 6.6 Viscosity–concentration curves for Vectra A/PA12 [27].

6.2.6 Vectra B950 in polycarbonate and polyethersulphone

Blends of Vectra B950 and two samples of PC were investigated by La Mantia *et al.* [29, 30]. Low concentrations of LCP remarkably reduced the viscosities of the blends, with non-newtonian behaviour becoming more evident with increasing LCP content. When the viscosity of the matrix material (PC) was higher than that of the LCP, the blend viscosities were intermediate between the two; when the viscosity of the two components were comparable, the viscosity of the blend showed a gradual minimum at about 40% LCP. For viscosity minima to occur, it is necessary for the viscosity of the LCP to be similar to or greater than that of the matrix polymer [30]. SEM micrographs indicated that the dispersed phase was spherical in shape and that the droplet size increased with decreasing shear rate and PC viscosity.

The present author and his co-workers [3, 4, 24, 31] found no minima when working with Vectra B950/PC blends and B950/PES using a die with $L/D = 20$ at extrusion temperatures of 270°C and 320°C or with blends involving polyethersulphone (PES) at 330°C. Under the experimental conditions, the viscosity of the LCP was less than that of the matrix polymer. In both cases, the LCP domains were elongated and aligned along the streamlines, with the $L:D$ ratio of the domain of 10 : 1, which seems less than reported elsewhere. It was supposed that the domains were insufficiently elongated to provide slip surfaces capable of giving a minimum. Migration of the LCP to the sheath region was not found. It was felt that the hydrogen bonding in the polyesteramide was sufficient to make the Vectra B950 so incompatible with PC or PES that it was difficult to generate a large interface, which was not the case for blends with PA6 and PBT [25–27]. However, Magagnini *et al.* [27, 32] found that blends of PES with Vectra B950 showed minima when the viscosities of the two components were similar; when that of the PES was the greater, the blend viscosities were intermediate between those of the components.

The present author and his co-workers [3, 4, 24, 31] examined the sheath/core morphology of the extrudates. This was done by examining the barrier properties of B950/PES extrudates to the ingress of water. Although the ingress of water into PES is not substantial, it is greatly reduced by the addition of small amounts of B950, as shown in Fig. 6.7. The water uptake is halved by adding as little as 6% LCP. This could suggest a migration of the LCP to the sheath region, although no morphological evidence was found to support this.

The second investigation concerned the ease by which the sheath region of an extrudate so readily peels away [2–4, 14, 16, 19, 31] in the same way as for extrudates of pure LCP; and it is known that some injection mouldings delaminate readily at the sheath/core interface. As little as 7% B950 in PES and PC will produce this effect. Figure 6.8 shows the darker sheath region surrounding the core in Fig. 6.8(b). This demarcation is not shown in Fig. 6.8(a), which was from an extrusion

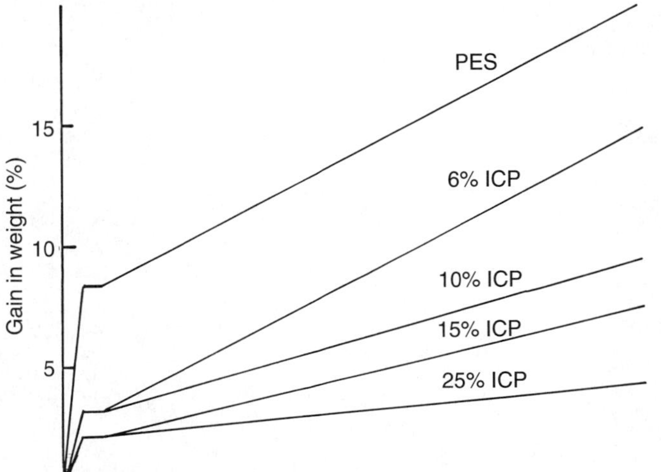

Figure 6.7 Water absorption versus time for Vectra B950/Polyethersulphone blends: 6%, 10%, 15% and 25% LCP in PES [3, 4, 24, 31].

(a)

(b)

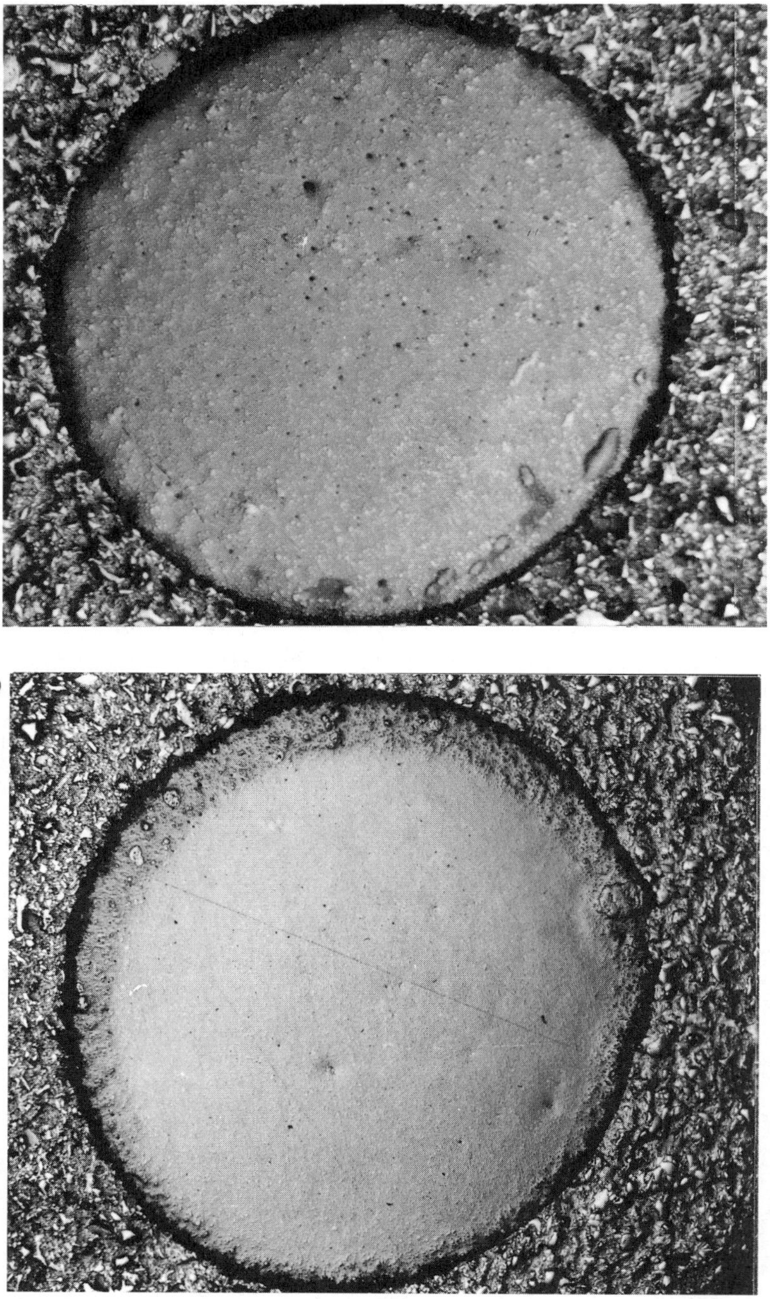

Figure 6.8 The position of the sheath/core interface in the extrudate of a 25Vectra B950/75PC blend: (a) below the transition temperature, 280°C, and (b) above it, at 310°C. Note the darker sheath region in (b).

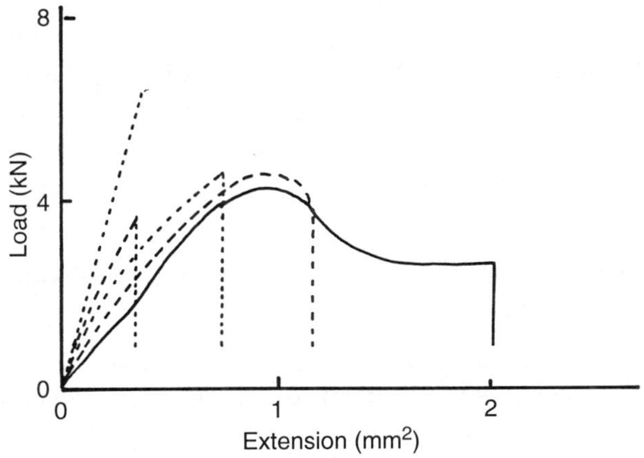

Figure 6.9 Tensile test results on Vectra B950/PES blends, showing increased brittleness and pure LCP response to stress as the LCP concentration increases [3, 4, 24, 31].

below the transition temperature (25Vectra/75PC at 280°C and 310°C). Measurements on a small number of extrudates showed that 16% of the blend lies outside this interface for a 25% LCP concentration. For 20% LCP in PC, there was a large variation in the results, with an average value of 19% lying outside in the sheath region, with a standard deviation of 2.5%. This indicates that migration is almost complete but the authors doubt this still.

Isayev and Modic [19] noted that the tensile strength of the extrudates increased when the $L:D$ ratio of the die was decreased. The present author and his co-workers [3, 4, 24, 31] carried out mechanical tests on blends to examine a different aspect, namely the effect of LCP concentration. Figure 6.9 shows results of tensile tests on B950/PES tensile specimens. As the LCP concentration increases, there is a marked shift towards brittle behaviour, characteristic of a pure LCP response. Again, as little as 15% LCP is needed to produce a marked departure from ductile PES behaviour. This could be attributed to the increase in the size of the sheath region of more oriented particles, which may be due to migration of LCP, but it is doubtful.

An explanation given by those authors is that migration of LCP is not a prerequisite, but that the LCP imposes a more aligned morphology on the isotropic polymer matrix by virtue of its presence, and this is what produces the improved barrier properties and the observed mechanical behaviour.

6.2.7 SBH/polycarbonate blends

La Mantia *et al.* [27, 33] used blends of a semi-rigid LCP (called SBH) in PC. This LCP was developed by Eniricerche and comprised sebacic acid (SA)/HBA/biphenyl (BP). In this case, the viscosity of the LCP is much lower than that of the PC, and the blend viscosities were intermediate between those of the constituents.

6.2.8 Wholly aromatic copolyester with polycarbonate and polyamide 66

Hoeck *et al.* [34] used a wholly aromatic copolyester with PC, whose viscosity was sometimes higher than, and sometimes lower than, that of the PC. When the LCP had the higher viscosity, which was at low shear rates, a minimum in viscosity of the blend was observed for an LCP content of 17.5%; when the flow curves of the two components intersected, the viscosity of the blends was intermediate between those of the components.

Hoeck *et al.* [34] also used blends of their LCP with PA66. The viscosity of the LCP was larger than that of the PA over the entire shear rate range, and viscosity minima appeared. Their results were very much dependent upon the way in which the original blends were mixed: whether co-rotating or counter-rotating extruders were used in compounding.

6.2.9 LCP blends with polyetherimide (PEI) and polysulphone (PSF)

The interesting part of this subject is the idea of blending LCPs with intractable or difficult polymers, which usually have a high service temperature and hence a high processing temperature. Regrettably, not much work has been carried out on these more difficult polymers. PES falls into this category but nevertheless has been lumped together for convenience and discussed with PC. PEI, in the commercial form of Ultem (General Electric) has few LCPs suitable for blending with it because of its high processing temperature.

Isayev and Swaminathan [35] and Acierno *et al.* [8, 36–38] blended PEI with Vectra A950 and K161 (Bayer) respectively. The K161 comprises HBA/isophthalic acid (IA)/TA/HQ/BP. This LCP is one of the few that can tolerate high processing temperatures. In all cases, the viscosities of the blends were reduced by the LCP additives but there were no minima. At high shear rates, an addition of about 5% LCP reduced the blend viscosity by about 30%. For both 5% and 10% LCP blends, the morphology contained spheroidal LCP domains of about

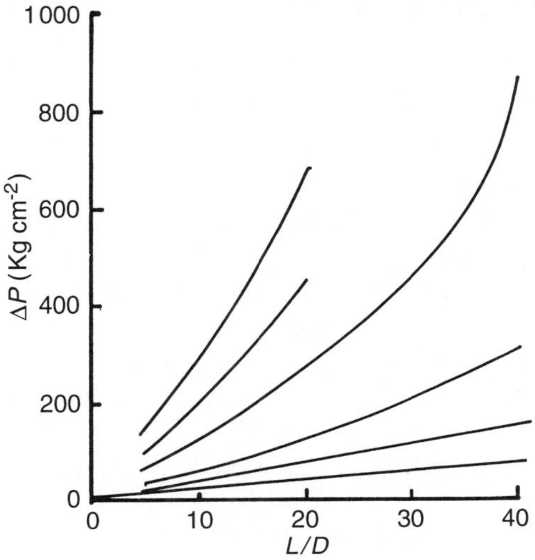

Figure 6.10 Bagley plots for polyetherimide at 330°C [37].

0.5 μm. For the 30% LCP blends, the initial LCP domains were of about 3 μm in size and, on extrusion at the higher shear rates, these deformed into long threadlike fibrils aligned along the streamlines. The results were for a die with L/D of 10, shear rate of $800\,s^{-1}$ at 330°C. The viscosity of Ultem and Ultem-rich blends were pressure dependent [37], giving nonlinear Bagley plots, as shown in Fig. 6.10. This sensitivity of amorphous polymers to hydrostatic pressure is well known [39–42]. Acierno *et al.* [8] corrected for entrance pressure using dies with L/D of 5 and 10, but later discovered that the entrance pressure was only about 10% the total pressure drop, and results were similar to the cone-and-plate values.

Kulichikhin *et al.* [43, 44] used Ultrax 4002 (BASF) as the LCP in PSF. Ultrax comprises BP/TA/IA/HQ. By changing the extrusion temperatures, these observers were able to modify considerably the ratio of the viscosities of the two components. This was due to their considerably different activation energies for flow. In all cases, the viscosity of the LCP was higher than the PSF although, at the lowest temperature, the viscosities were similar. These curves gave two minima at about 25% LCP and 75% LCP for 240°C, a single minimum at about 50% LCP at 260°C and a log additive response at 280°C. Kulichikhin, Shumskii and Semakov have presented their work on LCP/PSF blends with fillers in the previous chapter, and include a brief discussion on LCP blends to put their results in context in section 5.3.

6.3 SUMMARY OF RESULTS FOR CAPILLARY FLOW

The following points are generally agreed between the previous authors:

(a) LCP/isotropic polymer blends are generally NDBs, except for Vectra B950/PA6 at about 1% Vectra.
(b) Some, but not all, blends give viscosity minima.
(c) For viscosity minima to occur, the LCP viscosity must be higher than that of the matrix polymer or at least comparable to it; if the matrix polymer has the higher viscosity, the blend viscosities lie between those of the component polymers giving NDB.
(d) A sheath/core morphology is present even for low concentrations of LCP.
(e) A fibrillar LCP structure is much more likely for low values of die $L:D$ ratio.
(f) Elongation of the LCP dispersed phase is necessary for large viscosity reductions such as minima.
(g) The minimum viscosity may be due to interfacial slip.
(h) Some blends give rise to yield values, namely Vectra B950 with PA6 or PBT.

6.4 MODELS TO EXPLAIN VISCOSITY MINIMA

As mentioned earlier, viscosity minima are not restricted to LCP blends, although they seem to produce the most startling effects. Below are some of the suggestions used to explain viscosity minima in polymer blends.

6.4.1 Phase equilibria

The Hildebrand solubility parameter δ has been used to indicate the miscibility of polymer species or the miscibility of a polymer in a potential solvent or plasticizer. This parameter may be used when there is no specific interaction between the two species, such as hydrogen bonding. The difference between the solubility parameters of the two polymer species ($\Delta\delta = (\delta_1 - \delta_2)$) determines whether or not they will mix, because $\Delta\delta$ is related to the heat of mixing by

$$\Delta H = V_{\mathrm{T}} \Delta\delta \phi_{\mathrm{T}} \phi_2 \qquad (6.3)$$

where V_{T} is the total volume of the mixture and ϕ_1, ϕ_2 are the volume fractions of the two species. If $\Delta\delta$ lies between 0.4 and 0.8 $\mathrm{MJ^{1/2}\,m^{-3/2}}$, a two-phase blend will form with good interfacial adhesion [45].

In general, polymer molecules are immiscible, but the presence of specific interactions or the presence of one species in another in small quantities makes the behaviour of polymer pairs in a blend difficult to foresee, and the effect of flow or of the blend morphology on the solubility of the molecules is not fully understood. The Hildebrand solubility parameter is too simple to give satisfactory predictions but it may be used as an indicator. The method of a guided 'try and see' seems to be the correct approach.

The variation of viscosity with concentration of polyoxymethylene (POM) in cellulose acetate butyrate (CAB) was explained by Lipatov *et al.* [46] in terms of thermodynamic equilibrium by considering the heat of mixing ΔH. The way in which both ΔH and η vary with the concentration of POM are shown in Fig. 6.11. On increasing the concentration of POM in CAB, the heat of mixing decreases, showing compatibility

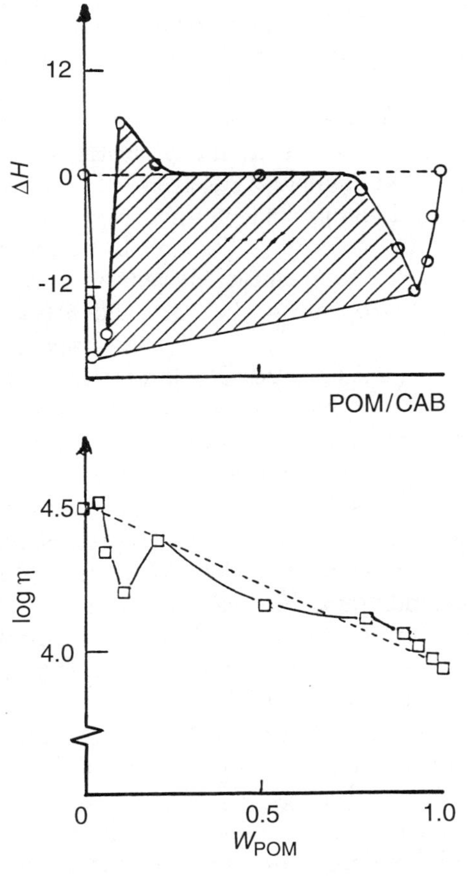

Figure 6.11 Variation of the heat of mixing ΔH and the viscosity η with the weight fraction of polyoxymethylene in cellulose acetate butyrate. The shaded area represents the region of immiscibility [45].

between the two species and a binodal (miscible) composition. This causes the viscosity to rise. As the concentration of POM is increased further the heat of mixing increases and the thermodynamic conditions become meta-stable in the region between the binodal and spinodal (immiscible) compositions. These two species are not particularly compatible and the increase in the interfacial area causes the viscosity to fall. A further increase in POM concentration gives rise to an increase in particle size, because the incompatibility of the two species encourages the particles to increase in size, rather than a further increase in the number of particles. Local perturbations in particle density lead to nucleation and growth of particles. The overall interfacial area may now decrease, giving less interfacial slip and an attendant increase in viscosity. Further increases in POM concentration cause a spontaneous phase separation as the spinodal regime is reached. From Fig. 6.11, the blend viscosity hovers around the log additive value. This explanation accounts for the appearance of viscosity minima in blends with one component in a great excess over the other. The degree of compatibility of the polymer pair may decide at what concentration the minimum occurs.

Limited compatibility occurs in many systems, one of which has already been mentioned, namely X7G in PC [13–15]. Wisniewski *et al.* [47] observed a limited compatibility between polystyrene (PS) and PC at low concentrations of PS (less than 5%). At these concentrations, the T_g of the PC in the blend was depressed by 10°C, with an intermediate transition appearing at 120°C. At 5% PS, a spontaneous phase separation set in, the T_g of the PC returned to 149°C and the intermediate transition disappeared. The intermediate transition was due to relaxations in the interphase region, which is a mixture of the two species, and disappears when this region reduces in volume. Limited compatibility and specific interactions may be investigated by DSC and DMTA, and appear as a shift of the usual relaxation peaks. Kulichikhin *et al.* [44, 48] reported work on blends of a small amount of Ultrax KR-4002 in polysulphone (PSF). The two-phase nature of the blend was shown by the individual T_g peaks (122°C for Ultrax and 183°C for PSF). The temperatures at which the β- and γ-peaks occurred also differed, −35°C and −87°C for Ultrax and PSF respectively. For small additions of Ultrax, the T_g of PSF lowered (although the initial T_g persisted), while for small additions of PSF to Ultrax the T_g of the latter rose. Adding a small amount of one of the polymers to the other caused a rise in both γ-processes.

The interpretation of the two values of T_g for PSF is that a more mobile form of PSF develops close to the surface of the dispersed Ultrax phase with a T_g of 177°C, while far away from this surface in the bulk of the PSF the T_g remains at 183°C. Close to the boundary separating the Ultrax particles from the PSF matrix, an increase is observed in the rigidity of the Ultrax molecules in the presence of the more flexible PSF molecules, giving a rise in the T_g of the Ultrax.

Unfortunately, not much of this type of work has been reported for blends of LCPs. It would be interesting to obtain results for blends of PA6 with Vectra B950. As it is, the explanation of the maximum in the viscosity–concentration curves for this blend near 1% may be attributed to a specific interaction between the two polymers at low Vectra B950 concentration. This may be due to hydrogen bonding between the amide part of the LCP and the matrix material but, when the number of LCP particles reaches a critical value spontaneous phase separation sets in, giving the more expected response.

From these kinds of considerations, the thermodynamics of the blend system must be suitable for providing large interfacial area, an area over which slip can occur. Compatibility and specific interactions are, therefore, generally undesirable, and the meta-stable region between the binodal and spinodal compositions may provide the best conditions for large slip surfaces. The problem encountered then is the effect this may have on the mechanical properties of the final product.

6.4.2 Droplet morphology

The occurrence of viscosity minima is assisted by the elongation of the dispersed phase. Plochocki [49] observed minima in polyethylene (PE)/polypropylene (PP) blends for a die with $L:D$ ratio of 33 : 1, but not for $L:D$ of 66 : 1, in which some of the droplet elongation was lost. Similar results have been noted for LCP blends [8, 19, 22]. The ability of LCPs to form fibres, and maintain them during the relaxation of the tensile stresses during capillary flow, accounts for their greater ability to give viscosity minima than blends involving isotropic polymers.

The conditions determining whether fibrils, and hence a large interface, or ellipsoidal domains occur are determined by λ, the ratio of the viscosities of the dispersed to the matrix phases, and λ_N, the ratio of the first normal stress differences of the dispersed to the matrix phases. Originally, it was believed that if $\lambda < 1$ a fibrillar structure would result, but this condition is not now believed to be sufficient [50, 51]. Ablazova *et al.* [52] and Tsebrenko *et al.* [53–55] widely investigated the formation of fibrillar morphologies. The optimum viscosity ratio range for the formation and retention of fibres during capillary flow is $0.76 < \lambda < 0.91$. Values of $\lambda \ll 1$ will be favourable for fibre formation but breakup will occur as the tensile stresses in the die relax. More recent research has shown that λ_N is important in influencing the driving force for domain deformation; for fibre formation, $\lambda_N \ll 1$. This is the situation in LCPs, which show small or negative first normal stress differences. Utracki [56] gives a more detailed coverage of this area.

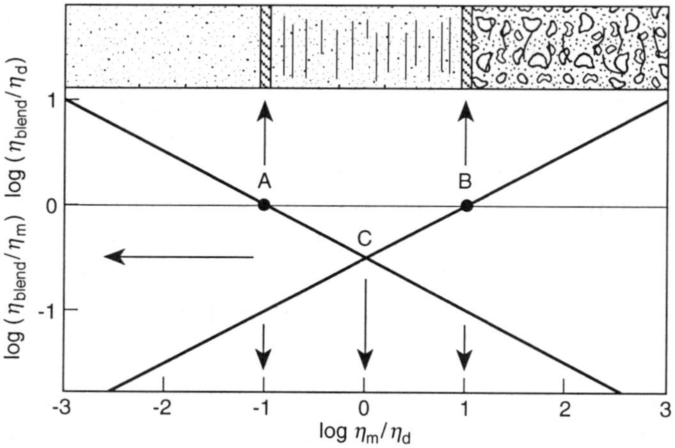

Figure 6.12 Dependence of (a) the viscosity of the blend normalized for that of the matrix and (b) the viscosity of the blend normalized for that of the dispersed phase on the viscosity ratio of the matrix to the dispersed phase [44, 57].

Kulichikhin *et al.* [44, 57] obtained viscosities of blends containing LCPs over a wide range of viscosity and concentration and proposed a model that used the initial viscosities of the blends and the constituents only to predict the morphology of the blends. This was based on data from 13 different polymer pairs including LCPs. Figure 6.12 gives the variation of η_{blend}/η_d and $\eta_{blend}\eta_m$ with $\xi = 1/\lambda = \eta_m/\eta_d$ on a log-log plot, where η_d is the viscosity of the dispersed phase, η_m is the viscosity of the matrix phase and η_{blend} is the viscosity of the blend. As ξ increases η_{blend}/η_m decreases in a linear manner and η_{blend}/η_d increases in a linear manner. The triangle ABC is of particular importance: the points A and B intersect the lines where $\eta_{blend} = \eta_d = \eta_m$; the point C corresponds to $\eta_d = \eta_m$. The characteristic points separate different morphological flow patterns shown schematically in the figure.

For very low ξ, the particles of the dispersed phase scarcely deform in the flow and the blend acts as a filled melt. At point A, $\eta_{blend} = \eta_m$, the presence of the dispersed phase hardly disturbs the flow of the matrix. At this point, asymmetric drops begin to form, which are converted into threads on passage from A to C on the graph. A further increase in ξ causes a fall in η_{blend} as a result of the presence of liquid threads of the dispersed phase and as a consequence of the formation of a skin of the dispersed phase at the edge of the flow. The formation of such a skin is completed when $\xi \simeq 8$–10, and this corresponds to the condition $\eta_{blend} = \eta_d$. LCP dispersed phases are particularly effective at forming this skin. At $\xi > 10$, the fibres are transformed into layers and the viscosity of the blend continues to fall as a result of a macrostructural plasticization of the blend.

This is a very interesting approach showing the decisive influence of the viscosity only in determining the morphology of the blends. Neither surface effects nor elastic properties are needed in this method. The interaction of the components and the surface phenomena, in particular, play an important role in the region of the edges of the triangle ABC, where experimental data suffered the greatest scatter.

6.4.3 Migration

The migration of the lower viscosity component to the die wall or to the regions of higher shear rate close to the wall is possible and would help to lubricate the flow. Collyer *et al.* [58–60], working with blends of polydimethylsiloxane (PDMS) and polyethersulphone (PES), sought evidence for the migration of the PDMS phases to the edges of extrudates using EDX by following the silicon in the PDMS. They did not succeed in finding migration but did not rule out the possibility. Verhoogt *et al.* [61], using blends of Vectra A900 and Kraton G1650 (Shell), which is a styrene/ethylene-butylene/styrene block copolymer elastomer, observed partial migration of the LCP to the sheath region, where the LCP fibres were much longer than in the core.

Han [62] discussed two mechanisms leading to the complete migration of one phase to the wall of a long capillary. This investigator found that migration of the less viscous phase to the wall occurs if $\lambda < 1$ and $\lambda_N = 1$. For $\lambda = 1$ and $\lambda_N < 1$, the more elastic phase will migrate to the die wall. Unfortunately, the LCP is the less elastic phase and migration may not be encouraged. Nevertheless, it is possible to obtain viscosity minima when a less viscous phase migrates to the wall region and it is instructive to see how this comes about.

Consider a co-extruded flow in which the variation of the viscosity of the two components is as given in Fig. 6.13. The important feature in this figure is the crossover point, after which material A becomes less viscous than material B. A viscosity minimum is possible without the need for interfacial slip if it can be arranged that material A occupies the high shear rate regimes, where $\eta_A < \eta_B$, and material B lies in the core region, $\eta_B < \eta_A$. This is illustrated in Fig. 6.14 in a plot of pressure gradient against concentration of material A. The shear rate close to the die wall must be greater than the shear rate at the crossover point in Fig. 6.13. When this is no longer true, the blend viscosity rises. If there were interfacial slip, the viscosity minimum would be much deeper.

The other observation of Verhoogt *et al.* [61], namely that the LCP fibres were much longer in the sheath region than in the core, may be

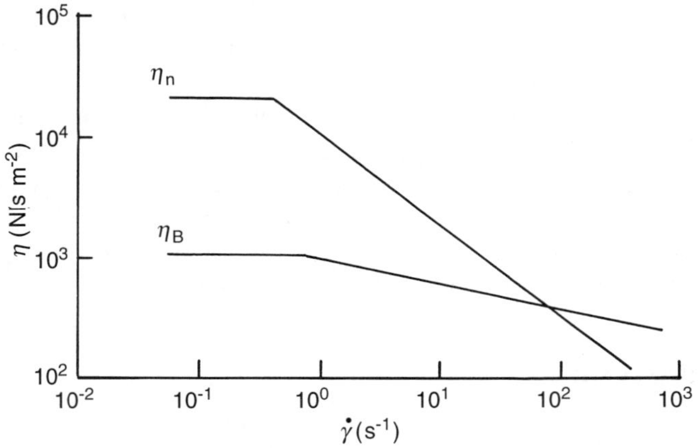

Figure 6.13 Viscosity against shear rate for two fluids whose viscosities cross over at a particular shear rate [62].

explained from a model proposed by Alderman and Mackley [63]. They stated that, outside a certain radius, there is an acceleration of the flow to permit the parabolic flow profile just before the die exit to change to a plane velocity profile in the free flow outside the die. This acceleration would lead to extensional forces that would elongate the LCP domains. Inside this radius, the domains would suffer deceleration, discouraging

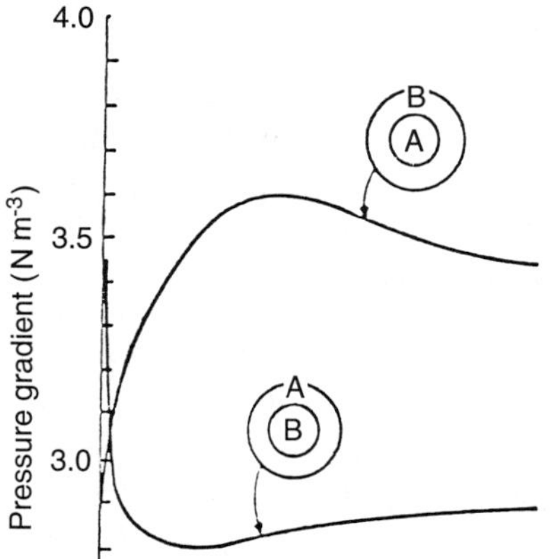

Figure 6.14 Theoretically predicted pressure gradient against volumetric flow rate for the two fluids in Fig 6.13 [62].

alignment and elongation. The simple theory they developed from this model was not very accurate in the prediction of the position of the sheath/core interface in pure LCP extrudates but, nevertheless, it was believed by Collyer [3] and Hawksworth *et al.* [4] that the basic model was correct.

Migration of the lower viscosity component to the high shear rate regions may not be important to the production of viscosity minima in LCP blends because, whereas dies of the larger $L:D$ ratios are more conducive to allowing migration, it is the dies of the lower $L:D$ ratios that encourage viscosity minima. Moreover, Han's theory favours migration of the more elastic component for components of comparable viscosity. Nevertheless, the co-extruded flow above does give rise to viscosity minima.

6.4.4 Interfacial slip

Several observers have attributed the viscosity minima to interfacial slip or slip at the die wall [3, 4, 8, 13, 14, 24, 31, 64–70]. Dispersed phases that maintain large elongations will provide large interfacial areas over which slip may occur. If this is accompanied by partial or complete migration, slip would occur in the higher shear rate regions and at the die wall. Migration and slip at the die wall give the lubrication of a plug flow by a low viscosity component in the extrusion of gun propellant cords [65–70]. In this process, nitrocellulose is extruded in the presence of a solvent (acetone) and nitroglycerine, which acts as a plasticizer. In experiments in which the nitroglycerine has been replaced by a poorer plasticizer of nitrocellulose, or in modelling experiments involving cellulose acetate/acetone/dibutyl phthalate, the poor compatibility between the solid and liquid components and the high hydrostatic pressures involved during the extrusion, causes a catastrophic phase separation, such that liquid was observed running out of the die with the solid cord. This gave considerable slip.

Lyngaae-Jørgensen *et al.* [71] demonstrated inter-layer slip in polystyrene/polymethylmethacrylate multi-layer samples and calculated the interphase viscosity. This was at least ten times smaller than that of the lower viscosity component. Such conditions would certainly give rise to minima, and it is believed that inter-layer slip is the main cause of these minima in LCP blends [3–6, 24–27, 31].

6.5 ELONGATIONAL FLOWS

There are few results involving the elongational flow of LCP blends because the high temperatures involved make measurements difficult. La Mantia *et al.* [27, 72–74] took non-isothermal data using a capillary viscometer equipped with a drawing module, which draws off an extruded rod at increasing speed. The force acting on the extrudate was monitored during the drawing to record the force at break. From this, the melt strength (MS) of the materials could be obtained, with the breaking stretch ratio (BSR). The materials used were PC/SBH (Eniricerche) and PA/Vectra B950. All results showed a slight increase of MS with increasing LCP content and a reduction of BSR. Results obtained by Baird and Sun [75] and Acierno *et al.* [36] were in agreement. The reduction in the BSR was expected on grounds of the incompatibility of the blend components, but the increase in the MS was not expected because of the reduction in the viscosity of these blends. However, it was noted [27] that the MS is a measure of the non-isothermal elongational viscosity so that the variation of viscosity with temperature may be important. In these blends, the LCP crystallizes at a higher temperature than the isotropic polymer, which means that the non-isothermal elongational viscosity of the blend can be higher than that of the matrix, enhancing the MS.

The morphologies involved in elongational flows consist of fibrils of the dispersed phase, but this requires relatively high draw ratios [76]. For PA/Vectra B950 blends, draw ratios of 50% are needed to give a fibrillar morphology [76]. By considering the rheological properties of the two components and the cooling conditions at the die exit, Blizard *et al.* [77] predicted the resultant morphology in the extrudates obtained in their experiments. Baird and Ramanathan [78] showed that the viscosity ratio is not the sole parameter affecting the fibrillation of the LCP phase. They demonstrated that two different LCP phases of similar viscosity used in the same matrix produced different sized droplet or fibrils, depending on the LCPs' viscosity–temperature dependence. Fibrils are formed and retained when the solidification behaviour of the LCP and the matrix polymer are similar, and fibrils are formed only when the LCP viscosity does not exceed that of the matrix.

6.6 DYNAMIC MEASUREMENTS

Results from oscillatory rheometry do not agree with those from capillary rheometry because of the lack of elongational flow, although occasionally minima have been found when the viscosity of the LCP is higher than, or comparable to, that of the isotropic polymer.

Dynamic viscosity results from blends of PA66 and X7G (Tennessee Eastman-Kodak) showed that the dynamic viscosity falls drastically with increased LCP content [12]. Sukhadia *et al.* [79] reported results of dynamic experiments with two different LCPs in PET, PET/60 HBA and PET/80 HBA. The complex viscosity of the blends was higher than that of the PET at low frequencies but not at high frequencies. The flow curves of the blends were not essentially different from that of pure PET, and only a blend with 20% LCP had a lower dynamic viscosity than that of PET. The behaviour was attributed to the relative compatibility between the two phases. It was noted by La Mantia [27] that the flow curves of blends were below that of the constituents when the viscosity of the LCP was higher than or equal to that of PET; the blend viscosities were intermediate when the viscosity of PET was greater than that of the LCP.

Blends of Vectra B950 copolyesteramide with a mixture of miscible components, 70PS/30PPE, showed a marked decrease in dynamic viscosity at high LCP content [80]. A maximum was observed at low frequency for 20% LCP, and this was attributed to the presence of a yield stress.

James *et al.* [81] reported that the dynamic viscosity of blends of PES/(HBA/HNA) copolyester was drastically reduced by the addition of about 2% LCP. At low LCP concentration, the flow curves of the blends were of similar shape to those of the PES; only at high frequencies was non-newtonian behaviour apparent. The flow curves for 2% LCP and 20% LCP were below those of the pure components at all frequencies.

Biing-Lin [82] used sweep strain tests on blends of chlorinated polyvinylchloride (CPVC)-PET/60HBA. Results showed that G^1 decreased as the strain increased. He attributed this to the existence of LCP microstructures in the blend. In frequency sweep tests, the blends and the pure materials showed remarkable shear thinning behaviour. The dynamic viscosity of the LCP was higher than that of CPVC and the blend viscosities were intermediate between those of the components, being higher than that of the isotropic matrix. Results on these blends from capillary experiments do not agree with the dynamic results and the Cox–Merz rule was not obeyed, as would be expected.

6.7 CONCLUSION (CAPILLARY FLOWS)

As mentioned in more detail by Kulichikhin *et al.* in Chapter 5, and therefore not mentioned before here, the apparent or 'bulk' viscosity in LCPs consists of at least two different coefficients, along and across the director and, in LCP domains that are long and thin, the viscosity in the direction of flow will be less than the bulk value; this will naturally lower the viscosity of the blend. Another possibility mentioned in section 5.1 is due to the matrix: in this model, the 'gaps' between droplets of the dispersed phase will give rise to locally high values of shear rate, and this will lead to a decrease in the apparent blend viscosity and may account for the shift in T_g. This applies when the viscosity of the dispersed component is higher than that of the matrix component. Both are discussed more fully in Chapter 5.

Thus, various phenomena could assist in giving the minimum turning point in the viscosity–concentration graphs for LCP blends in capillary flows at low concentrations of LCP. Many of these phenomena also occur in blends of isotropic polymers but the effect in LCP blends is often more marked.

6.7.1 Model to explain the viscosity minimum in capillary flows

The model proposed by Collyer *et al.* [3, 4, 24, 31] to explain the minimum at low LCP concentrations indicates that, in these circumstances, the thermodynamic conditions of the blend are in a meta-stable regime between the binodal and spinodal. This gives rise to numerous small domains. If there is poor compatibility between the two phases, the interfacial slip may permit the low concentration phase to lubricate the matrix phase, giving a shift to lower temperatures of the T_g of the matrix phase and perhaps an intermediate peak; this will lower the viscosity of the blend. As the concentration of the LCP is increased, nucleation and growth occurs and more domains will be created, increasing the interfacial area over which slip can occur and thereby decreasing the blend viscosity. Eventually, spontaneous phase separation sets in and the turning point occurs when further increases in LCP content lead to an increase in the size of the original domains rather than to the creation of new ones; although there is more LCP, the interfacial area has reached its maximum value (the concentration at which this occurs may be influenced by the degree of compatibility between the constituents). Thereafter, further increases in LCP content will give an increase in domain size and many domains may coalesce. This will reduce the surface over which slip can occur and the viscosity will rise towards the log additive value.

The ease with which LCP domains are oriented and elongated and retain their elongation during passage through the die is the reason for the enhanced effect over isotropic polymer blends. The effect seems to be greatest when $\lambda \simeq 1$, which is in agreement with practice, and for LCPs $\lambda_N \ll 1$, which again assists the formation of fibrils [52–56]. In the theory of Kulichikhin *et al.* [44, 57], the best morphology for giving minima occurs when $\xi = 1/\lambda \simeq 1$, which is in agreement with practice.

Results of the present author and his co-workers [3, 4, 24, 31] on the barrier properties on extrudates of LCP blends indicated that a small amount of LCP was needed to reduce greatly the uptake of water. This was attributed to the presence of the LCP fibrils in the sheath region having enforced a more aligned structure on the isotropic polymer than it would normally adopt. This would reduce the blend viscosity, adding to the effects of domain elongation and interfacial slip [3, 4, 24, 31].

The tensile stresses in a capillary die decrease along its length towards the exit, and the domains of the LCP may become less elongated, giving a less prominent minimum or a blend viscosity intermediate between those of the constituents. This is even more apparent in isotropic polymer blends, where the domains are less able to maintain their elongation. Results show that an increase in the die $L:D$ ratio leads to a reduction in the size of the viscosity minimum.

Migration of the dispersed LCP phase, if it occurs, would reduce blend viscosity. Some observers have found migration of the LCP domains to the walls of the die [61]. According to Han's theory [62], for $\lambda = 1$ and $\lambda_N < 1$, the more elastic phase will migrate to the die wall, which will mean that the LCP will tend not to migrate.

6.7.2 Explanation of the viscosity maximum in blends of Vectra B950/PA6

La Mantia *et al.* [25–27] reported the maximum in the viscosity–concentration curve for Vectra B950/PA6 at about 1% concentration of the LCP. A similar explanation to the above may be used to model the findings [3, 4, 24, 31]. At low concentration of the LCP, many small domains are formed and the hydrogen bonding from the amide part of the Vectra B950 links across the interfaces to the PA6. In this way, the usual slip is replaced by a specific reaction, which causes an increase in viscosity. It would be interesting to discover whether this is accompanied by a change in the T_g of the PA6. As the LCP content is increased, the number of domains increases until there is sufficient LCP material for the bonding to link between LCP molecules, and then the interface will be devoid of hydrogen bonding across it, and slip may occur. The maximum in the viscosity–concentration curve occurs just before there is spontaneous phase separation, after a phase of nucleation and growth. This may be at a lower concentration than with other systems because the hydrogen bonding in the Vectra B950 may prefer to be intermolecular.

6.7.3 The yield value

La Mantia *et al.* [25–27] also observed yield values in blends of Vectra B950 in PA6 and PBT and details of similar observations of Kulichikhin *et al.* are given in Chapter 5 and discussed in sections 5.3.1 and 5.4.

6.7.4 Suitable LCPs for blending

Although the conditions for obtaining minima in capillary flow are not fully understood, there are some guidelines that can be followed. When choosing an LCP as a processing aid for use with a high viscosity isotropic polymer, the three important choices to make are to ensure that:

1. the processing range and continuous service temperature of the LCP are appropriate to that of the isotropic polymer;
2. that in this range, or at least at the intended process temperature, the viscosity of the LCP is higher or similar to that of the isotropic polymer; and
3. that the two polymers are incompatible and show no specific interaction such as with PA6/Vectra B series.

Table 6.1 gives processing and service data for some of the high temperature and speciality plastics.

Of the commercially available LCPs, Xydar (Amoco) has the highest processing temperature (400–425°C) and upper service limit. K161 (Bayer) has been used up to 350°C in polyetherimide [8, 36–38]. The

Table 6.1 Processing and continuous service data for high temperature and speciality plastics [83, 84]

Isotropic polymer	*Melt temperature (°C)*	*Continuous service temperature (°C)*
Polyetherketone	385–450	260
Polyamideimide	315–360	250
Polyamideimide/glass	315–360	–
Polyetheretherketone	350–390	250
Polyetheretherketone/glass	350–390	250
Polyetheretherketone/carbon	350–390	250
Polyethersulphone	320–380	180
Polyethersulphone/glass	320–380	180
Polyethersulphone/carbon	360–380	–
Polyetherimide	340–420	170
Polyetherimide/glass	340–420	180
Polyetherimide/carbon	340–420	–
Polysulphone	310–390	160
Polysulphone/glass	310–390	–
Polysulphone/carbon	310–390	–
Polyphenylenesulphide/glass	315–360	200–220
Polyphenylenesulphide/mineral	315–360	240
Polycarbonate	280–320	120
Polycarbonate/glass	290–330	135
Polycarbonate/carbon	290–330	–
Polybutyleneterephthalate (PBT)	230–270	120
PBT/glass	240–270	140
PBT/carbon	240–270	–
Polyarylate	290–330	–
Polyarylate/glass	275–350	–

processing temperature range of the Vectra (Hoechst-Calanese) materials is 285–325°C, but has been used up to 330°C with PES [3, 4, 24, 31] and other commercial LCPs have a similar processing range to Vectra. From Table 6.1, it can be seen that a polymer such as polyetherketone would have to be processed with Xydar, with K161 being used with polyetherketone (perhaps) and polyamideimide. The remainder in the table could be used with the Vectra series and other similar commercial LCPs.

6.8 CONCLUSIONS (OTHER FLOWS)

The rheological behaviour of LCP blends is quite complex in capillary flows whereas, in Couette and cone-and-plate flows, the morphology is fairly uniform, consisting of spherical droplets, irrespective of shear rate, but no fibrils. By increasing the shear rates, the droplets become more constant and uniform. The viscosity of the blends are intermediate between those of the constituent polymers and are usually NDB.

In non-isothermal elongational flows, the dispersed LCP phase is in the form of fibrils but, to achieve this, relatively high draw ratios (approximately 50) are required [27, 76]. Increasing the LCP content increases the MS and decreases the BSR. Blizard *et al.* [77] developed a model based on the rheological parameters of the constituents and the cooling conditions at the die exit that predicted the morphology in the extrudates.

In dynamic experiments, some minima have been reported provided that the dynamic viscosity of the LCP is greater than or equal to that of the isotropic polymer [27, 79]. Yield stresses were also reported [80]. When results are compared with those from capillary measurements the Cox–Merz rule does not apply, as would be expected, because of the lack of an elongational flow [82].

Capillary flows have been much more extensively studied and have produced some fascinating results. Perhaps future work may focus more on elongational flows and oscillatory rheometry, as well as on DMTA. It may be that these approaches will give a better understanding to the flow of LCPs and their blends in extrusion dies.

ACKNOWLEDGEMENTS

Figures 6.2, 6.3, 6.4, 6.5, 6.7, 6.9, 6.10, 6.11, 6.13 and 6.14 were reproduced by courtesy of ChemTech Publishing, Toronto-Scarborough, Canada and Fig. 6.1 was reproduced by courtesy of The Institute of Materials, London.

REFERENCES

1. Cogswell, F.N., Griffin, B.P. and Rose, J.B. (1983) US Patent 4,386,174, 31 May.
2. Siegmann, A., Dagan, A. and Kenig, S. (1985) *Polymer*, **26**, 1330.
3. Collyer, A.A. (1991) PPS European Regional Meeting, Palermo, 15–18 Sept., Paper 5.01 kn, 145.
4. Hawksworth, M., Hull, J.B. and Collyer, A.A. (1993) in *Processing and Properties of Liquid Crystalline Polymers and LCP Based Blends*, D. Acierno and F.P. Mantia (eds), ChemTech Publishing, Toronto-Scarborough, 65–89.
5. Weiss, R.A., Huh, W. and Nicolais, L. (1987) *Polym. Eng. Sci.*, **27**, 684.
6. Weiss, R.A., Huh, W. and Nicolais, L. (1988) in *High Modulus Polymers*, Marcel Dekker Inc. Publishers, New York, 145.
7. Jung, S.H. and Kim, S.C. (1988) *Polymer J.*, **20**, 73.
8. Acierno, D., Nobile, M.R., Nicolais, L. *et al.* (1990) in *Polymer Rheology And Processing*, A.A. Collyer and L.A. Utracki (eds), Elsevier Applied Science Publishers, London and New York, Ch. 4.
9. Blizard, K.G. and Baird, D.G. (1986) *Polymer News*, **12**, 44.
10. Blizard, K.G. and Baird, D.G. (1986) *ANTEC*, 311.
11. Blizard, K.G. and Baird, D.G. (1987) *Polymer News*, **12**, 173.
12. Blizard, K.G. and Baird, D.G. (1987) *Polym. Eng. Sci.*, **27**, 653.
13. Acierno, D., Amendola, E., Carfagna, C. *et al.* (1987) *Mol. Cryst. Liq. Cryst.*, **153**, 553.
14. Nobile, M.R., Amendola, E., Nicolais, L. *et al.* (1989) *Polym. Eng. Sci.*, **29**, 244.
15. Zhuang, P., Kyu, T. and White, J.L. (1989) *Polym. Eng. Sci.*, **28**, (17), 1095.
16. Nishimura, T. and Sakai, H. (1988) *Kobunshi Ronbunshu*, **45**, 401.
17. Brydson, J.A. (1980) *Flow Properties of Polymer Melts*, Iliffe.
18. Cogswell, F.N. (1980) *Polymer Melt Rheology*, George Goodwin.
19. Isayev, A.F. and Modic, M. (1987) *Polymer Composites*, **8**, 158.
20. Malik, T.M., Carreau, P.J. and Chapleau, N. (1990) *Polym. Eng. Sci.*, **29** (9), 600.
21. Beery, D., Kenig, S. and Siegmann, A. (1991) *Polym. Eng. Sci.*, **31** (6), 451.
22. Kohli, A., Chung, N. and Weiss, R.A. (1989) *Polym. Eng. Sci.*, **29**, 573.
23. Calundann, G., Jaffe, M., Jones, R.S. *et al.* (1988) in *Fibre Reinforcements for Composite Materials*, A.R. Bunsell (ed.), Elsevier, Amsterdam, Ch 5, 211–48.
24. Hawksworth, M. (1993) PhD Thesis unpublished, Sheffield Hallam University, UK.
25. La Mantia, F.P., Valenza, A., Paci, M. *et al.* (1989) *Rheol. Acta.*, **28**, 417.
26. La Mantia, F.P., Valenza, A., Paci, M. *et al.* (1990) *Polym. Eng. Sci.*, **30**, 7.
27. La Mantia, F.P. (1993) *Liquid Crystal Polymer Blends*, Technomic, Lancaster, USA, Ch 4.
28. Chung, T.-S. (1988) in *Advances in Polymer Blends and Alloys Technlogy*, M.A. Kohudic (ed.), Lancaster PA: Technomic Publishing Co. Inc., 175–88.
29. Valenza, A., La Mantia, F.P., Paci, M. *et al.* (1991) *Int. Polym. Processing*, **6** (3), 247.

30. La Mantia, F.P. and Valenza, A. (1992) *Makromol. Chem. Macromol. Symp.*, **56**, 151.
31. Hull, J.B., Hawksworth, M. and Collyer, A.A. (1991) Irish Materials Forum, Limerick, 4–6 Sept.
32. Magagnini, P.L., Paci, M., Pracella, M. *et al.* (1992) (in press).
33. La Mantia, F.P., Magagnini, P.L. and Pedretti, U. (1992) *Polymer Networks and Blends*, **2** (1), 13.
34. Hoeck, P., Brinkmann, Th. and Michaeli, W. (1990) *Proc. of Compalloy Europe 90, 8–9 Oct. 1990*, Brussels, Belgium, 281–91.
35. Isayev, A.I. and Swaminathan, S. (1987) *Proc. Third Ann. Conf. on Advanced Composites, Detroit, 15–17 Sept., 1987*, Detroit, Michigan, 259–67.
36. Nobile, M.R., Acierno, D., Incarnato, L. *et al.* (1990) *J. App. Polym. Sci,* **41** (11/12), 2723.
37. Nobile, M.R., Acierno, D., Incarnato, L. *et al.* (1990) *J. Rheol.*, **34** (7), 1181.
38. Acierno, D., Nobile, M.R. and Marino, G. (1991) PPS European Regional Meeting, Palermo, Italy, 15–18 Sept., Paper 5.03.
39. Penwell, R.C. and Porter, R.S. (1971) *J. Polym. Sci.*, **A2** (9), 463.
40. Toelke, G.A., Gogos, C.G. and Biesenberger, J.A. (1967) *SPE 13th ANTEC*, 98.
41. Casale, A., Penwell, R.C. and Porter, R.S. (1971) *Rheol. Acta.*, **10**, 412.
42. Crowson, R.J., Folkes, M.J. and Bright, P.F. (1980) *Polym. Eng. Sci.*, **20**, 925.
43. Kulichikhin, V.G., Vassil'eva, O.V., Litvinov, I.A. *et al.* (1991) *J. App. Polym. Sci.*, **42** (2), 363.
44. Kulichikhin, V.G. and Platé, N.A. (1991) *Polym. Sci. USSR*, **33** (1), 1–37.
45. Stehling, F.C., Huff, T., Speed, C.S. *et al.* (1981) *J. App. Polym. Sci.*, **216**, 2693.
46. Lipatov, Yu.S., Shumskii, A.N., Gorbatenko, A.M. *et al.* (1983) *Fiz. Khim. Mekh. Disp. Struktur.*, 117.
47. Wisniewski, C., Marin, G. and Monge, Ph. (1985) *Eur. Polym. J.*, **21** (5), 479.
48. Semakov, A.V., Kantor, G.Ya., Vasil'eva, O.V. *et al.* (1991) *Polym. Sci. USSR*, **A33**, 1.
49. Plochocki, A.P. (1972) in *Polymer Blends*, D.R. Paul and S. Newman (eds), Academic Press, New York, Ch 2.
50. Lyngaae-Jørgensen, J. and Valenza, A. (1990) *Makromol. Chem., Macromol. Symp.*, **38**, 43–60.
51. Lyngaae-Jørgensen, J., Utracki, L.A., Valenza, A. *et al.* (1991) PPS European Meeting, Palermo, Italy, 15–18 Sept., P5.06, 159.
52. Ablazovo, T.I., Tsebrenko, M.V., Yudin, A.V. *et al.* (1975) *J. App. Polym. Sci.*, **19**, 1781.
53. Tsebrenko, M.V., Yudin, A.V., Ablazovo, T.I. *et al.* (1976) *Polymer*, **17**, 831.
54. Tsebrenko, M.V., Benzar, A.I., Yudin, A.V. *et al.* (1979) *Vysokomol. Soed.*, **A21**, 830.
55. Tsebrenko, M.V., Rezanova, M.N. and Vinogradov, G.V. (1982) *Nov. Reol. Polim., 11th Mater. Vses. Simp. Reol.*, **2**, 136.
56. Utracki, L.A. (1990) in *Polymer Blends and Alloys*, Hanser Publishers, Munich, Vienna, New York.

57. Borisenkova, Ye.K., Kulichikhin, V.G. and Platé, N.A. (1990) *Dokl. Akad. Nauk. SSSR*, **314**, 7.
58. Collyer, A.A., Clegg, D.W., France, G.H. *et al.* (1985) IXth Int. Congr. On Rheol., Acapulco, 8–13 Oct., 1984; Elsevier Applied Science Publishers, London and New York, Vol. 3, 543–50.
59. France, G.H., Collyer, A.A., Blackmore, D.L. *et al.* (1988) *Proc. Xth Int. Congr. On Rheol., Sydney, 14–19 Aug.*, Vol. 1, 234–6.
60. Collyer, A.A., France, G.H., Bott, K. *et al.* (1989) The 1989 Int. Chem. Conf. of the Pacific Basin Countries, Honolulu, Hawaii, USA, 17–22 Dec., Paper O7J-98.
61. Verhoogt, H., Willems, C.R.J., Van Dam, J. *et al.* (1991) PPS European Regional Meeting, Palermo, Italy, 15–18 Sept. , P5.02, 149–50.
62. Han, C.D. (1981) in *Multiphase Flows in Polymer Processing*, Academic Press, New York, 348–52 and 386–94.
63. Alderman, N.J. and Mackley, M.R. (1985) *Faraday Discussions, Chem. Soc.*, **79**, 149–60.
64. Kann, R. and Shaw, M.T. (1981) 52nd Ann. Meeting of Soc. Rheol., Williamsburgh, Virginia, USA, 23–28 Feb., Paper D8.
65. Baker, F.S., Carter, R.E. and Warren, R.C. (1980) *Rheology*, Vol. 3, Plenum Press, New York, 591.
66. Baker, F.S. and Carter, R.E. (1982) *Propellants, Explosives and Pyros.*, **7**, 139.
67. Baker, F.S. and Privet, G.J. (1987) *Polymer*, **28**, 1121.
68. Collyer, A.A., Tsang, C.N., Clegg, D.W. *et al.* (1988) *Proc. Xth Congr. on Rheol., Sydney, Australia, 14–19 Aug.*, Vol. 1, 282–4.
69. Collyer, A.A., Tsang, C.N., Clegg, D.W. *et al.* (1989) The 1989 Int. Chem. Congr. of Pacific Basin Countries, Honolulu, Hawaii, 17–22 Dec., Paper O75–354.
70. Collyer, A.A., Clegg, D.W., Tsang, C.N. *et al.* (1991) PPS Seventh Ann. Meeting, Hamilton, Ontario, Canada, 21–24 April, 52–3.
71. Lyngaae-Jørgensen, J., Dahl Thomson, L., Rasmussen, K. *et al.* (1988) *Int. Polym. Proc.*, **2**, 123.
72. La Mantia, F.P., Valenza, A. and Magagnini, P.L. (1992) *J. App. Polym. Sci.*, **44**, 1257.
73. La Mantia, F.P., Valenza, A., Magagnini, P.L. *et al.* (1991) *Proc. of Compalloy Europe 91, Luxembourg, 24–25 Oct.*, 263–84.
74. La Mantia, F.P. and Acierno, D. (1985) *Polym. Eng. Sci.*, **25** (5), 279.
75. Baird, D.G. and Sun, T. (1990) in *ACS Symp. Ser. 435*, C.K. Ober and R.A. Weiss (eds), ACS Washington, 416–38.
76. La Mantia, F.P., Saiu, M., Valenza, A. *et al.* (1990) *Europ. Polym. J.*, **26** (3), 323.
77. Blizard, K.G., Federici, C., Federici, O. *et al.* (1990) *Polym. Eng. Sci.*, **30** (22), 1442.
78. Baird, D.G. and Ramanathan, R. (1990) in *Contemporary Topics in Polymer Science*, Vol. 6, B.M. Culberson (ed.), Plenum Press, 73–93.
79. Sukhadia, A.M., Done, D. and Baird, D.G. (1990) *Polym. Eng. Sci.*, **30** (9), 519.
80. Crevecoeur, G. and Groeninckx, G. (1990) in *Polymer Based Molecular Composites*, D.W. Schaefer and J.E. Mark (eds), Pittsburgh PA: Materials Research Society, 166–70.
81. James, S.G., Donald, A.M. and MacDonald, W.A. (1987) *Mol. Cryst. Liq. Cryst.*, **153**, 491.

82. Biing-Lin, L. (1988) *Polym. Eng. Sci.*, **28** (17), 1107.
83. Collyer, A.A. (1989) *Prog. in Plastics and Rubber Technol.*, **5** (1), 36–87.
84. Collyer, A.A. (1990) *A Practical Guide to the Selection of High-temperature Engineering Thermoplastics*, Elsevier Science Publishers, Oxford.

7

Processing of liquid crystal polymers and blends

J.B. Hull and A.R. Jones

7.1 INTRODUCTION

In recent years, many areas of research relating to the development of new and improved commercial products made from liquid crystal polymers (LCPs) and LCP blends have been engineered. LCPs have the ability to form partially ordered melts, and the focus of the research has been to investigate the rheological characteristics of these LCPs and their blends, and to determine how such characteristics affect both the final structure of finished products, and the operational processing requirements. The development of ultra-high modulus and high tensile strength Kevlar fibres by the Du Pont Company during the 1970s [1–5], focused the attention of polymer processing engineers on the potential of LCPs as either commercial materials in their own right, or as unique processing aids for other polymers, and initiated widespread research interest in a range of LCP systems. LCP concepts have now been extended to encompass a large number of homo-polymer and copolymer compositions which exhibit lyotropic or thermotropic behaviour.

In addition to their attractive mechanical properties, some LCPs possess desirable optical and physical properties, which have promoted their use as promising materials for components in the electrical/ electronic, chemical, transportation and aerospace industries, and are used in nonlinear optical devices, optical data storage items, and as

Rheology and Processing of Liquid Crystal Polymers
Edited by D. Acierno and A.A. Collyer
Published in 1996 by Chapman & Hall
ISBN 0 412 59640 7

'orienting carriers' for conducting polymers. In particular, main chain nematic, thermotropic liquid crystalline polymers, which feature anisotropic (mesomorphic) melts within a given temperature range, have received considerable attention due to their low melt viscosity and ease of melt processability in the liquid crystalline state. Using conventional techniques such as injection moulding, extrusion and melt spinning, they are melt processable and attain a high degree of molecular orientation under conditions of elongational flow [6–12]. The high melting point of the basic homopolymers ($T_m \simeq 600°C$) inhibits melt processing and their processing temperature requirements tend to be in a range where rapid chemical decomposition takes place. However, a substantial decrease in processing temperature requirements has been achieved through the development of random copolymers with small mesogenic units and random copolymers which include flexible spacers. As a consequence, much lower melting points ($T_m \simeq 275°C$) can be attained. The lack of periodicity along a rigid chain effectively reduces the possibility of crystallization by the copolymer.

Among main chain nematics, the thermotropic copolyesters have been found to be extremely versatile. Examples of commercial resins are Vectra^{TM}, developed by Hoechst-Celanese, and Xydar^{TM} marketed by Amoco. As in the case of lyotropic nematic polymers, thermotropic copolyesters orientate easily in an elongational flow field, producing highly oriented uniaxial structures with moduli as high as 140 GPa [13]. Long melt relaxation times, low melt viscosities, negative first normal stress differences and little or no die swell are major parameters which influence the processing of thermotropic copolyester. Other thermotropic materials which have received lesser interest include polyesteramide and polyesteramide based LCPs.

Following melt solidification, moulded articles possess a high degree of molecular anisotropy in the direction of flow, and the resulting mechanical properties are equivalent to, or can supersede, those of corresponding glass reinforced composites [14, 15]. Moulded articles appear 'layered', with local orientation of the hierarchy of fibres reflecting the local orientation of the moulded part. *In situ* reinforcement of conventional thermoplastics, using minor LCP component additions, often results in highly elongated fibrous structures. These not only improve blend properties, but also reduce equipment wear and fibre breakage, as experienced with conventional short fibre composites. A number of groups have studied the rheology, morphology and mechanical properties of LCP/thermoplastic blends [16–28]. The studies have been conducted on both crystalline and amorphous materials, such as polyethyleneterephthalate, polyamides, polycarbonate, polyethersulphone and polystyrene. The process of forming an elongated thermotropic LCP fibrous structure depends on factors such as composition, interfacial tension between the constituent materials, the viscosity ratio, and the deformation flow field involved in the processing procedures [29].

In this chapter, the behaviour of LCPs and polyblends of LCPs with common thermoplastics during different processing operations is discussed, and the important interrelationships between the processing conditions and the resulting structure and mechanical properties exhibited by moulded products are examined.

7.2 STRUCTURE OF LCPs

7.2.1 Structural order

In polymer systems, the microstructure can exist at many different levels of order, with the highest level being crystallinity. Crystalline polymers possess chains arranged in a three-dimensional lattice network. Lower levels of order include two-dimensional smectic and one-dimensional nematic systems. Liquid crystalline materials exist with all of the different levels of order. However, most systems of interest have some type of nematic structure . In liquid crystalline systems, it is often necessary to specify the degree of local structural order [30, 31]. One method of approach adopts the specification of an order parameter S mentioned in Chapter 1, defined as:

$$S = 0.5(3\,\overline{\cos^2\theta} - 1) \tag{7.1}$$

where θ is the angle between the long axis of a molecule and the axis of preferred orientation. This definition of S derives from the work of Zocher [32]. In low molecular weight liquid crystalline systems, S has a value between 0.3 and 0.8. S has a value of 0 for completely isotropic systems and a value of 1.0 if uniaxial orientation exists. The degree of order in liquid crystals has been determined by nuclear magnetic resonance and infra-red dichroism [31].

Cholesteric liquid crystalline systems exhibit a superposed pitch on the nematic structure, which is of the order of the wavelength of visible light [33, 34], producing a myriad of colours. In some systems, a much larger pitch is evident, resulting in spherulites or grains similar to metals [34].

7.2.2 Orientation and its role in processing

Industrially, as mentioned in the introduction, major research effort has focused on main chain thermotropic LCPs because of their inherent ease of processing and versatility, combined with the high thermal and

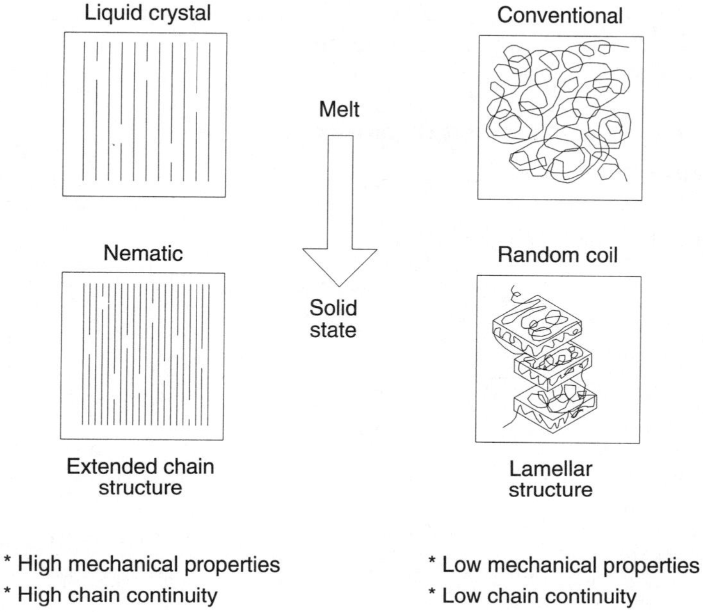

Figure 7.1 Schematic comparison of the structures formed during the uniaxial processing of LCP and conventional, crystallizable, random coil polymers.

mechanical properties of the resulting moulded products. A schematic comparison of the structures of an LCP with a nematic and crystallizable conventional polymer, shown in Fig. 7.1, illustrates the reason for the ease of processability of LCPs. Processing capability originates from the relative ease with which the molecular 'rods' can slide past one another, while the high thermal and mechanical properties emanate from the extended chain morphology present in the solid state.

Almost all of the LCPs of interest are based on a basic para-phenylene chain structure. In the case of polyesters, the ester groups act as links in the para position as shown in Fig. 7.2.

Copolyesters, are produced by chemical modification with comonomers, chain substitution, molecular swivels, etc. which drives the

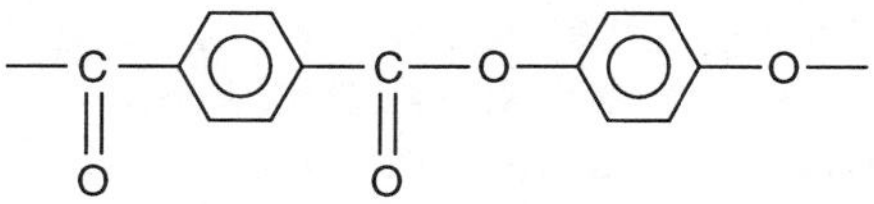

Figure 7.2 Basic structure in liquid crystal polyesters.

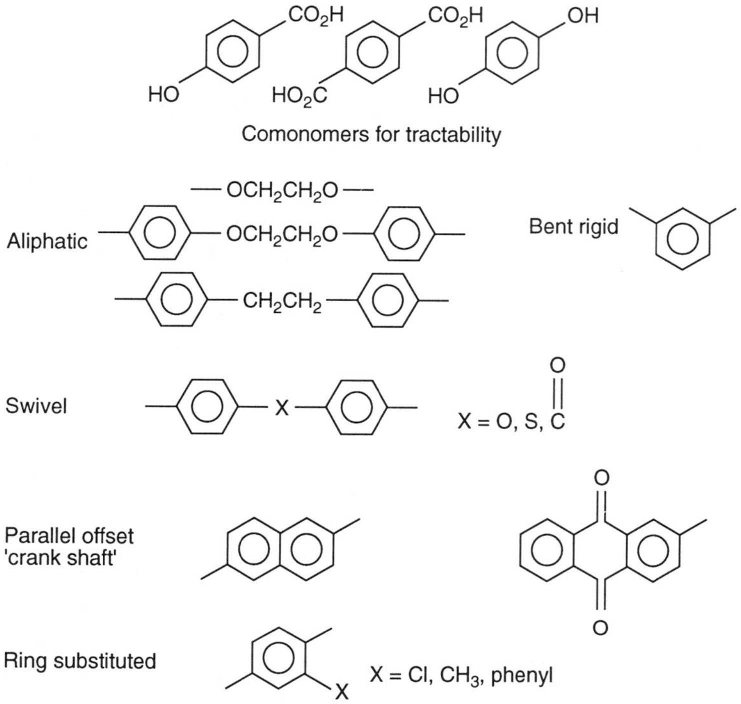

Figure 7.3 Copolymer approaches to tractible, thermotropic polyesters based on para-phenylene.

mesogenic transition temperature of the polymer far enough below the temperature range where decomposition takes place to allow melt processing. Copolymerization imparts flexibility to inherent rigidity of the chain backbone of the homopolymers, while maintaining chain linearity. A summary of the molecular approaches adopted is shown in Fig. 7.3.

Investigations of the structure of liquid crystal copolyesters, including studies of three-dimensional order on solidification have been well reported in the literature [35–44]. X-ray analysis of the structure of melt spun fibres of different compositions of a HBS/HNA aromatic LCP indicated that only minimal block structure was evident in these materials. However, the possibility of some three-dimensional order was recognized even if the formation of large crystallites was precluded. Crystals of different sizes from 3 nm to 30 nm have been indicated by different authors [37, 38, 40–43] and in some cases the evidence of the formation of small crystals was confirmed by thermal analysis [41, 45].

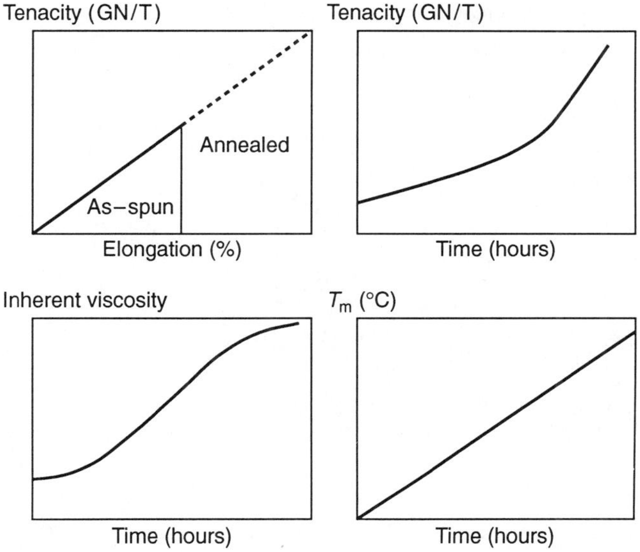

Figure 7.4 Structure–property response of highly orientated thermotropic copolyester fibre to annealing close to the crystal to nematic transition temperatures; after Jaffe [13].

The effects of heat treatment on enhancing the structure of LC copolyesters have also been investigated [41, 46], and it was concluded that although heat treatment did not produce development of macro-crystals in block form, for the HBA/HNA copolyester under investigation, some transformation to increased lateral ordering did take place [46]. Differential scanning calorimetry (DSC) analysis of annealed 73/27 HBA/HNA samples has indicated that high melting crystals ($T_m \simeq 315°C$) can form at temperatures as high as 280–295°C [47]. This effect was most noticeable in stretched rather than as-moulded samples and it has been realized that these phenomena could affect the rheology and processing of such LCPs. One effect relating to the presence of high melting crystals was found to be the flow instability resulting in melt fracture, observed at low shear rates, during processing of HBA/HNA LCPs at temperatures within the range 320–335°C [48–50]. Similar observations have been made with other LCP systems [51]. Jaffe [13] reported property improvement in LC copolyester extrudates achieved by heat treatment with tensile strength increases of five to six times that of as-spun fibres (approximately 3 GPa) obtained by annealing close to the melting point. This process appears to combine structural improvements with solid state polymerization and the changes obtained through heat

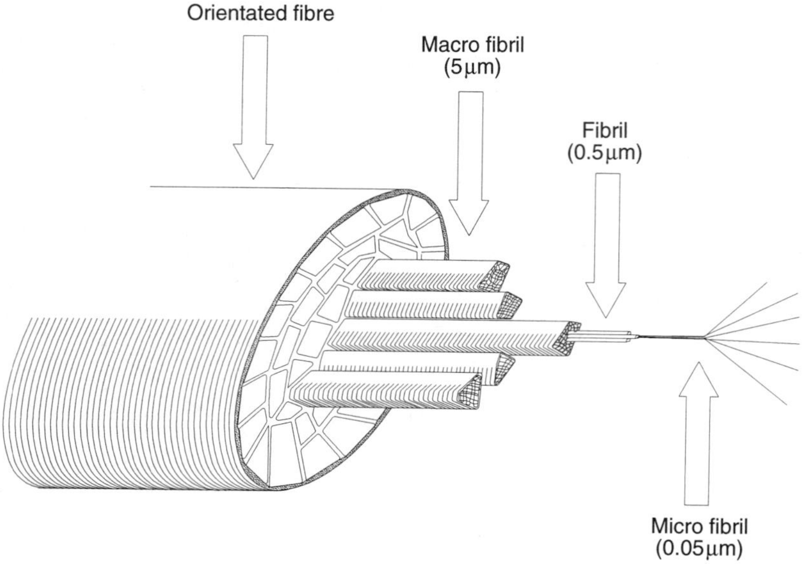

Figure 7.5 Hierarchical structure of uniaxially orientated thermotropic copolyester extrudate; after Sawyer and Jaffe [54].

treatment are illustrated in Fig. 7.4. Heat treatment has been observed to improve elongation, melting point, chemical stability and retention of thermal properties, as well as increasing tensile strength. However, some property 'drop-offs' at elevated temperatures resulting from secondary transitions have also been observed [13]. The molecular origins of such processes have been defined by Yoon and Jaffe [52] and Green *et al.* [53].

The morphology of fibres has been identified by Sawyer and Jaffe [54], who indicated that the fibres are composed of a hierarchy of fibrillar structures covering several orders of magnitude in diameter. This is shown schematically in Fig. 7.5. Fibres exhibit relatively poor properties in compression. The cause of this characteristic, whether molecular or morphological in origin, has led to vigorous debate among researchers [55].

The potential for high-modulus thermotropic LCPs to be processed industrially by conventional melt processing techniques such as injection moulding, extrusion and melt spinning is easily recognized. These materials offer a balance of properties unobtainable with other polymers, including controllable melting capability, low melt viscosity, ease of processing, ease of orientation, low die swell and low shrinkage.

7.3 PROCESSING OF LCPs

7.3.1 Injection moulding

Around 40% of all polymeric components are manufactured by injection moulding. The injection moulding process is, however, one of the most complex processing techniques for polymers, more complex for example than extrusion. During processing, in addition to the agitated processing action of the screw in the barrel, non-isothermal conditions are in operation, and the melt is subjected to both shear and elongational flows in the nozzle and in the mould. Moulded components are often characterized by the distribution of micro and macro orientation which are dependent upon the induced flow deformation in shear and elongational flows. The wide range of operating variables in injection moulding necessitates that the process operations are properly sequenced if high quality products are to be manufactured with consistent repeatability. The main operating variables are:

- distance of screw travel
- injection speed
- screw dwell time
- moulding freeze time, including gates and runners
- injection pressure and its variation throughout the injection cycle
- number of impressions in the mould
- mould packing
- barrel and nozzle temperatures
- type, position, size and number of gates in the mould
- mould locking force
- type, size and layout of the runners in the mould
- mould temperature and the variation of mould temperature
- overall cycle time

Although current equipment is operated under computer control, with a relatively good 'apparent' repeatability of the moulding cycle (nominal variation of approximately 0.6% during operation for one day), rapid transients in the rheological behaviour of polymers during processing, which are beyond the ability of most commercial data acquisition and control equipment to detect and correct, often have a marked influence on the final properties of moulded products [56]. For example, such variability can be caused by small changes in stock polymer, from charge to charge. Unlike lyotropic LCPs such as Kevlar, thermotropic LCPs can be processed by injection moulding in the same manner as commercial thermoplastics. When these materials are injection moulded, the moulded components exhibit high mechanical properties without additional fibre reinforcement. The components also show high anisotropy of physical properties, measured parallel to the flow direction and across the flow direction [57–59]. However, this anisotropic variation has been observed to reduce as the thickness of a moulded article increases [57]. This observation was considered to be an indication of the presence of layered structures in the final moulded products.

A major feature of thermotropic LCPs is their relatively low thermal expansion coefficient, which is around one tenth of the value for conventional thermoplastics. This property, combined with their low shear viscosity, makes thermotropic LCPs ideal materials for injection moulding. However, the final properties of moulded articles depends on the large number of interacting thermo-mechanical processing parameters. Because LCPs inherently exhibit anisotropic properties, the study of structure, as related to processing conditions, is of fundamental importance in the understanding of the mechanical and physical properties of mouldings.

Structural characterization of mouldings

Detailed structural characterization of injection moulded parts was initiated in the 1980s. Ophir and Ide [60] have studied the injection moulding of an aromatic polyester melt polymerized from p-acetoxybenzoic acid (60%), terephthalic acid (20%), and naphthalene diacetate (20%), invented by Calundann. They concluded that the rheological behaviour and orientation development during injection moulding was similar to the characteristics of short fibre filled resins. Short fibre orientation distribution in injection moulded components can be highly complex, due to the extreme sensitivity of fibre alignment to elongational flow fields [60]. For example, when the cross-section of the flow path contracts (converging flow), the fibres tend to align along the flow direction, but when the cross-section expands (diverging flow), the fibres tend to align along the transverse direction. Moulded parts sometimes exhibit a transversely oriented core surrounded by a longitudinally oriented skin.

Joseph *et al.* [12] conducted a study of the morphological characteristics of polyethyleneterephthalate/polyhydroxybenzoicacid based LCP under different deformation flow histories. They observed a typical skin/core structure with a higher concentration of the PHB in the skin layers. The microstructure of the injection moulded samples was shown to be dependent on the thickness of the mouldings, varying from three to five layers. An unorientated layer was observed to be sandwiched between either set of orientated layers with thin mouldings. As the thickness of mouldings increased, two sets of sandwiching layers were observed.

Following observations on extruded and injection moulded parts, Sawyer and co-workers [54, 61, 62] have proposed a fibrillar hierarchical structural model. The orientated fibres and extrudates appear to be formed by a hierarchy of structure ranging from micro-fibrils (approximately 0.05 μm in diameter) through fibrils (approximately 0.5 μm in diameter) to macrofibrils (approximately 0.5 mm in diameter) in a highly orientated skin. Injection moulded parts exhibit a much lower degree of orientation.

Another hierarchical model for injection moulded HBA/HNA LCPs has been proposed by Weng *et al.* [63] and a schematic diagram of the suggested model is shown in Fig. 7.6.

The model consists of a skin macro-layer and core region. The skin layer comprises three sub-layers. Described from the surface inwards, there is a top layer, and a thin (approximately 20 μm thick) highly orientated fibrillar layer. Below the top layer, there were considered to be several sub-layers (30–50 μm in thickness), of varying degrees of orientation leading to a less orientated inner zone. The sub-layers were themselves divided into stacks of smaller units around 0.5 μm in thickness, and about 30 μm in width. In the core region, molecular orientation was found to be represented by localized flow patterns, perpendicular to the direction of injection.

Thapar and Bevis [64, 65] have carried out micromorphological studies on injection moulded tensile bars of a random copolyester LCP (70%

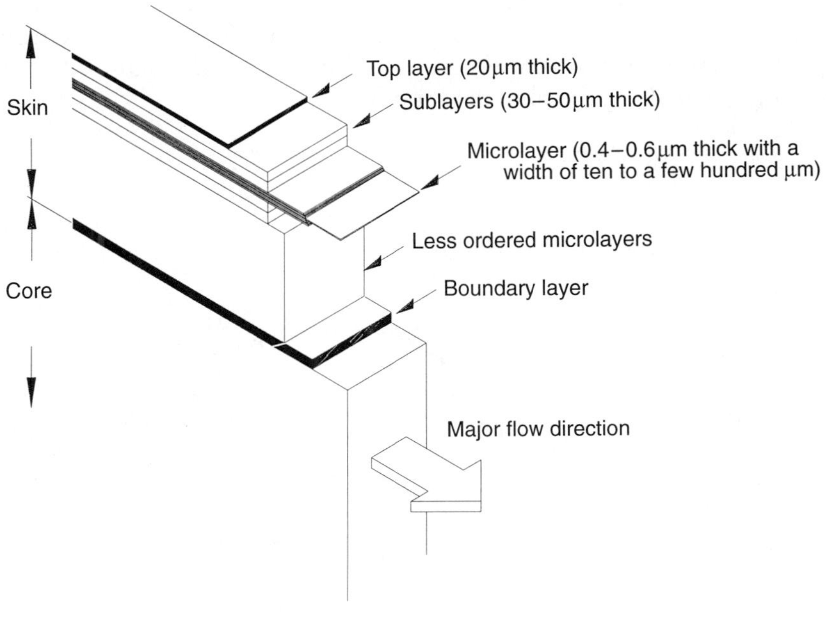

Figure 7.6 Schematic of the hierarchial model proposed by Weng *et al.* [63].

para-acetoxy-benzoic acid and 30% para-acetoxy-naphthoic acid). They observed a five-layer structure within the moulded components including skin, nodular, nodular/fibrillar (intermediate layer), fibrillar and core regions. The core was composed of layers with a tendency to align along the flow direction. Microhardness measurements confirmed the existence of four distinct zones. The first three zones, from the skin edge inwards were observed to be highly anisotropic, while the fourth zone was isotropic to the indenter. Qualitative observations of X-ray diffraction patterns indicated slight changes in the fibre axis orientation through the depth and the patterns were found to be consistent with etched photomicrographs. Changes in azimuthal spread and width were also apparent from the diffraction patterns. The azimuthal spread was observed to be lowest for tightly packed shear flow induced fibrillar micromorphology. X-ray intensity measurements of equatorial reflections in the region of shear flow induced fibrils were high. Low intensities, observed in some regions, were thought to be caused by reduced volumes of suitably aligned material. The existence of a second phase based upon the observation of a very low intensity Debye ring was also proposed. This was considered to be interfibrillar or intrafibrillar in nature. Sheet-like structures with fibrillar material as part of the macrostructure were observed in both skin and core regions with the sheets lying parallel to the injection direction in the skin and perpendicular to the injection direction in the core. The high mechanical property values reported for the polymer (see below) were considered to arise from the existence of aligned microfibrils inherently composed of nematic domains with high axis alignment along the fibril direction.

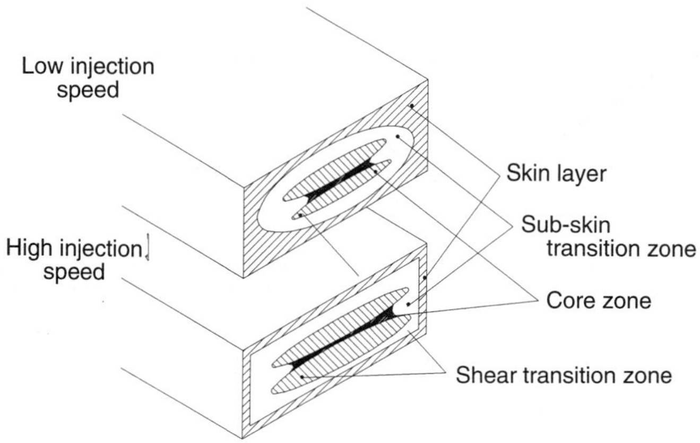

Figure 7.7 Schematic of a four-layer model of injection moulded LCP at low and high injection speeds as proposed by Hsiung *et al.* [66].

Hsiung *et al.* [66] have also carried out a thorough investigation of structural layering phenomena after injection moulding a wholly aromatic thermotropic copolyester (ULTRAX KR 4002–BASF). Dumb-bell shaped tensile bars were moulded at three injection speeds (1.5, 5.2 and 23.2 $cm^3 s^{-1}$) and at different mould temperatures (15°C, 80°C and 120°C). A four-layer structure was found to prevail in moulded samples, which is broadly in accord with the work of Thapar and Bevis [65]. The observed layers were a skin layer, a sub-skin transition zone, a shear layer and a core region, as shown schematically in Fig. 7.7.

The effect of annealing on the orientation of the skin region of a PET/60 PHB sample injection moulded at 275°C has been investigated by Joseph *et al.* [12]. A microtomed skin layer was heat-treated at 250°C for 30 minutes. WAXS scattering analyses carried out before and after heat treatment indicated that relaxation of the anisotropic structure did not take place, and that the structure was extremely stable, even at a temperature close to the melting point of the polymer.

Barrel temperature and injection speed

Initial studies of the properties of injection moulded copolyesters were carried out in the 1970s by Jackson and Kuhfuss [57], using a plunger injection moulding machine. Various compositions of a copolymer of PET/PHB were prepared and analysed in terms of the moulding conditions, and the effects of these moulding conditions on mechanical properties. For a polymer containing 60 mol% PHB, the effect of moulding temperature was reported in terms of the cylinder or barrel temperature, as this temperature was varied between 210°C and 280°C. Typical variations observed in the properties of the LCP over the range are shown in Table 7.1.

In general, it was observed that the mechanical properties increased with increasing processing temperature over the range, due to an increase in polymer orientation in mouldings resulting from a decreased melt viscosity and higher injection speed. However, processing at 280°C resulted in a lower level of properties, which was attributed by the authors to lower orientation caused by partial relaxation. Copolyesters with higher PHB contents (up to 80 mol%) were processed at temperatures up to 400°C. However, the resultant mechanical properties of mouldings made with 80 mol% PHB were found to be much lower than those of the 60 mol% PHB samples. Repetition of the investigation employing a screw injection machine, rather than the original plunger injection machine, indicated that the materials were susceptible to changes in properties when subjected to a different shear history. Corresponding samples exhibited a marked improvement in notched

Table 7.1 Effects of mould temperature as reported by Jackson and Kuhfuss [57]

	Barrel temperature			
Property	*210°C*	*240°C*	*250°C*	*260°C*
Inherent viscosity	0.62	0.61	0.63	0.62
Tensile strength (MPa)	150	201	206	229
Elongation (%)	8	9	20	20
Flexural modulus (GPa)	10.2	10.8	13.0	12.3
Izod impact strength–notched ($J\,m^{-1}$)	152	242	298	416
Izod impact strength–unnotched ($J\,m^{-1}$)	693	800	1120	1493
Rockwell hardness (L)	50	36	45	42
Heat-deflection temperature (°C)	–	–	66	64

impact properties (approximately +74%) with a relatively smaller loss in tensile strength (approximately −21%).

Thapar and Bevis [65] have carried out a detailed study of the effect of varying injection moulding conditions on a thermotropic copolyester of 70% para-acetoxybenzoic acid and para-acetoxynaphthoic acid. Tensile bars ranging in thickness from 1 mm to 20 mm were prepared, and the effects of barrel temperature and injection speed on mechanical properties were investigated. The combination of process conditions investigated is shown in Table 7.2. The bars were characterized by etching/SEM analysis, by WAXS and by X-ray microradiography.

Mechanical testing was carried out by subjecting the tensile bars to tensile loads at 21°C using a test machine with a constant cross-head speed of $0.05\,mm\,min^{-1}$. A strain gauge was used to measure Young's modulus at 0.1% to 0.2% strains. In addition, microhardness of the samples was investigated because of the sensitivity of this property to structural changes [67, 68]. This technique can detect changes in preferred orientation through the thickness of injection moulded semi-crystalline thermoplastics, and can also detect a variety of

Table 7.2 Comparison of processing conditions used by Thapar and Bevis [65]

Barrel temperature (°C)	*Injection speed*
295	Fast
295	Slow
315	Fast
315	Slow

Table 7.3 Injection moulded properties of a wholly aromatic copolyester as reported by Ophir and Ide [60]

	Melt temperature (°C)					
	320	*340*	*340*	*340*	*340*	*340*
Mould temperature (°C)	40	40	40	100	100	100
Packing pressure (MPa)	28	21	39	39	28	0
Injection speed	Fast	Fast	Slow	Slow	Fast	Fast
Tensile strength (MPa)	192	173	189	208	176	153
Tensile elongation (%)	1.8	1.3	1.5	1.6	2.5	1.5
Tensile modulus (GPa)	17.2	20.0	19.3	18.6	15.2	16.5
Flex strength (MPa)	178	174	175	175	170	176
Flex modulus (GPa)	15.2	15.2	15.2	14.5	13.1	13.1
Notched impact ($J\,m^{-1}$)	395	427	230	283	283	347

morphological and textural changes in polymers.

An increase in barrel temperature from 275°C to 315°C resulted in increased tensile strength and Young's modulus. A decrease in injection speed resulted in mechanical property increases at all of the processing temperatures. This observation is in accord with conclusions drawn by Ophir and Ide [60], who have carried out a detailed study of the injection moulding characteristics of a wholly aromatic copolyester polymerised from p-acetoxy-benzoic acid (60 mol%), terephthalic acid (20 mol%) and naphthalene diacetate (20 mol%). A strong interrelationship between injection moulding variables and properties of moulded samples was evidenced as shown in Table 7.3.

Thapar and Bevis [65] observed that greater increases in tensile strength and flexural modules were achieved when the barrel temperature was increased from 275°C to 295°C (+80 MPa and +3.14 GPa, respectively), than corresponding increases in mechanical properties when the barrel temperature was increased from 300°C to 320°C (+46 MPa and +1.4 GPa, respectively). The drop in shear viscosity observed from 280°C to 300°C was large in comparison with the change in shear viscosity between 300°C and 310°C at similar shear rates or shear stresses. The shear viscosity at 280°C was around two orders of magnitude greater than that at 310°C and the indications were that the melt did not flow suitably to allow high preferred orientation in mouldings produced at 280°C or less. Melt fluidity and consequently orientation was improved when the processing temperature was in excess of 290°C. This observation was substantiated by the DSC evidence provided by Windle *et al.* [41]. This work demonstrated that the number of crystalline regions which were molten increased as the temperature was raised above 280°C and consequently, processability was improved.

Table 7.4 Effect of injection speed on Young's modulus and tensile strength as reported by Thapar and Bevis [65]

Injection speed (mm/s)	*Young's modulus (GPa)*	*Tensile strength (GPa)*
17	16.25	184
63	11.73	174
105	11.29	170

Table 7.5 Effect of changes in injection speed on Young's modulus through the depth of a 20-mm-thick specimen [65]

Zones	*Injection speed* ($mm\ s^{-1}$)	*Young's modulus (GPa)*
Skin	21.43	12.6
		6.8
		2.5
Core		1.9
Skin	42.86	8.6
		4.9
		3.5
Core		2.9
Skin	150	8.5
		5.2
		2.7
Core		4.0

The effects of changes in injection speed on the mechanical properties of 3-mm-thick samples as recorded by Thapar and Bevis [65] are summarized in Table 7.4. Property changes with injection speed through the depth of 20 mm sample are shown in Table 7.5.

The property reduction with increases in injection speed has been claimed to be caused by a decrease in net molecular orientation [60]. However, Thapar and Bevis pointed out that by comparison with thermoplastics, the molecular orientation in the LCP mouldings was solely due to the flow process. The melts did not exhibit significant die swell, and had unusually long relaxation times. Hence, the resulting decrease in properties with increased injection speed could not be explained by such effects as molecular relaxation or recoil after cessation of flow. The observed reductions in mechanical properties were shown to relate to a reduction in skin orientation with increased injection speed. The preferred orientation in other regions through the depth was affected correspondingly. These observations were correlated to microhardness measurements through the depth of the test bars and, in general, it was found that the microhardness properties of the samples were improved with higher processing temperatures and lower injection speeds.

Flow within moulds

In general the extent of the die swell in LCPs is relatively low when compared with conventional thermoplastics [69]. Hence, it is important when designing and siting gates to ensure that the melt touches the mould walls in order to prevent 'jetting' problems. Side gates or fan gates have been found to reduce such phenomena [69]. However, core orientation of polymer injected through fan gates is often considerably more transverse than that of polymer injected through edge gates [70]. In addition, because moulded LCP parts exhibit relatively poor resistance to failure at weld lines (formed by the impingement of two or more flow fronts), it is important to ensure that weld lines are not formed in critical areas. Some improvements can be achieved through the employment of conventional approaches such as fast injection, and high melt and mould temperatures [69]. As expected, fibre orientation at weld lines is induced by localized flow fields which are often perpendicular to the main flow direction. Photomicrography has clearly demonstrated that the flow pattern and orientation are preserved at weld lines [69].

Hedmark *et al.* [71] have shown that the morphology developed during processing can depend on the thickness of moulded bars (i.e. mould design). In the case of thin specimens (2.9 mm), an oriented skin, a less oriented layer and a core were detected. However, when the thickness was increased to 5.8 mm, five different layers could be distinguished. In particular, the existence and dimensions of a non-oriented zone close to the skin, which was not observed in thin specimens, apparently depended upon sample thickness. In addition, the thickness of unoriented layers was found to increase with distance from the mould gate and with mould temperature.

Garg and Kenig [72] have attempted to analyse the behaviour of LCPs during mould filling. The analysis was based on a model of dilute short fibre suspensions [73] with LCP molecules seen as rigid rod-like articles. They proposed that four deformation flow systems were responsible for orientation:

1. spreading radial flow in the vicinity of the gate;
2. converging flow due to melt levelling following radial flow distribution;
3. fountain flow; and
4. shear flow at a distance from the mould wall.

They assumed that the different effects could simply be superimposed in order to model the complex flow patterns in a mould and, hence, that

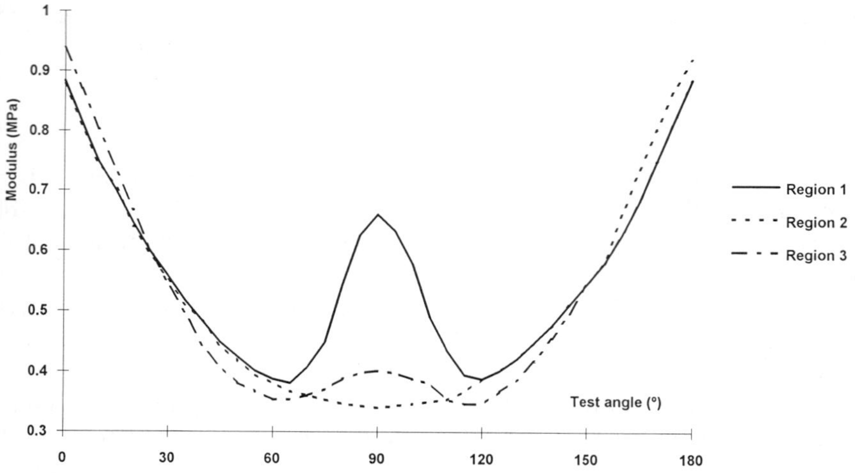

Figure 7.8 Variation in modulus with test angle in three regions of moulded bars at various distances from the gate (region 1 is closest to the gate) [73].

the contribution of each flow field to the orientation pattern could be analysed individually. Analysis of the radial flow indicated that the LCP molecules gradually orientated in a direction transverse to the principal flow path. Once the radius of the melt front exceeded half the width of the mould, the melt front levelled off, and became a convergent flow. This resulted in a consequent increase in the orientation of the polymer in the flow direction, as distance from the gate increased. The fountain flow pattern resulted in maximum elongation by elements at the surface of the advancing front and, at the cool mould walls, the onset of solidification formed a skin which froze in the orientation of the polymer. The authors suggested that if the concepts of dilute short fibre solutions could be applied, then the orientation of the skin was strictly related to the orientation parameter of the polymer. Finally, shear induced orientation was considered to take place at some distance from the wall, at a position of maximum shear rate.

A comparison of these predictions with experimental investigations of the orientation and properties of moulded bars of a LCP which possess a high orientation parameter was carried out. Three different regions of moulded bars were examined, and the variations in moduli with different test angles and between regions were recorded, as shown in Fig. 7.8.

Region 1 was closest to the gate, and exhibited a high orientation both in the flow direction and in a transverse direction at 90° to the flow path. This indicated that the region closest to the gate was subjected to spreading radial flow during moulding. Higher levels of anisotropy were

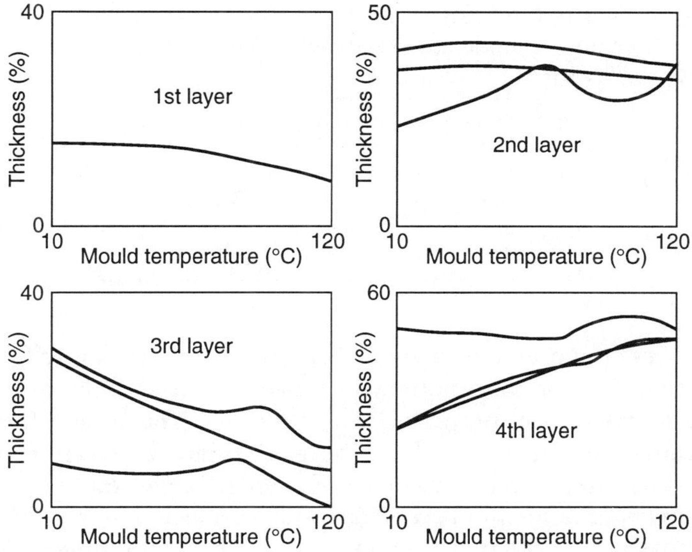

Figure 7.9 Changes in layer dimensions with temperature variation; after Suokas [74].

evident in regions 2 and 3, which was in good agreement with the predictions of the flow model. In addition, it was observed that, in the layered structure, the skin was characterized by a high modulus, while minimum modulus values were obtained at some distance from the surface. As the distance from the surface increased further, a peak in modulus value was recorded and finally the modulus decreased towards the centre with a small increase at the onset of the core region. The large peak corresponded to the region of elevated shear due to melt shearing against a solidification front.

Hsuing *et al.* [66] observed that increasing mould temperature from 115°C to 120°C resulted in a slight decrease in the thickness of the skin layer of moulded samples of ULTRAX KR 4002, and also that slight changes in the condition of layer boundaries were apparent. However, the formation of the multi-layer structures observed was significantly affected by injection speed; see section 3.1.2. Studies of the effects of mould temperature by other authors [74–76] have indicated that an increase in temperature expands the dimensions of the core region. This was clearly evident in data presented by Suokas [74] as shown in Fig. 7.9.

The variations in thickness could be related to corresponding trends in changes of mechanical properties. The tensile modulus, in the flow direction, increased with decreasing mould temperature and decreasing injection pressure. A correlation between the mechanical properties and

Table 7.6 Effects of moulding thickness on Young's modulus as reported by Thapar and Bevis [65]

Thickness (mm)	*Young's modulus (GPa)*
4	11.2
3	12.2
2	21.3
1	29.6

the thickness of unoriented zones against oriented zones was self-evident. As the thickness of the unoriented zones increased, a corresponding decrease in mechanical properties was observed. Thapar and Bevis [65] also clearly demonstrated the interrelationship between moulding thickness and mechanical properties as shown in Table 7.6.

The effects of different processing conditions including mould design, gate geometry and the use of an exit cavity have been investigated by Boldizar [77].

7.3.2 Extrusion of LCPs

A particularly important feature of polymers, which has contributed greatly to the considerable growth in usage of these materials, is their relative ease of melt processing. Screw extrusion lies at the heart of melt processing of polymeric materials, because the bulk of polymers pass through some form of extruder during their life cycle, whether compounded and pelletized, following polymerization in a reactor, or fabricated into semi-finished/finished products such as pipes, profiles, films, fibres or coatings. Extrusion costs tend to be dominated by raw materials costs and, because polymers are not cheap materials, the need to avoid out of specification product, and to reduce wastage, is a necessary requirement. This factor is particularly important in the extrusion processing of relatively expensive liquid crystalline materials. Moreover, because the properties of LCP products are highly anisotropic and strongly dependent on the level and distribution of molecular orientation, some understanding of the interrelationship between domain orientation and the type of flow occurring during processing is of prime importance. However, a lack of true real-time process monitoring, using specialized instrumentation, has hampered investigations, and has sometimes resulted in contradictory conclusions concerning the overall effects of shear rate and shear strain on the behaviour and properties of LCPs chosen for different studies.

Ide and Ophir [78] have investigated the effect of extrusion variables on the structure and properties of an LCP with mole composition 60% p-acetoxy hydroxybenzoic acid, 20% naphthalene diacetate and 20% terephthalic acid. Extrudate sections exhibited a characteristic skin/core morphology, with the greatest orientation within the surface skin. Studies of shear rate and shear deformation on mechanical properties indicated that the tensile modulus (approximately 3.5 GPa) and the tensile strength (approximately 100 MPa) were low, and were independent of shear rate and shear strain. Other workers [79] have reported that, for a PET/60PHB LCP, enhancement of orientation with increasing shear rate was apparent through a significant increase in Young's modulus (5 GPa to 18 GPa over the two decades of shear rate $10^2\ s^{-1}$ to $10^4\ s^{-1}$). However, both Muramatsu and Krigbaum [80] and Kenig [81] maintain that orientation developed in LCP extrusion does not depend on shear rate but only on shear deformation. The shear fields created during extrusion of LCPs are complex and non-uniform. Acierno and Nobile [82] have attempted to explain the discrepancies in observed mechanical property changes with shear rate by different investigators in terms of simple Poiseuille flow, considering shear rate to be a maximum at the die wall and zero along the centre of the die. They postulate that the total effect of increasing shear rate results in only a minor increase in orientation due to averaging effects created by the flow behaviour, as indicated by the Poiseuille model. They conclude that the only situation in which orientation should be sensitive to shear rate is at very low shear rates. This is in direct opposition to the observations of Ide and Ophir [78]. In addition, Acierno and Nobile [82] suggest that the interdependence of orientation and shear strain could be associated with die length. Longer capillaries could enable molecular flow to approach steady state, resulting in a significant increase in the total orientation. In shorter dies, steady-state conditions are never realized during the time of shearing within the die and, hence, orientation changes are less marked. It is clear that no simple model will suffice to produce a generalized explanation of the interrelationship between rheological parameters and the structure/property characteristics of extruded LCPs. More detailed investigations, employing modern developments in real-time on-line rheometry, may provide a better understanding of the behaviour of LCPs during extrusion.

An unusual feature of extruded LCPs is the observation of small or negligible die swell, even in the presence of high melt elasticity [83, 85]. La Mantia and Valenza [85], working on Vectra B950, explain the apparent discrepancy in elastic properties by using two different models to describe the different effects. Low die swell was considered to be due to the lack of deformability of the rigid LCP chains resulting in only a small amount of stored energy recovery at the die exit. High apparent elasticity, measured as a function of reduced entrance pressure losses, was attributed to the high energy necessary to orientate the LCP molecules at a capillary die inlet.

A recent development of particular interest to manufacturers and users of flight balloons is the employment of aromatic LCP resins in the reinforcement of thin polyethylene films [86, 87]. Small additions of LCP (around 10%) to standard polyethylene LDPE/HDPE blend resins can produce a modulus enhancement of approximately 400% (modulus > 1 GPa) in blown films, when specialized film extrusion equipment is used to produce a strong elongational flow field [87]. A typical set-up employing a 16-mm mini-extruder is shown schematically in Fig. 7.10.

The research has identified the effect of varying temperature profiles, take-up rates (v/V) and chill roller/die separations (x). Strong elongational flow fields were created through the use of die adapters with different converging angles which modify the flow pattern. Uniform film with minimal variation in thickness was produced employing die temperatures in the region 215–230°C which was close to, or higher than, the extruder zone temperature.

Film with a suitable LCP fibrillar structure rather than discrete droplet morphology meets the higher modulus specification of long duration flight balloons, while maintaining good low temperature properties and ease of fabrication, normally associated with current low modulus (approximately 250 MPa) polyethylene blends. Hsu and Harrison [87] point out that although balanced film properties are generally desirable, with LCP reinforced films, it is necessary to concentrate on improving the dispersion and distribution of LCP domains, while increasing the fibre aspect ratio ($L:D$) via the creation of strong elongational flow fields during processing. In general, the films were observed to exhibit an apparent wood grain surface pattern when extruded as dry-mixed, single pass through the extruder. More uniform films, in terms of thickness

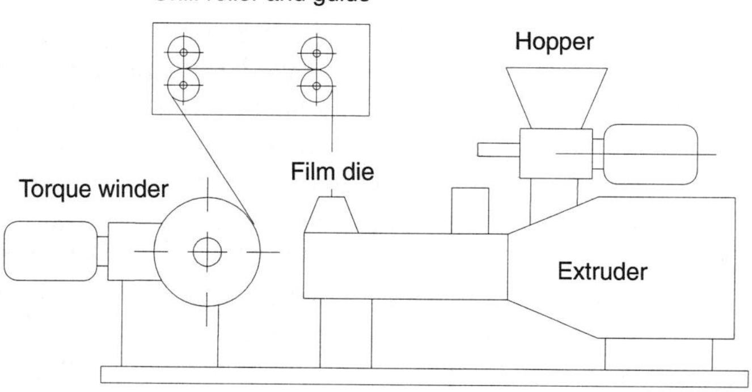

Figure 7.10 Schematic of typical 16 mm mini extruder.

and surface finish, were produced after two to three passes. However, sometimes distinct parallel lines, probably due to the high orientation of LCP in the skin, were visible along the machine direction. As expected, it was found to be undesirable to hold the blends for any period above the LCP melting temperature, without the presence of a strong flow field. Discrete droplets were observed where the LCP fibres had begun to reduce to droplet form and coalesce. Small droplets are commonly observed in cast films; these can result from either jet breakage during processing [88, 89] or LCP domain remelting.

7.8.3 LCP fibre spinning

Spinning is a process for the production of fibres by the extrusion of polymer melts or solutions through a metal die containing a number of symmetrically arranged small holes to form a corresponding number of continuous 'fluid' strands. Following this extrusion, the strands are subjected to post-die treatment which includes both stretching and cooling (for melts) and cold drawing. These drawn fibres are smaller in diameter than the original strands and are extremely anisotropic. High-modulus high-strength polymer fibres can be obtained through the retention of a high degree of chain orientation and extension. The mechanical properties and morphological features produced depend mainly on the drawing technique (whether tensile drawing or die drawing is employed) and the drawing conditions, typically the draw ratio.

For the development of a spinning process, the determination that a fluid is potentially fibre forming is a necessary, but not solely exclusive, condition. In particular, the rheological characteristics of the extrudate in the post-extrusion region must be such that a coherent filament can be drawn into the quenching or coagulating region. For any material, there is an upper limit to the extrusion rate or a lower limit to the length of the drawing region and beyond these limits, the liquid stream or filament 'breaks down'. Hence, a principal problem with all-spun fibres is the establishment of the optimum processing conditions. Even so, as a result of kinks, folds and chain entanglements, the theoretical modulus limits for flexible polymer fibres are never reached. For example, although the theoretical modulus of polyethylene fibres is 300 GPa, the highest modulus obtainable with melt spun fibres is around 70 GPa. Solution spinning techniques have been employed to produce polyethylene fibres with a modulus of about 100 GPa.

The inherent properties of thermotropic LCPs make them prime candidates for fibre manufacture, and crystalline fibres of relatively high modulus and strength can be obtained by simple melt spinning. Hence, in recent years, significant attention has been applied to the manufacture, mechanical properties and related orientation and structure of liquid crystalline fibres [90–100]. Most of the studies of LC copolyesters have centred on PET/60PHB and HBA/HNA. Acierno and Nobile [101] have carried out an excellent in-depth review of this literature, and interested readers are advised to search out this publication. An interesting extension to the ongoing work in this area has been the recent interest in the manufacture and properties of LCP polyblend fibres.

The production of fibres from mixtures of two or more polymers was first demonstrated by Cates and White in the 1950s [102]. They manufactured fibres from a mixture of cellulose acetate and polyacrylonitrile. Since then, many approaches have been developed in attempts to make polyblend fibres with superior properties to the individual constituent polymers. According to Hersh [103], the addition of a dispersed phase of higher modulus than the continuous phase, generally increases the modulus and strength of the blend or fibre, while the addition of a low modulus filler is sometimes employed to increase impact resistance and elongation to break. A number of researchers have studied polyblends of LCPs and conventional materials where the LCPs are employed as reinforcing components [17, 23, 104, 105].

Qin *et al.* [106, 107] have investigated the manufacture and properties of polypropylene/Vectra A900 polyblend fibres with different weight ratios (100 : 2.5, 100 : 5, 100 : 10, 100 : 15). Test samples were produced by the melt extrusion of both the base polypropylene and the LCP/polypropylene polyblends, followed by hot stretching the as-spun fibres on a small-scale drawing unit comprising of two pairs of advancing rollers and a hot plate. Fibres were produced by drawing at temperatures within the range 120–160°C. At temperatures above 160°C, drawing was not possible, because the fibres tended to stick to the hot plate. Both single-stage and multi-stage drawing were employed with the single-stage drawing producing fibres with a slightly higher initial modulus than polypropylene, but with a significantly lower tenacity. After single-stage drawing, hot stage photomicrography revealed that the well-oriented thin LCP fibrils, observed in the as-spun fibres, were broken into short pieces with an aspect ratio of around 10 : 1. It has been suggested that a large aspect ratio (greater than 40 : 1) is necessary for an effective reinforcement [108]. The mechanical properties exhibited by fibres of polypropylene and the polyblend in

Table 7.7 Tensile properties of the as-spun fibres [107]

	Polypropylene	*PP/LCP*
Fibre (Tex)	44.87	49.27
Yield stress (mN Tex^{-1})	28.50	36.53
Initial modulus (N Tex^{-1})	0.93	1.38

Table 7.8 Draw ratios and tensile properties of the single-stage drawn fibres [107]

	Polypropylene	*PP/LCP*
Draw ratio	12.13	11.46
Fibre (Tex)	3.70	4.30
Tenacity (N Tex^{-1})	0.960	0.794
Elongation (%)	23.6	20.6
Initial modulus (N Tex^{-1})	8.32	8.46

Table 7.9 Drawing conditions and tensile properties of the three-stage drawn fibres [107]

	1st stage at 120°C	*2nd stage at 150°C*	*3rd stage at 160°C*
Draw ratio	5.84	1.83	1.58
Overall draw ratio	5.84	10.71	15.94
Fibre (Tex)	8.42	4.60	3.09
Tenacity (N Tex^{-1})	0.387	0.750	0.946
Elongation (%)	31.6	20.9	14.2
Initial modulus (N Tex^{-1})	3.44	8.15	13.13

the as-spun condition are shown in Table 7.7, and after single-stage drawing in Table 7.8.

It was observed that the as-spun fibres were reinforced by introduction of the LCP component. However, major improvements in the tensile properties of samples were only achieved through the use of multiple drawing operations (optimum of three stages). Depending on processing conditions and the amount of LCP addition to the polyblend, multiple drawing produced increases of the order of 55% in tensile modulus and 19% in tenacity over the corresponding properties of fibres produced by single-stage drawing [20]. Table 7.9 shows typical property improvements achieved using an extrusion rate of 25 m min^{-1} followed by a three-stage drawing operation at three different temperatures in the processing range (120°C, 150°C and 160°C).

Lee and Di Benedetto [109] have studied the properties of thermotropic LCP/LCP fibre blends. A wholly aromatic copolyester K161 (Bayer AG) and an aliphatic containing LCP PET/PHB 60 (Kodak Tennessee Eastman) were melt blended at 345°C in a Brabender for 30 minutes. The mixture was then quenched to room temperature and fibres were subsequently drawn from an extruder at a temperature of 260°C. Five different compositions of the blend were prepared, in the range 10–90% K161. In addition to establishing the mechanical properties of the blends, the main objectives of the study were to examine and characterize the adhesion, the interphase and possible interactions between the two LCP phases.

The Young's modulus of the fibre blends as a function of K161 content was estimated from single fibre tensile tests carried out at a strain rate of 5 cm min^{-1}. A rule of mixture behaviour was apparent with the tensile

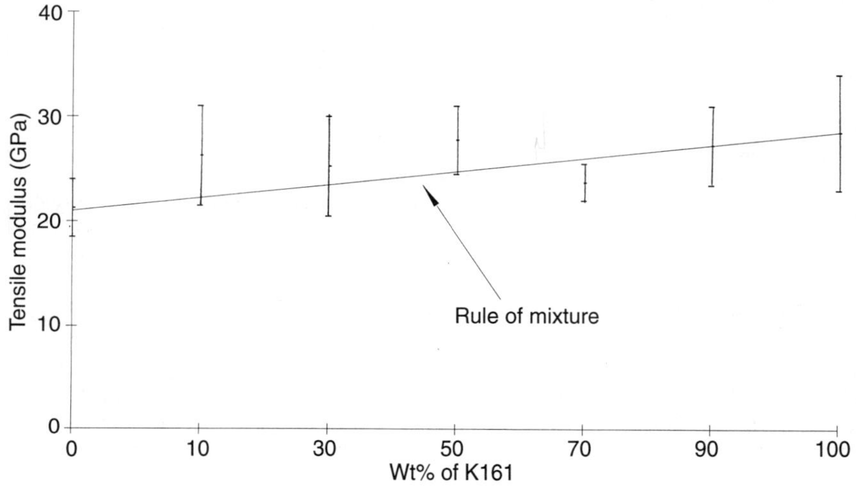

Figure 7.11 Young's modulus of the fibre blends as a function of K161 content; after Lee and Di Benedetto [109].

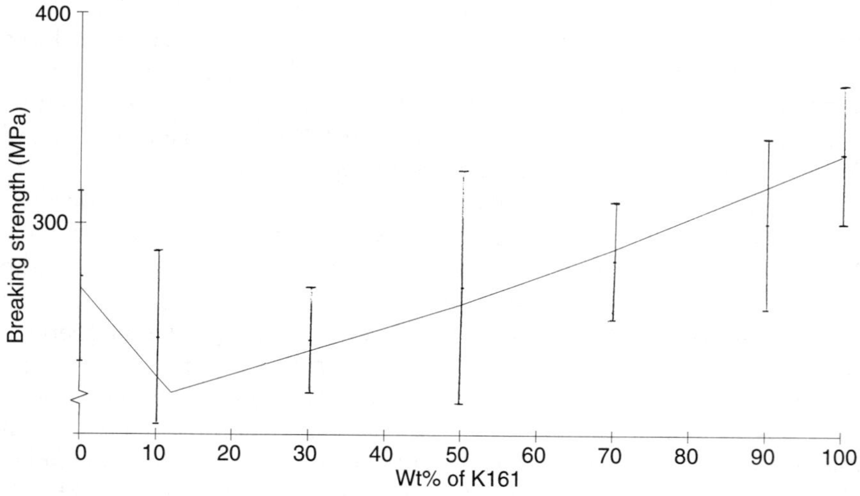

Figure 7.12 Ultimate strength of the fibre blends as a function of K161 content; after Lee and Di Benedetto [109].

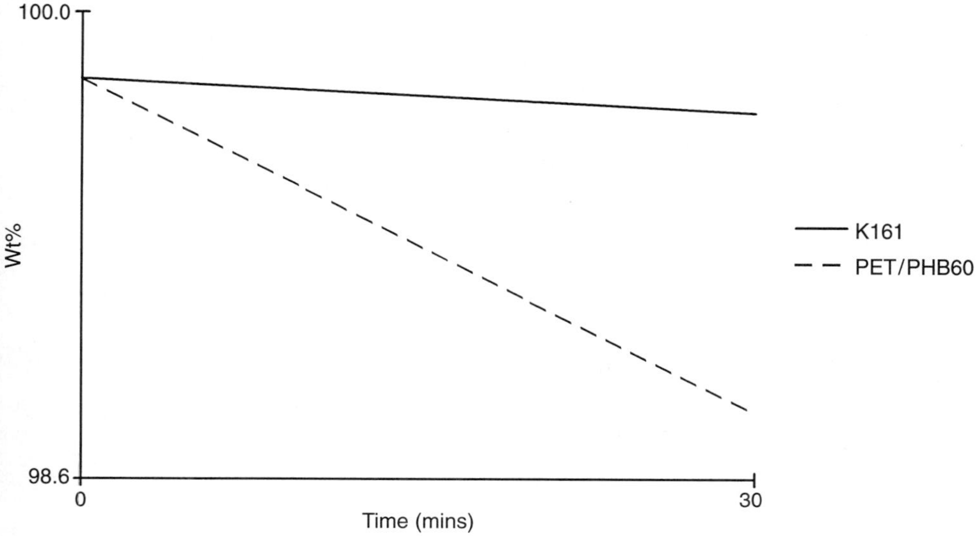

Figure 7.13 Thermogravimetric analysis of both PET/PHB60 and K161 at 345°C for 30 minutes; after Lee and Di Benedetto [109].

moduli in the range 20–30 GPa, as shown in Fig. 7.11. The blends were about two to three times stiffer than conventional nylon or PET fibres.

The corresponding tensile strengths of the blends are shown in Fig. 7.12. The trend appears to indicate that a simple composite model applies, with a reinforcing effect starting at around 40–50% K161 content.

Following standard specimen preparation procedures, the morphology of the fibre blends was studied using both scanning electron microscopy and transmission electron microscopy. The authors report evidence of an obvious two-phase morphology in samples of 30K161/70PET(PHB) from observations both on fracture surfaces and TEM microtomed fibres. It was considered likely that the embedded phase was either K161 or a K161-rich phase within a PET/PHB60 or PET/PHB60-rich matrix. A strong bonding between the two phases was evident in all microscopic studies.

The processing conditions employed during melt blending were observed to have a strong effect on the structure of the PET/60PHB, whereas the K161 appeared to remain stable during processing. Both thermogravimetric analyses of the base LCPs (Figure 7.13), and dynamic viscosity analyses of the PET/60PHB before and after thermo-mechanical treatment (Figure 7.14) indicated that chain scission of the aliphatic containing material had taken place during processing.

It was considered that the nature of the interface observed in electron micrographs strongly indicated that an interface reaction between the

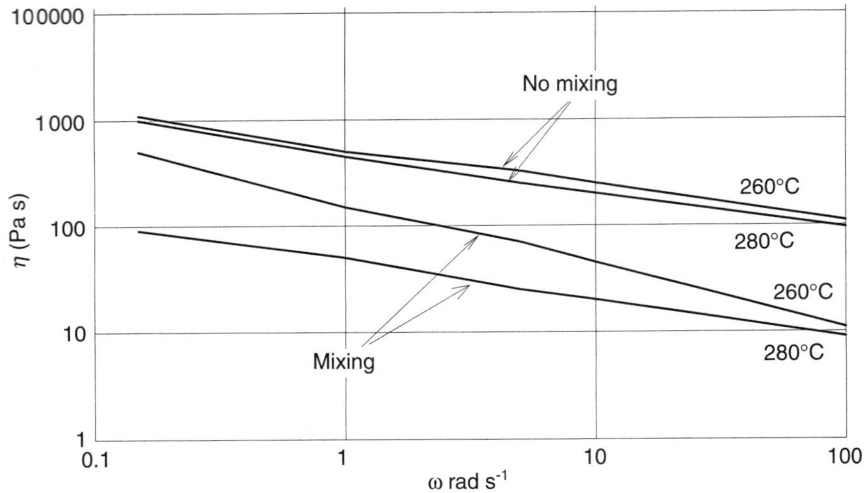

Figure 7.14 Comparison of the dynamic viscosity of PET/PHB60 before and after thermomechanical treatment; after Lee and Di Benedetto [109].

two constituent materials and/or phases had taken place during blending. The possible causes were considered to be either the production of radicals due to chain scission of the PET/60PHB, which initiated a chemical reaction, or an improved miscibility of the two components brought about by a reduction in molecular weight of the PET/60PHB.

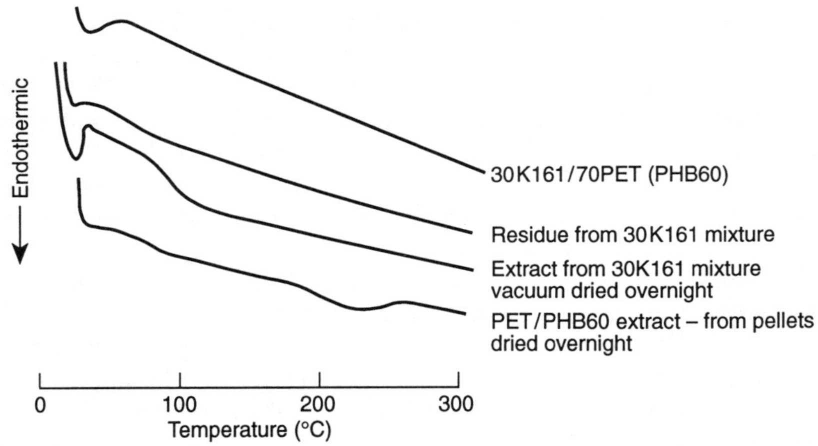

Figure 7.15 DSC thermograms of the extracted and the insoluble fractions from the 30K161/70PET(PHB60) mixture in comparison with the as-received PET/PHB60 and the original mixture; after Lee and Di Benedetto [109].

Table 7.10 Summary of the relative intensity changes in absorptions from Proton NMR Spectra; after Lee and Di Benedetto [109]

As-received PET/PHB60	*Theoretical ratio*	*Soluble fraction from 30K161/70PET(PHB60)*
1.0/1.0	1.0 : 1.0	7.4/1.0
2.7/1.0	3.0 : 1.0	8.4/1.0
2.7/1.0	3.0 : 1.0	1.1/1.0

DSC analyses (Figure 7.15), backed up by evidence derived from NMR spectroscopy (Table 7.10), supported the theory of a chemical interaction.

The authors suggested that the work should form the basis of further studies to explore the development of *'in situ'* coupling of thermally stable thermotropic LCPs and incompatible thermoplastic matrices using a melt processable LCP as an intermediate phase.

7.4 OTHER DEVELOPMENTS

In recent years, considerable attention has been devoted to the study of the rheological characteristics and properties of blends of commercial thermoplastics with LCPs. In addition to the reinforcing action of the dispersed LCP phase, improved processability is observed, resulting from the considerable reduction in melt viscosity caused by the presence of the LCP. This strong reduction in melt viscosity is particularly apparent at high shear rates even with small LCP additions (less than 5%). In the direction of melt flow, considerable improvements in tensile modulus and tensile strength over that of the base thermoplastic are also evident, provided that processing conditions promote fibrillation of the dispersed LCP phase. However, the transverse strength of blends is often observed to be much lower than expected [110].

Adhesion between phases in an LCP/thermoplastic blend with a fibrillar LCP morphology tends to be poor because of the intrinsic incompatibility between the constituent materials. This lack of adhesion greatly reduces both the transverse and shear strengths of a blend, especially when the dispersed LCP phase is highly orientated. Approaches utilized to improve adhesion between *in situ* formed fibrils include modification of the fibre geometry to provide mechanical locking with the matrix [111], and chemical grafting of flexible polymers to the fibre surface [112]. The feasibility of employing a second thermotropic LCP as a coupling agent to improve adhesion between incompatible phases has also been investigated [113]. The objective of this approach is to produce a composite LCP/LCP fibre blend in which the dispersed phase is thermally stable during moulding operations and in which the fibrous matrix is processable and compatible with both the dispersed phase and the base thermoplastic matrix. This approach, using a partially compatible and/or reactive additional thermotropic additive to the blend, is similar in nature to the use of compatibilizers in other polymer blends, such as polyethylene/polystyrene and other blends [114, 115]. Evidence suggests that the latter approach has high potential for future utilization. It has been demonstrated that a highly aromatic copolyester LCP (K161, Bayer AG, KU-9211) can be finely dispersed and well bonded to an inherently incompatible PET matrix through the presence of a second aliphatic LCP (PHB60).

The inclusion of a second LCP in an incompatible LCP/thermoplastic blend not only produces a ternary system with improved inter-phase adhesion, but also provides the potential of enhanced processability, as an additional advantage. Melt blending of an LCP with a thermoplastic requires the processing temperature range of the two materials to be similar, so that the thermoplastic is in the molten state while the LCP is in the anisotropic melt state. Attempts to melt blend materials with dissimilar melt characteristics tend to result in both the degradation of the lower melting constituent, usually the thermoplastic, and poor overall dispersion. This requirement limits the choice of LCP/thermoplastic for blending. The copolyester-based LCP Xydar has outstanding thermal stability and fibre-forming properties. However, this material does not melt until 400°C, which makes it difficult to blend with many engineering thermoplastics. Blending a thermally stable, reinforcing LCP, such as Xydar, with a second lower melting point LCP coupling phase, could conceivably allow a tailored processing temperature window to be attained.

REFERENCES

1. Blades, H. (1972) US Patent 3,767,756.
2. Morgan, P.W. (1976) *Macromolecules*, **10**, 1381.
3. Kwolek, S.L., Morgan, P.W., Schaefgen, J.R. *et al.* (1976) *Macromolecules*, **10**, 1390.
4. Bair, T.I., Morgan, P.W. and Killian, F.L. (1976) *Macromolecules*, **10**, 1396.
5. Panar, M. and Beste, L.F. (1976) *Macromolecules*, **10**, 1401.
6. Wissburn, K.F. (1980) *Brit. Polym. J.*, Dec., 163.
7. Wissburn, K.F. (1981) *J. Rheol.*, **25**, 619.
8. Wissburn, K.F. and Griff, A.C. (1982) *J. Polym. Sci. Polym. Phys. Ed.*, **20**, 1835.
9. Ide, Y. and Ophir, Z. (1983) *Polym. Eng. Sci.*, **23**, 261.
10. Ide, Y. and Chung, T.S. (1984–85) *J. Macromol. Sci. Phys.*, **4–6**, 497.
11. Baird, D.G. (1983) *Polymeric Liquid Crystals*, A. Blumstein (ed.), Plenum Press, New York.

12. Joseph, E.G., Wilkes, G.L. and Baird, D.G. (1985) *Polym. Eng. Sci.*, **25**, 377.
13. Jaffe, M. (1991) *J. Stat. Phys.*, **62**, 985.
14. Calundann, G.W. and Jaffe, M. (1982) *Proc. R.A. Welch Conf. on Chemical Research, Synthetic Polymers*, Texas.
15. Prevorsek, D.C. (1982) in *Polymer Liquid Crystals*, A. Ciferri, W.R. Krigbaum and R.B. Meyer (eds), Academic Press, New York.
16. Huh, W., Weiss, R.A. and Nicolais, L. (1983) *Polym. Eng. Sci.*, **23**, 779.
17. Siegmann, A., Dagan, A. and Kenig, S. (1985) *Polymer*, **26**, 1325.
18. Joseph, E.G., Wilkes, G.L. and Baird, D.G. (1985) in *Polymeric Liquid Crystals*, A. Blumstein (ed.), Plenum Press, New York, pp. 197–216.
19. Isayev, A.I. and Modic, M.J. (1986) *SPE ANTEC Tech. Papers*, **32**, 573.
20. Chung, T.S. (1987) *SPE ANTEC Tech. Papers*, **33**, 1404.
21. Blizard, K.G. and Baird, D.G. (1987) *Polym. Eng. Sci.*, **27**, 653.
22. Ramanthanan, R., Blizard, K.G. and Baird, D.G. (1987) *SPE ANTEC Tech. Papers*, **33**, 1399.
23. Kiss, G. (1987) *Polym. Eng. Sci.*, **27**, 410.
24. Ramanthanan, R., Blizard, K.G. and Baird, D.G. (1988) *SPE ANTEC Tech. Papers*, **34**, 1123.
25. Zhuang, P., Kyu, T. and White, J.L. (1988) *Polym. Eng. Sci.*, **28**, 1095.
26. Ko, C.U. and Wilkes, G.L. (1989) *J. Appl. Polym. Sci.*, **37**, 3063.
27. Hull, J.B., Hawksworth, M. and Collyer, A.A. (1992) in *Key Eng. Mats.*, M. Buggy and S. Hampshire (eds), Trans. Tech. Publications, IMF 8, Limerick, pp. 627–34.
28. Hawksworth, M., Hull, J.B. and Collyer, A.A. (1993) in *Processing and Properties of Liquid Crystalline Polymers and LCP Based Blends*, D. Acierno and F.P. La Mantia (eds), Chem. Tech. Publishing p. 65.
29. Sukhadia, A.M., Done, D. and Baird, D.G. (1990) *Polym. Eng. Sci.*, **30**, 519.
30. Brown, G.H. (1969) *Anal. Chem.*, **29A**, 41.
31. DeGennes, P.G. (1974) *The Physics of Liquid Crystals*, Clarendon Press, Oxford.
32. Zocher, H. (1933) *Trans. Faraday Soc.*, **29**, 945.
33. Werbowyi, R.S. and Gray, D.G. (1976) *Mol. Liquid Crystal Lett.*, **97**, 43.
34. Onogi, Y., White, L. and Fellers, F. (1980) *J. Polym. Sci. Polym. Phys. Ed.*, **18**, 663.
35. Blackwell, J., Gutierrez, G.A. and Chevers, R.A. (1984) *Macromolecules*, **17**, 1219.
36. Blackwell, J., Biswas, A., Gutierrez, G.A. *et al.* (1985) *Faraday Discussions of the Chem. Soc.*, **79**, 79.
37. Blackwell, J., Biswas, A. and Bonart, R.C. (1985) *Macromolecules,*, **18**, 2126.
38. Blundell, D.J. (1982) *Polymer*, **23**, 359.
39. Butzbach, G.D., Wendorff, J.H. and Zimmermann, H.J. (1985) *Makrom. Chem. Rapid Commun.*, **6**, 821.
40. Butzbach, G.D., Wendorff, J.H. and Zimmermann, H.J. (1986) *Polymer*, **27**, 1337.
41. Windle, A.H., Viney, C., Golombok, R. *et al.* (1985) *Faraday Discussion Chem. Soc.*, **79**, 55.
42. Hanna, S. and Windle, A.H. (1988) *Polymer*, **29**, 207.
43. Spontak, R.J. and Windle, A.H. (1990) *Polymer*, **31**, 1395.

44. Shiwaku, T.A., Nakai, H., Hasegawa, H. *et al.* (1990) *Macromolecules,* **23**, 1590.
45. Cheng, S.Z.D. (1988) *Macromolecules,* **21**, 2475.
46. Kaito, A., Kyotani, M. and Nakayama, K. (1990) *Macromolecules,* **23**, 1035.
47. Lin, Y.G. and Winter, H.H. (1988) *Macromolecules,* **21**, 2439.
48. Krigbaum, W.R., Liu, C.K. and Yang, D.K. (1988) *J. Polym. Sci. Phys. Ed.,* **26**, 1711.
49. Yang, D.K. and Krigbaum, W.R. (1989) *J. Polym. Sci., Phys. Ed.,* **27**, 1837.
50. Yang, D.K. and Krigbaum, W.R. (1989) *J. Polym. Sci., Phys. Ed.,* **27**, 819.
51. La Mantia, F.P., Valenza, A.L., Pedretti, U. *et al.* (1990) *Proc. Comp '90, Patras (Greece), 20–24 August*, 33.
52. Yoon, H. and Jaffe, M. (1984) *Polymer Preprints.*
53. Green, D.J., Davies, G.R., Ward, I.M. *et al.* (1990) *Polym. Adv. Technol.,* **1**, 41.
54. Sawyer, L.C. and Jaffe, M. (1986) *J. Mater. Sci.,* **21**, 1897.
55. Lee, C.Y.C. and Wang, C.S. (1990) *Proc. Polym. Mater. Sci. Eng.,* **62**, 81.
56. Speight, R.J. (1993) PhD Thesis, Bradford.
57. Jackson, W.J., Kuhfuss, H.F. and Grey, T.F. (1975) *Soc. Plastics Ind. Tech. Conf.,* **17-D**.
58. Jackson, W.J. and Kuhfuss, H.F. (1976) *J. Polym. Sci. Polym. Chem. Ed.,* **14**, 2043.
59. McFarlane, F.E., Nicely, V.A. and Davis, T.G. (1977) *Cont. Topics in Polym. Sci.,* **2**, 109.
60. Ophir, Z. and Ide, Y. (1983) *Polym. Eng. Sci.,* **23**, 792.
61. Sawyer, L.C. (1984) *J. Polym. Sci. Polym. Lett. Ed.,* **22**, 347.
62. Sawyer, L.C. (1985) *Intern. Fibre J,* **27**.
63. Weng, T., Hilmer, A. and Baer, E. (1986) *J. Mater. Sci.,* **21**, 744.
64. Thapar, H. and Bevis, M. (1983) *J. Mater. Sci. Lett.,* **2**, 733.
65. Thapar, H. and Bevis, M. (1989) *Plastics and Rubber Processing Applications,* **12**, 39.
66. Hsuing, C.M., Tian, J. and Cakmak, M. (1993) *Intern. Polym. Process.,* **2**, 164.
67. Bowman, J. and Bevis, M.J. (1977) *Colloid and Polym. Sci.,* **255**, 954.
68. Bowman, J. and Bevis, M.J. (1976) *Plastics and Rubber: Materials and Applications,* **1**, 177.
69. Ophir, Z. and Ide, Y. (1982) *National Tech. Conf. on Plastics,* **88**, 164.
70. Goettler, L.A. (1970) *Mod. Plast.,* **48**, 140.
71. Hedmark, P.G., Lopez, J.M.R., Westdahl, M. *et al.* (1988) *Polymer Eng. Sci.,* **28**, 1248.
72. Garg, S.K. and Kenig, S. (1988) in *High Modulus Polymers,* A.E. Zachariades and R.S. Porter (eds), Marcel Dekker Inc., New York, pp. 71–103.
73. Kenig, S. (1987) *Polym. Eng. Sci.,* **27**, 887.
74. Suokas, E. (1989) *Polymer,* **30**, 1105.
75. Hsiung, C.M. and Cakmak, M. (1991) *SPE ANTEC Tech.,* **37**, 1063.
76. Kennache, G. and Kamal, M.R. (1991) *SPE ANTEC Tech.,* **37**, 1075.
77. Boldizar, A. (1988) *Plastic and Rubber Processing Applications,* **10**, 73.
78. Ide, I. and Ophir, Z. (1983) *Polym. Eng. Sci.,* **23**, 261.

79. Takeuchi, Y., Yamamoto, F. and Yamakawa, S. (1984) *Polym. J.*, **16**, 579.
80. Maramatsu, H. and Krigbaum, W.R. (1986) *J. Polym. Sci., Phys. Ed.*, **24**, 1695.
81. Kenig, S. (1989) *Polym. Eng. Sci.*, **29**, 1136.
82. Acierno, D. and Nobile, M.R. (1993) in *Thermotropic Liquid Crystal Polymer Blends*, F.P. La Mantia (ed.), Technomic, Lancaster, USA, p. 67.
83. Jerman, R.E. and Baird, D.G. (1981) *J. Rheol.*, **25**, 275.
84. Sugiyami, H., Lewis, D.N., White, J.L. *et al.* (1985) *J. Appl. Polym. Sci.*, **30**, 2329.
85. La Mantia, F.P. and Valenza, A. (1989) *Polym. Eng. Sci.*, **29**, 717.
86. Lichkus, A.M. and Harrison, I.R. (1992) *SPE ANTEC '92*, **38**, 2257.
87. Hsu, T.C. and Harrison, I.R. (1993) *SPE ANTEC '93*, **39**, 1183.
88. Pracella, M., Chiellini, E., Galli, G. *et al.* (1987) *Mol. Crystl. Liq. Cryst.*, **153**, 525.
89. Sukhadia, A.M., Datta, A. and Baird, D.G. (1991) *SPE ANTEC '91*, **37**, 913.
90. Acierno, D., La Mantia, F.P., Polizotti, G. *et al.* (1982) *Macromolecules*, **15**, 1455.
91. Maramatsu, H. and Krigbaum, W.R. (1986) *Macromolecules*, **19**, 2850.
92. Tealdi, A., Ciferri, A. and Conio, G. (1987) *Polym. Comm.*, **28**, 22.
93. Jenkis, J.C. and Jenkis, G.M. (1987) *J. Mat. Sci.*, **22**, 3784.
94. Itoyama, K. (1988) *J. Polym. Sci. Phys. Ed.*, **26**, 1845.
95. Chen, G.Y., Cucolo, J.A. and Tucker, P.A. (1988) *J. Polym. Sci. Polym. Phys. Ed.*, **26**, 1677.
96. Nobile, M.R., Amendola, E., Nicolais, L. *et al.* (1989) *Polym. Eng. Sci.*, **29**, 244.
97. Di Benedetto, A.T., Nicolais, L., Amendola, E. *et al.* (1989) *Polym. Eng. Sci.*, **29**, 153.
98. Sarlin, J. and Tormala, P. (1990) *J. Appl. Polym. Sci.*, **40**, 453.
99. Acierno, D., Incarnato, L., Nobile, M.R. *et al.* (1991) *Proc. of X Convegno Italiano Di Scienza e Technologia Delle Macromolecole*, Ferrara, Italy, pp. 869–72.
100. Fujiwata, K., Takahashi, M. and Masuda, T. (1991) *Int. Polym. Process.*, **6**, 232.
101. Acierno, D. and Nobile, M.R. (1993) in *Processing and Properties of Liquid Crystal Polymers*, D. Acierno and F.P. La Mantia (eds), Chem. Tech. Publishing, pp. 59–103.
102. Cates, D.M. and White, H.J. (1956) *J. Polym. Sci.*, **20**, 155.
103. Hersh, S. (1985) in *High Technology Fibres*, L. Lewin and J. Preston (eds), Marcel Dekker Inc., New York.
104. Isayev, A.I. and Modic, M.J. (1987) *Polym. Comp.*, **8**, 158.
105. Chung, T.S. (1987) *Plast. Eng.*, **43**, 39.
106. Qin, Y., Brydon, D.L., Mather, R.R. *et al.* (1992) *Textiles and Composites*, **133**, 199.
107. Qin, Y., Brydon, D.L., Mather, R.R. *et al.* (1993) *Polymer*, **34**, 1203.
108. Huh, W. (1986) PhD Thesis, Univ. of Connecticut.
109. Lee, W. and Di Benedetto, A.T. (1992) *Polym. Eng. Sci.*, **32**, 400.
110. Valenza, A., La Mantia, F.P., Paci, M. *et al.* (1991) *Intern. Polymer Process.*, **3**, 247.
111. Shaw, M.T. and Rueda, G. (1989) *Polym. Prepr.*, **30**, 538.

112. Datta, A., De Souza, J.P., Sukhadia, A.M. *et al.* (1991) *SPE ANTEC Tech. Papers*, **37**, 913.
113. Lee, W. and Di Benedetto, A.T. (1993) *Polymer*, **34**, 684.
114. Paul, D.R., Barlow, J.W. and Keskkula (1988) *Encyclopaedia of Polym. Sci. and Eng.*, Vol. 12, Wiley, New York.
115. Fayt, R., Jerome, R. and Teyessie, P. (1989) in *Multiphase Polymers*, L.A. Utracki and R.A. Weiss (eds), Series 395.

8

Time-dependent effects in lyotropic systems

P. Moldenaers

8.1 INTRODUCTION

8.1.1 Liquid crystallinity

The term 'liquid crystals' describes a state of matter which is intermediate between that of amorphous liquids and that of crystalline solids [1]. Such materials possess some degree of orientational ordering but limited or no positional order. Hence anisotropy is a key feature of liquid crystals. As a result of their particular structure, the mechanical behaviour of liquid crystals often resembles that of fluids whereas their optical characteristics reflect their crystal-like nature. A general introduction to liquid crystalline materials is presented by Collyer in Chapter 1 of this book.

Liquid crystallinity was first identified in low molecular weight materials by Reinitzer in 1888. Somewhat over half a century ago, the same type of behaviour was also reported to exist in polymeric materials. Suspensions of tobacco mosaic virus [2] and solutions of synthetic polypeptides [3] were the first polymeric liquid crystalline systems studied in some detail. Apart from the already mentioned division between low molecular weight and polymeric liquid crystals (PLCs), other classification schemes exist. A first one is based on the type of ordering the materials exhibit and consequently can be associated with their optical properties. According to this classification, one can distinguish between three major types of structures: **nematics**, **cholesterics**

Rheology and Processing of Liquid Crystal Polymers
Edited by D. Acierno and A.A. Collyer
Published in 1996 by Chapman & Hall
ISBN 0 412 59640 7

and **smectics**. A second classification scheme distinguishes between systems in which the isotropic/anisotropic transition is caused by changes either in concentration or in temperature. The corresponding classes of liquid crystals are referred to as **lyotropics** and **thermotropics** respectively. As the latter division is important for the material covered in this chapter, we will discuss it in more detail in the next section. It should also be noted that PLCs can be divided into main chain and side chain PLCs depending on the location of the mesogenic units. This chapter deals only with the main chain type.

8.1.2 Lyotropics versus thermotropics

Lyotropic liquid crystals are formed by mesomorphic molecules in a non-mesomorphic solvent; see Chapter 1. At a given temperature, the nature of the solution is determined by the concentration of the mesomorphic molecules. At low concentrations, an isotropic solution is formed. Such solutions become biphasic at intermediate concentrations. At still higher concentrations, a fully anisotropic solution or mesophase results. The phase transitions can be detected by various methods, such as viscometric or optical ones. One of the most striking features of these materials is the non-monotonic increase of the viscosity with concentration. The concentration range, around which the viscosity curve displays a maximum, corresponds to the transition from an isotropic to an anisotropic solution. Several theories attempt to predict the critical conditions for the formation of a lyotropic liquid crystalline structure. Of the older theories, the one by Flory [4] has attracted the most attention. This theory was based on purely geometrical considerations. More recently, Doi [5] presented a molecular theory which takes into account the interactions between neighbouring molecules. In the limit of no flow (equilibrium case), the theory predicts the concentrations at which the material becomes biphasic and fully liquid crystalline. These theories assume rigid rod-like molecules, the critical concentration for phase transition then depends on the aspect ratio ($L : D$) of the molecules. In thermotropic systems, temperature plays a key role. A liquid crystalline phase appears upon heating the crystalline solid above its melting temperature. Further increase of the temperature eventually causes a transition into an isotropic liquid. Viscosity can again be used to detect the transition points.

This chapter discusses time effects in PLCs, with the emphasis on lyotropic systems. This limitation to solutions results from the practical consideration that most of the reliable information available in the literature at present has been obtained on lyotropics. This is not surprising as PLCs show time effects that are much slower than those encountered in regular isotropic polymeric systems. The experimental determination of slow transients does not really pose problems when using lyotropics, provided one can limit solvent evaporation. Prolonged experiments on thermotropic systems suffer, however, from many experimental difficulties, most of them being related to the high melting temperature of these systems (typically around 300°C). Problems include recrystallization, polymerization and chemical changes such as hydrolysis and transesterifications. They have been discussed in the literature by various research groups [6–9] and limit drastically the time over which reliable data can be obtained on thermotropics. These problems are probably also the source of much of the disagreement that exists between published data. Problems related to experimentation with thermotropics are discussed in detail by Cogswell and Wissbrun in Chapter 4 of this book.

In addition to the above-mentioned stability, lyotropics have the extra advantage over thermotropics in that they are relatively transparent. This makes them amenable to microscopic and rheo-optical investigations such as birefringence, dichroism and small angle light scattering (SALS). These techniques have been applied to lyotropics only relatively recently, but they have contributed substantially to our insight into the mechanisms underlying the rheological phenomena. Rheological, microscopic and rheo-optical results on lyotropics will be discussed systematically in the next sections.

Once the characteristics of the flow behaviour of lyotropics have been identified, it becomes of course much easier to verify whether they also apply to thermotropics. Only a limited number of well-chosen experiments are then required for such a validation. This approach is not unlike the one used many years ago for isotropic polymeric systems where much of our insight into the behaviour of melts resulted from the understanding of concentrated polymeric solutions in the first place.

8.2 COMMONLY INVESTIGATED LYOTROPICS

8.2.1 Types of polymers

When reviewing the literature, it is surprising that our understanding of the behaviour of lyotropics is based on investigations of a relatively limited number of materials. The most studied polymers are polybenzylglutamates (PBG), hydroxypropylcellulose (HPC) and aromatic polyamides. All three form main chain liquid crystals because of their rigid molecular structure in the appropriate solvents. For convenience, their basic properties are described below as they will be referred to frequently in this chapter. In addition to these three types of polymers, other materials have sometimes been used for special purposes, such as the generation of monodomains.

Polybenzylglutamates (PBGs)

The PLC system that has attracted most attention over the past 15 years is certainly the solution of PBG in m-cresol. This material is mainly of academic interest [10–37]. PBG is a synthetic polypeptide, with a basic repeat unit as shown in Fig. 8.1. PBG forms an α-helix in several solvents, of which m-cresol has been the most popular but other ones, such as dimethylformamide, have also been used [38]. The α-helix provides the rigid rod-like structure which is necessary to generate a liquid crystalline phase. When a single enantiomer (PBLG or PBDG) is dissolved, a cholesteric structure develops, whereas a racemic mixture of the L and D forms will give rise to a nematic structure. The cholesteric structure of the single enantiomer solutions transforms, however, into a nematic one during flow [10].

Although PBG solutions have no industrial importance as such, there are various reasons why they are useful as reference materials. First of all, the flexibility of the α-helix is limited, meaning that the ratio persistence length/molecular length of the rod is relatively close to 1. This implies that PBG solutions can approach the 'ideal' liquid crystal. This is especially important from a modelling point of view where an ideal rigid rod-like structure is assumed to exist [5, 39, 40]. Indeed, one has only recently tried to incorporate some degree of molecular flexibility into the molecular models (see Chapter 2 of this book). Liquid crystalline PBG solutions can also be prepared with different degrees of intermolecular interactions by varying the concentration. Moreover, different molecular weights of PBG are commercially available, making it possible to study the effect of the intrinsic molecular parameters on the flow behaviour and the structure. The relatively high level of polydispersity encountered in commercial samples, however, hampers such studies to some extent [41]. An additional advantage of the use of PBGs is that, once dissolved, the solutions turn out to be very stable. Limitations inherent to the use of

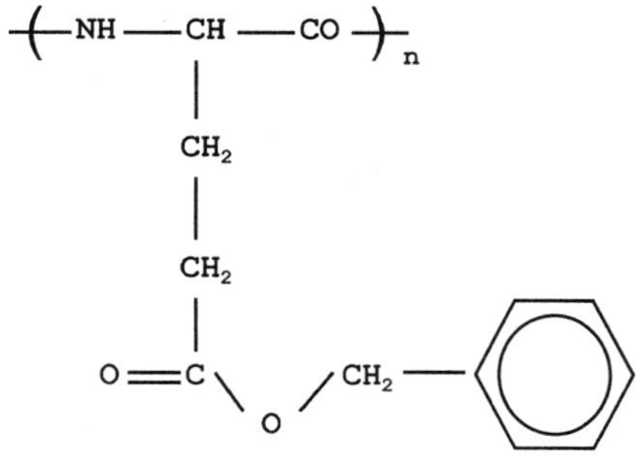

Figure 8.1 Basic repeat unit of poly-benzylglutamate (PBG).

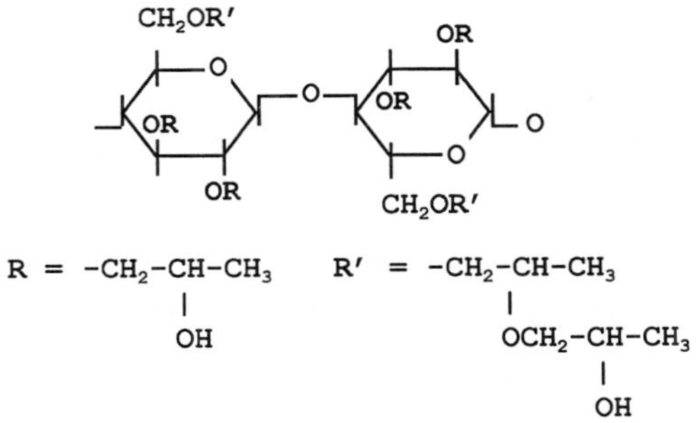

Figure 8.2 Basic repeat unit of poly-hydroxypropylcellulose (HPC).

PBG solutions are the cost of the polymer and the toxicity and volatility of the solvents.

Hydroxypropylcellulose (HPC)

A second lyotropic system that has been widely studied consists of HPC solutions, water being the most commonly used solvent [17, 21, 34, 42–50]. The chemical structure of HPC is depicted in Fig. 8.2. Contrary to the rather ideal rigid rod-like structure of PBG, the HPC molecules display much more flexibility. This results in a persistence length, much shorter than the typical molecular length. Although this degree of flexibility hampers the modelling of these solutions, it probably reflects more closely the characteristics of commercially relevant PLCs such as the aromatic polyamide solutions and the thermotropics. An additional advantage of using HPC rather than PBG is that it is readily available in large quantities and at low cost. This makes it possible to use HPC solutions in experiments that require an appreciable amount of sample, e.g. spinning [51] or mould filling [52] experiments.

Aromatic polyamides

The third category of relatively well studied lyotropics consists of solutions of aromatic polyamides such as poly-p-phenyleneterephthalamide (PPTA), poly-p-benzamide (PBA) and poly-4,4′-benzanalidyleneterephthalamide (DABT) in concentrated sulphuric acid [53–58]. Figure 8.3 shows the molecular structure of PPTA, the most widely used aramid polymer. Solutions of aromatic polyamides can be spun into high-modulus high-strength fibres that are better known by their trade names Twaron® (Akzo–Nobel) and Kevlar® (Du Pont). It is mainly their industrial importance that has stimulated research on this type of lyotropic solutions.

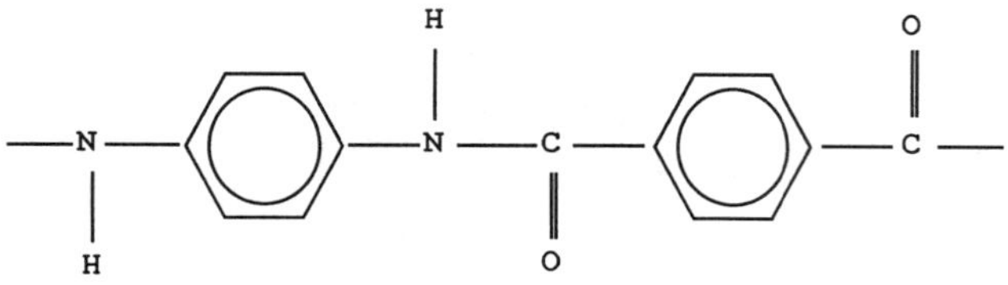

Figure 8.3 Basic repeat unit of poly-p-phenyleneterephthalamide (PPTA).

8.2.2 Behaviour in steady-state flow

This section briefly describes the generally observed flow behaviour of lyotropics under steady-state shear conditions. The steady-state behaviour serves to identify the specific ranges of the viscosity curve. In 1980, Onogi and Asada [59] suggested the three-region flow curve (Fig. 8.4) for the dependency of the viscosity on shear rate for liquid crystals. This well-known curve consists of a shear thinning region (region I) at low shear rates, a region of constant viscosity (region II) at intermediate shear rates and a second shear thinning region (region III) at still higher shear rates. The intermediate region of constant viscosity is commonly referred to as 'Newtonian' although it does not satisfy the complete definition of Newton's law.

This type of flow behaviour has been reported for many lyotropic systems although not all three regions were always observed [60]. For the PBGs for example, regions II and III have been reported for various concentrations, molecular weights and temperatures [10, 11, 12, 15, 17]. Region I seems to appear only when the concentration of PBG is rather

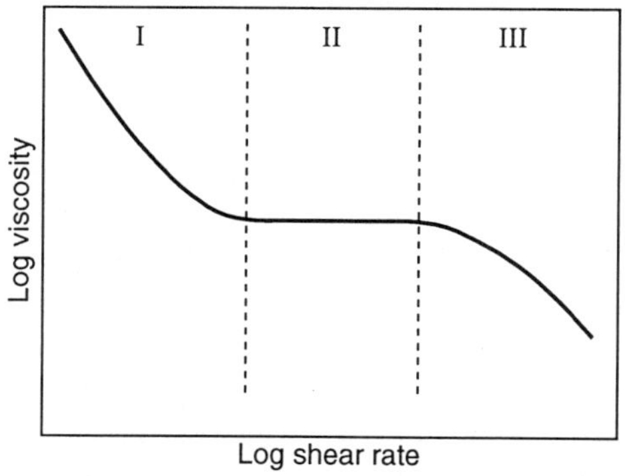

Figure 8.4 Shear viscosity versus shear rate for PLCs, showing the characteristic three-region flow behaviour; reprinted with permission [59].

high [13, 30, 37]. In addition, the viscosity curve in region I might consist of two branches [30, 37]. It can, of course, not be excluded that region I exists in the PBG systems with relatively low concentrations at immeasurably low shear rates. In isotropic polymeric liquids, the shear rate range of constant viscosity represents the limiting behaviour of the material at low shear rates. This does not hold for PLCs where the constant viscosity region (region II) occurs at intermediate shear rates. It will be demonstrated in sections 8.3 and 8.4 that the microstructure is not constant in this region, resulting in quite complex transients.

For solutions of HPC in water, a clear three-region curve has been recorded; the viscosity at intermediate shear rates is often not perfectly constant, hence it is better referred to as pseudo-Newtonian [43, 44, 46, 50]. Sigillo and Grizzuti recently suggested that, for their HPC solutions, region I could be preceded by another Newtonian region (region 0) at very low shear rates [47]. It remains, however, to be verified whether these solutions were fully anisotropic [61]. For other lyotropics, such as PPTA, only a single shear thinning region could be measured, so it is not clear from the viscosity data whether it actually represents region I or region III [62]. This is not unlike the case of many thermotropes where, often, only a single shear thinning range is observed [7, 63]. However, for the thermotropic polyester Vectra A900, a full three-region flow curve has been recorded; see the discussion in Chapter 4.

In ordinary polymers, the first normal stress difference (N_1) is always positive. Lyotropics, on the other hand, often display a very typical pattern for N_1 as a function of shear rate. At low shear rates, positive values for N_1 are recorded which increase with increasing shear rate. N_1 then goes through a maximum and subsequently decreases to reach negative values. When the shear rate is increased further, the values of N_1 become positive again. This behaviour, reported for the first time on PBGs in the pioneering work by Kiss and Porter [10, 11], was later confirmed by other research groups [14, 15, 23, 25]. A typical response of N_1 for a PBG solution is shown in Fig 8.5. The steady shear viscosity has also been added in this figure. It is interesting to note that the viscosity curve displays a 'kink' in the shear rate range with negative N_1 values. This distortion is not accidental; it correlates with the transition from tumbling to flow aligning, as was shown by molecular models (see Chapter 2). Recently, it has been shown that the region of negative N_1 values at intermediate shear rates might disappear if highly concentrated solutions are used [27]. In addition to the change in sign of N_1, it has been reported that the second normal stress difference also changes sign twice with increasing shear rate. This occurs at about the same values for the shear rate as where N_1 changes sign [23]. Hence N_1 and N_2 have opposite

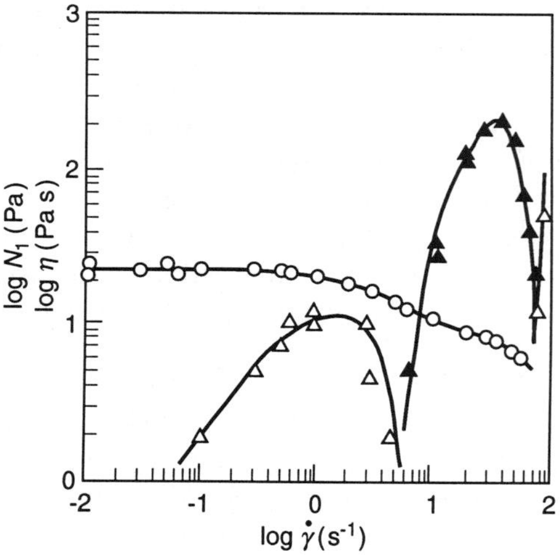

Figure 8.5 Steady shear flow results for a 12% PBLG solution at 293 K (Mw = 250 000) (○: viscosity; △: positive N_1; ▲: negative N_1); reprinted with permission [16] © 1987 by Gordon and Breach Science Publishers.

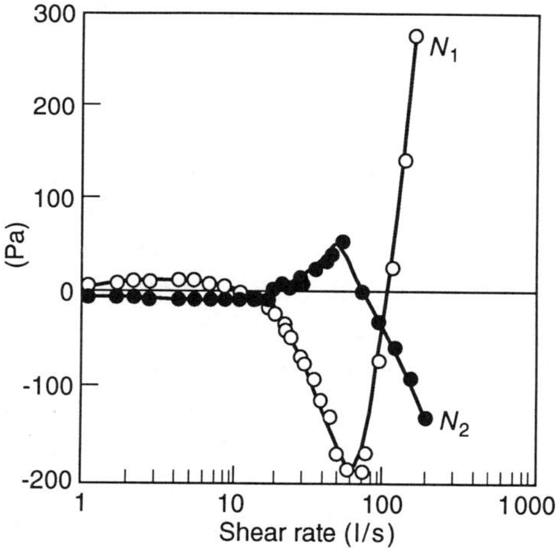

Figure 8.6 First (N_1) and second (N_2) normal stress difference measured for a 12.5% PBLG solution (Mw = 238 000); reprinted with permission [23] © 1991 by American Chemical Society.

signs at most shear rates. This is illustrated in Fig. 8.6 for a PBLG solution of very similar characteristics to the one shown in Fig. 8.5.

With HPC in water or m-cresol, a similar sequence of positive–negative–positive values has been obtained for the N_1-shear rate curve [42–46]. Baek *et al.* [45] also investigated N_2 curves for solutions of HPC in m-cresol. Both the N_1 and N_2 curves depend on shear rate, in a way similar to that obtained with PBG, including the double sign change for N_1 and N_2. As for the PBGs, the region of negative N_1 disappears at high concentrations [48].

From a theoretical point of view, the steady-state flow behaviour of lyotropics is well understood now. An excellent overview of the state of the art was recently published by Marrucci and Greco [64]. In this book (Chapter 2), Marrucci also deals with the modelling of PLC flow. Present-day theoretical predictions are based mainly on the molecular theory for concentrated solutions of rod-like molecules by Doi [5]. Marrucci and Maffettone [39] calculated the two-dimensional predictions of the Doi theory, omitting the closure approximation used by Doi, whereas Larson performed similar three-dimensional calculations [40]. The viscosity–shear rate curves resulting from these models consist of a Newtonian region at low shear rates and a shear thinning region at higher shear rates. For relatively concentrated systems, the direction of the average molecular orientation (the **director**) is of a tumbling nature at low shear rates and flow-aligning at high shear rates. At intermediate shear rates, wagging of the director results from these calculations. The negative N_1 is predicted to occur in the range of shear rates encompassing the transition from tumbling to flow-alignment. The experimentally observed sign inversion in N_2 is also reproduced by this theory. Region I behaviour is not described by these monodomain theories. It can be generated by more complex polydomian models in which the shear flow affects the polydomain texture [65, 66, 67].

8.3 TIME-DEPENDENT EFFECTS DURING SHEAR FLOW

Lyotropic PLCs exhibit a wide range of time-dependent responses that reflect their molecular and textural characteristics. Transient data can be generated when steady-state shear conditions are suddenly changed, including an instantaneous cessation of the shear flow. This section focuses on the transient phenomena as recorded during flow; the time-dependent behaviour upon cessation of flow is dealt with in section 8.4.

Various kinds of transient experiments can be used to study time-dependent characteristics of a material during flow: stress growth, stepwise changes in shear rate or shear stress, flow reversal experiments and intermittent shear flow. The kinematics of such experiments are shown schematically in Fig. 8.7. Although all the shear histories of Fig. 8.7 basically represent an instantaneous change of the shear rate, their usefulness for deriving scaling relations and gaining insight into structural mechanisms is rather different.

In principle, one could follow the evolution in time of both the shear stress and the normal stress differences. In practice, it turns out that

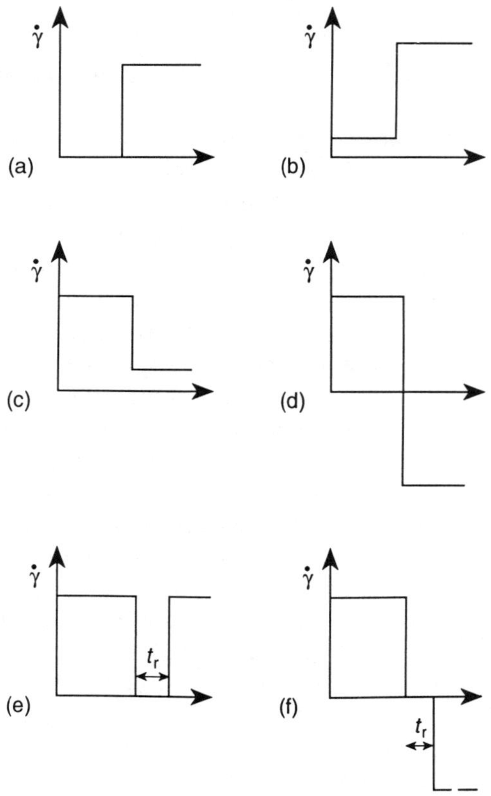

Figure 8.7 Kinematics of some transient shear flow experiments: (a) stress growth; (b) stepwise increase in shear rate; (c) stepwise decrease in shear rate; (d) flow reversal; (e) intermittent forward flow; and (f) intermittent flow reversal.

reliable transient normal stresses are very difficult to obtain and the majority of the experimental results in the literature are limited to shear stress transients. In addition to rheological properties, rheo-optical properties have also been recorded under similar kinematic conditions as those used in the rheological experiments. The latter is extremely useful to elucidate the structural mechanisms at the origin of the mechanical phenomena.

8.3.1 Stress growth

In this kind of experiment, the response of the material to the inception of a steady shear flow is recorded (Fig 8.7(a)). This is an easy experiment if the instrument can apply the desired shear rate quasi instantaneously and if it is capable of fast data collection. Stress growth experiments have been used extensively when studying isotropic polymeric systems. Provided one applies a shear rate, low enough to be in the limiting low shear rate range, the transient shear stress curves contain the necessary information to describe the linear, viscoelastic characteristics of the material [68].

Stress growth experiments have also been used to some extent to investigate the flow behaviour of lyotropic materials. Figure 8.8 displays a typical example of the response of a liquid crystalline solution. The material under consideration in this particular case was a solution of an aromatic polyamide in sulphuric acid [55]. Several characteristics that are unlike the response of an isotropic polymeric material can be observed in this figure. First of all, a damped oscillatory pattern with multiple oscillations is recorded for the shear stress. It only reaches steady-state conditions after a considerable time. In addition, the period of the oscillations decreases with increasing shear rate. However, if one replots the data from Fig. 8.8 as a function of applied deformation ($\dot{\gamma}.t$) rather than as a function of time, a relatively constant period is obtained. These features have been measured on other lyotropic systems, including PBGs [16] and HPC [44, 51], and turn out to be a general feature of lyotropic PLCs. With thermotropes, the oscillations are generally less pronounced

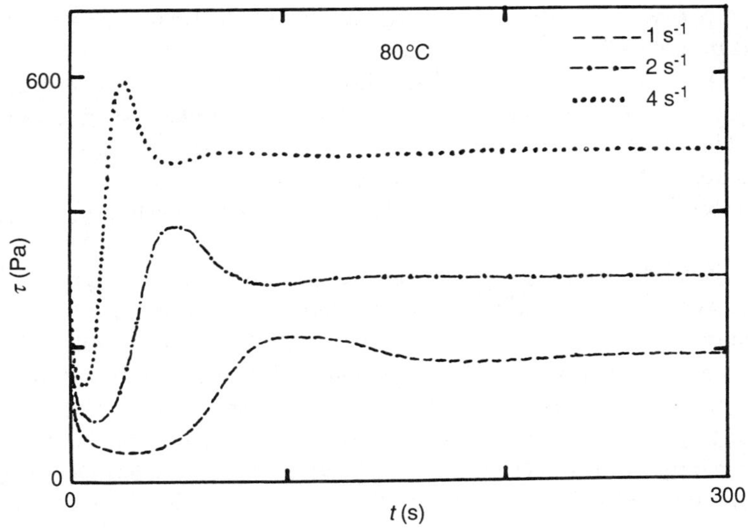

Figure 8.8 Shear stress versus elapsed time upon inception of steady shear flow at three different shear rates for an aromatic polyamide solution; reprinted with permission [55] © 1987 by Gordon and Breach Science Publishers.

but the trends are the same as the ones obtained with lyotropics [8, 9, 63, 69].

Stress growth experiments on PLCs turn out to be very sensitive to the previous shear history. This has to be taken into account in order to perform meaningful experiments. Obviously, it will also affect the comparison between theory and experiment. Reproducible and well-defined initial conditions can be generated in various ways. An external field, electrical [70] or magnetic [22], is often used to induce a particular orientation throughout the sample.

A flow field can also be used to achieve reproducible initial conditions for a stress growth experiment. At low shear rates, PLCs will show director tumbling, whereas they are flow aligning at high shear rates. After shearing for a sufficiently long time at a given shear rate in the tumbling region, a well-defined and reproducible polydomain structure is obtained. This provides a suitable, albeit complex, initial condition for further experimentation. The result will not be a start-up experiment but a stepwise change in shear rate. The latter can be performed in two different ways, depending on the relative direction of the flow before and after the stepwise change. When only the magnitude of the shear rate is changed without altering the direction of flow, one has the normal experiment of stepwise change in shear rate (Figs. 8.7(b) and 8.7(c), see section 8.3.2). When, on the contrary, the direction of the flow is changed, one has a flow reversal experiment (Fig. 8.7(d)). Here (see section 8.3.3), only flow reversals where the magnitude of the shear rate is kept constant are discussed. It is also possible to incorporate a rest period between the previous shearing and the stress growth experiment, resulting in an intermittent shear flow experiment (Figs. 8.7(e) and 8.7(f)). Long rest periods would result in the standard start-up experiment. However, this might take hours for PLCs and even then the result would depend on the final director orientation in the sample, which might be affected by the nature of the walls. Hence, such experiments are rarely performed.

Nearly all the stress growth experiments reported in the literature have been taken in the tumbling region. Both the molecular Doi [5] and continuum mechanics Leslie–Ericksen [71] theories can predict director tumbling. This results in oscillating stresses, the period of which depends on the material parameters. In the Leslie–Ericksen theory, the viscosity ratio $\alpha_3 : \alpha_2$ is the governing factor. In practice, tumbling nematics do not maintain a monodomain structure. As predicted theoretically, the director starts to change orientation along the velocity gradient direction. The growing gradients cause director 'winding up' and growing elastic stresses [72]. They result in the formation of defects and domains. As a result, there might still be local director rotation but the average orientation over the sample becomes constant.

From the previous discussion, it will be clear that it is difficult to do experiments in conditions that can be compared with theory. Stress growth data are available on samples that were electrically oriented in the velocity gradient direction [70]. At least, the initial part of the curves should be useful for such a comparison. The shear stresses reached overshoot values which were several times the steady-state value. The normal stresses went through a negative minimum before the shear stresses reached their first maximum. All these results are in agreement with the theoretical predictions. The oscillation dampens out, which is attributed to the breaking up of the monodomain. Application of an electrical field during the stress growth experiment can give rise to a monotonous increase in stress rather than to an overshoot [70].

Modelling the polydomain structure is extremely complex. As a first approximation, one can assume an assembly of non-interacting domains. Starting with a random distribution of domain orientations, the results of stress growth experiments can be computed. This results again in oscillating stress signals in stress growth experiments but with a much smaller amplitude than for monodomain structures. The overshoot seems to occur at comparable times for both types of texture [70]. To describe the damping of the stress oscillation, as seen in Fig. 8.8, some kind of interaction between the domains has to be incorporated [67, 73]. The Frank elasticities provide a suitable physical mechanism for this purpose, but they have to be incorporated in a phenomenological manner in the models. Notwithstanding the complex microstructure and the difficulty of developing an adequate theory, it has been possible, as will be discussed below, to derive some precise scaling laws for such a behaviour.

8.3.2 Stepwise changes in shear rate

The kinematics of experiments in which one applies a stepwise increase or a stepwise decrease in shear rate are shown schematically in Figs. 8.7(b) and 8.7(c). Obviously, the initial conditions of the material are well defined in these experiments, provided the preshearing has been applied for a sufficiently long period. A typical stress response to a stepwise increase in shear rate is given in Fig. 8.9. In this case, the material under investigation is a PBG solution. In Fig. 8.9(a), the shear stress transients are shown for two step-up experiments. Figure 8.9(b) contains the corresponding normal stress transients. It should be noted that the shear rates in this experiment were restricted to the Newtonian region (region II). The data in Figs. 8.9(a) and 8.9(b) display the typical damped oscillatory pattern. For isotropic polymeric materials, one would not

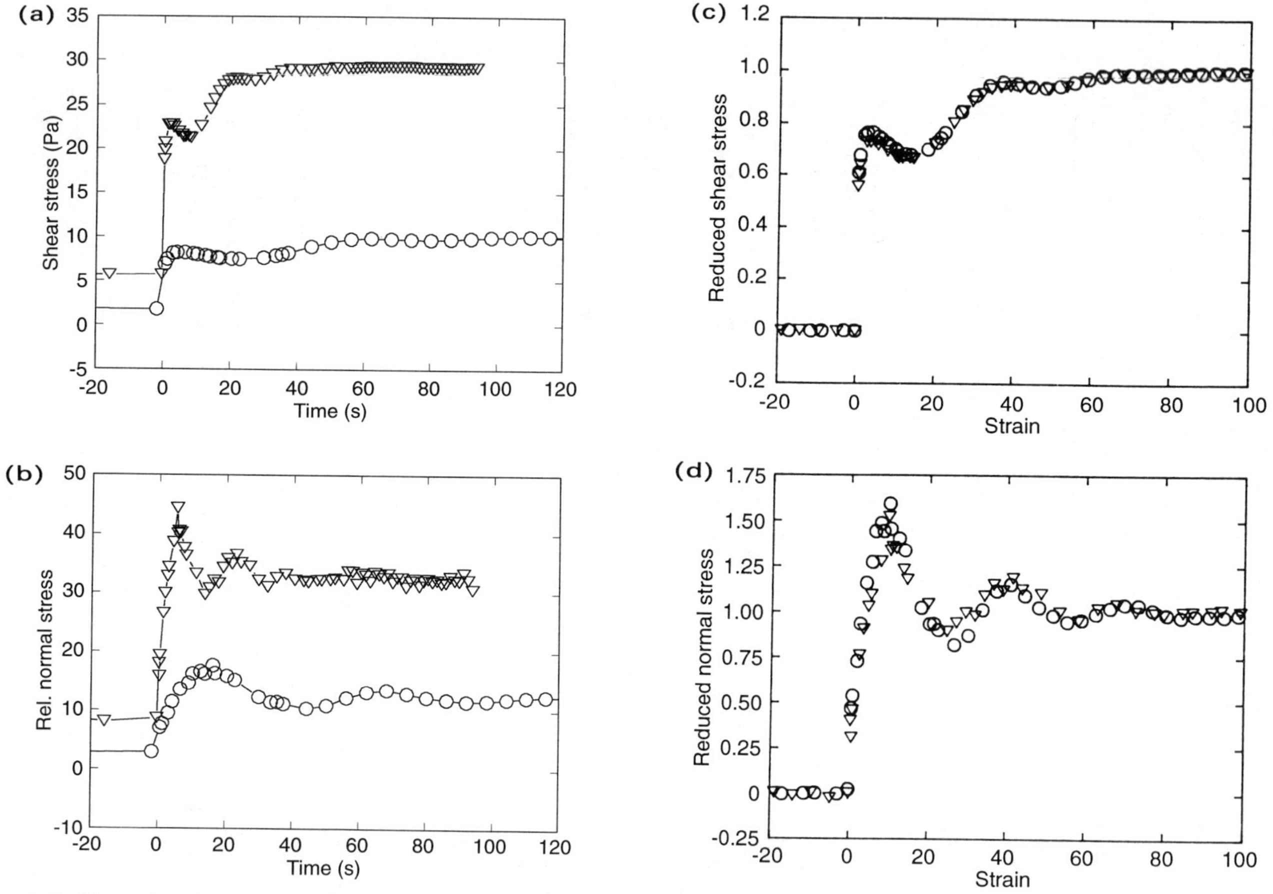

Figure 8.9 Stepwise increase in shear rate experiment for a 25% PBG solution at 293 K (○: $\dot{\gamma}_{\text{initial}} = 0.1\,\text{s}^{-1}$, $\dot{\gamma}_{\text{final}} = 0.6\,\text{s}^{-1}$; ▽: $\dot{\gamma}_{\text{initial}} = 0.3\,\text{s}^{-1}$, $\dot{\gamma}_{\text{final}} = 1.8\,\text{s}^{-1}$): (a) shear stresses versus time; (b) normal stresses versus time; (c) reduced shear stresses versus strain; (d) reduced normal stresses versus strain; (c), (d) reprinted with permission [34].

expect such an oscillatory response. Moreover, in the Newtonian region, the time dependency would be independent of shear rate (linear viscoelastic behaviour).

To compare the transients from different experiments, the stresses from Figs. 8.9(a) and 8.9(b) have been rescaled between their initial and their final values as is shown in Figs. 8.9(c) and 8.9(d). In addition the transients are plotted versus strain rather than time. It is clear from Fig. 8.9(c) that such a scaling causes a superposition of the shear stress transients for experiments with the same ratio between final and initial shear rate. The same holds for the corresponding normal stress transients (Fig. 8.9(d)). It is also worthwhile to note that the reduced curves do not depend on temperature although the absolute stress levels are rather sensitive to this parameter [74].

From a theoretical point of view the scaling in Figs. 8.9(c) and 8.9(d) is not unexpected. As discussed by Marrucci in Chapter 2, dimensional arguments lead to the following relation for the reduced transient stresses in the Newtonian region:

$$\sigma_{\text{red}} = f\left(\dot{\gamma}_{\text{f}}.t, \frac{\dot{\gamma}_{\text{i}}}{\dot{\gamma}_{\text{f}}}\right) \tag{8.1}$$

in which $\dot{\gamma}_{\text{i}}$ is the initial shear rate and $\dot{\gamma}_{\text{f}}$ is the final one. Hence, experiments with a constant ratio between the initial and final shear rate should render a single curve for the stress transients if they are plotted versus the total deformation at the final shear rate. The data from Fig. 8.9 correspond to two experiments with a sixfold increase in shear rate. The scaling represented in equation (8.1) is also valid for other materials without an internal time-scale as was shown by Doi and Ohta [75]. For LCPs, it has also been verified experimentally that the scaling is not only valid for an increase in shear rate but also for stepwise decreases [34]. However, it does not hold any longer outside the Newtonian region [32, 24]. In that case, the stress still goes through extrema at similar values of the strain, but the magnitude of the oscillations depends on the shear rate.

8.3.3 Flow reversal

In this type of experiment, the direction of the flow is reversed at time zero, the magnitude of the shear rate being the same in the two directions (Fig. 8.7(d)). For flow reversal experiments in the Newtonian region equation (8.1) can be rewritten as:

$$\sigma_{\text{red}} = f(\dot{\gamma}t) \tag{8.2}$$

In Fig. 8.10, the validity of equation (8.2) is illustrated for a 25% PBG solution at 293 K for five shear rates in the Newtonian shear rate range. As in Figs. 8.9(c) and 8.9(d), the data are plotted as reduced stresses versus total deformation. The flow reversal transients can be seen to superimpose quite accurately in this representation. It can be noted in Fig. 8.10(b) that the transient first normal stress difference initially decreases and goes through a (negative) minimum at about 6–7 strain units. A minimum in the N_1 transients is not a unique feature of liquid crystals. It is also expected to occur in isotropic solutions of rigid molecules as was recently demonstrated by Chow *et al.* [76]. Steady-state

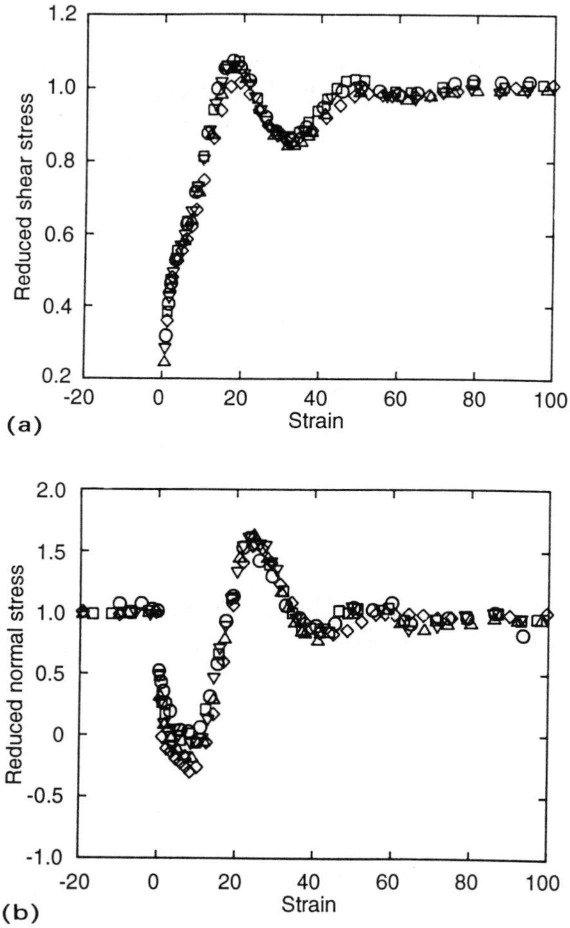

Figure 8.10 Reduced stresses: (a) shear stress; and (b) first normal stress difference after flow reversal for 25% PBG at 293 K (○: 0.3 s^{-1}; ▽: 0.6 s^{-1}; □: 1.0 s^{-1}; △: 3 s^{-1}; ◇: 6 s^{-1}); reprinted with permission [34].

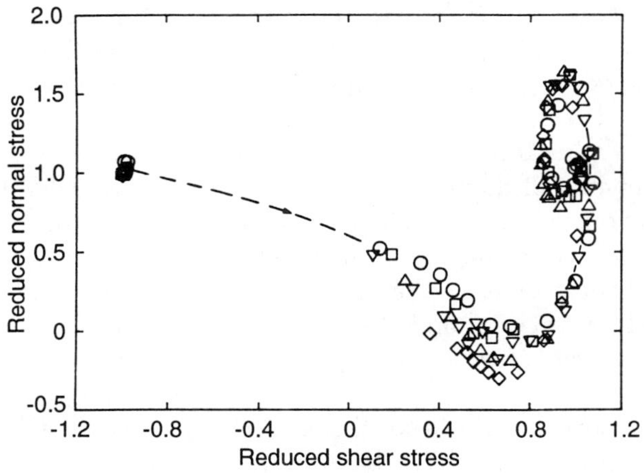

Figure 8.11 Stress paths for the data of Fig. 8.10 (symbols as in Fig. 8.10); reprinted with permission [34].

conditions are reached in Fig. 8.10 after approximately 60–80 deformation units. It can also be observed that the shear stress and the normal stress transients are phase shifted by roughly 90°.

The data from Figs. 8.10(a) and 8.10(b) could also be presented in an alternative way by plotting the normal stress versus the shear stress. In this representation, one eliminates time as a parameter, as is illustrated in Fig. 8.11. In this figure, the steady-state flow in the initial direction is represented by the point (−1, 1). Upon reversing the flow direction, the

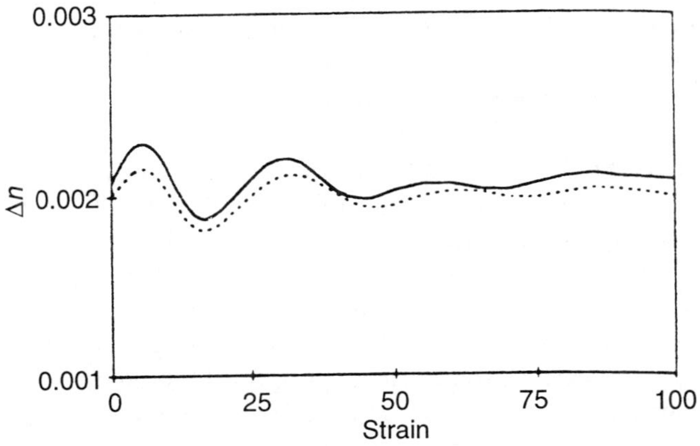

Figure 8.12 Birefringence versus shear strain following flow reversal for a 13.5% PBG solution. Shear rate: - - - : $0.2\ s^{-1}$; ——— : $1\ s^{-1}$); reprinted with permission [29] © 1993 by American Chemical Society.

material evolves to steady-state flow in the other direction along a counterclockwise spiral. The steady state at the end of the experiment is given by the point (1, 1). The shape of such stress paths is very sensitive to the detailed relative evolution of the two stress components. Therefore, it could be used as a critical test to evaluate model predictions [34]. Unfortunately, a full theoretical analysis is not yet available for the low shear, polydomain region. Hence, the data of Fig. 8.11 cannot be compared with modelling results at this stage. On the other hand, Larson and Doi [67] could describe the observed behaviour, at least for the shear stresses, in flow reversal and stepwise changes in shear rate quite well with a mesoscopic, phenomenological domain model for textured liquid crystals.

To gain more insight into the underlying structural mechanisms, rheo-optical experiments have been performed under similar kinematic conditions as in the rheological experiments. Both birefringence [29] and dichroism [18] have been used for this purpose. Figure 8.12 shows an example of the transient birefringence upon reversing the flow direction for a PBG solution at two shear rates in the Newtonian region. The birefringence describes a similar damped oscillatory pattern upon reversing the flow direction as does the shear stress and exhibits the same scaling with shear strain. The same conclusions hold for the dichroism transients [18].

In the high shear (i.e. flow-aligning) regime, the material exhibits a spatially uniform texture. Under these circumstances, predictive theories are reasonably well developed and their predictions can be compared with experimental observations. Figure 8.13 shows the results of two flow reversal experiments ($\dot{\gamma} = 30\,\mathrm{s}^{-1}$ and $\dot{\gamma} = 50\,\mathrm{s}^{-1}$) in the flow-aligning regime for a 12% PBG solution at 283 K. The 'universal' curve as obtained in the tumbling low shear regime is added for reference in Fig. 8.13 (solid curve). It can also be noted in Fig. 8.13(b) that the normal stress is negative at $30\,\mathrm{s}^{-1}$ and positive at $50\,\mathrm{s}^{-1}$.

From Fig. 8.13, it can be concluded that the scaling law, as written in equation (8.2), no longer holds outside the Newtonian region. In the non-Newtonian region maxima and minima occur at different locations and differ widely in magnitude as compared with the curve that was recorded in the low shear region. The stress paths for the flow reversal experiments of Fig. 8.13 are presented in Fig. 8.14. For every experiment, the stress path can essentially be described by a counterclockwise spiral, as was the case in the tumbling region. However, the results for the flow aligning region should be described by any suitable monodomain model. It turns out that apparently proper versions of the Doi model generate stress paths that evolve clockwise [32].

The discrepancy between theory and experiment provided strong evidence for a flaw in the analysis and required a closer look at the theory.

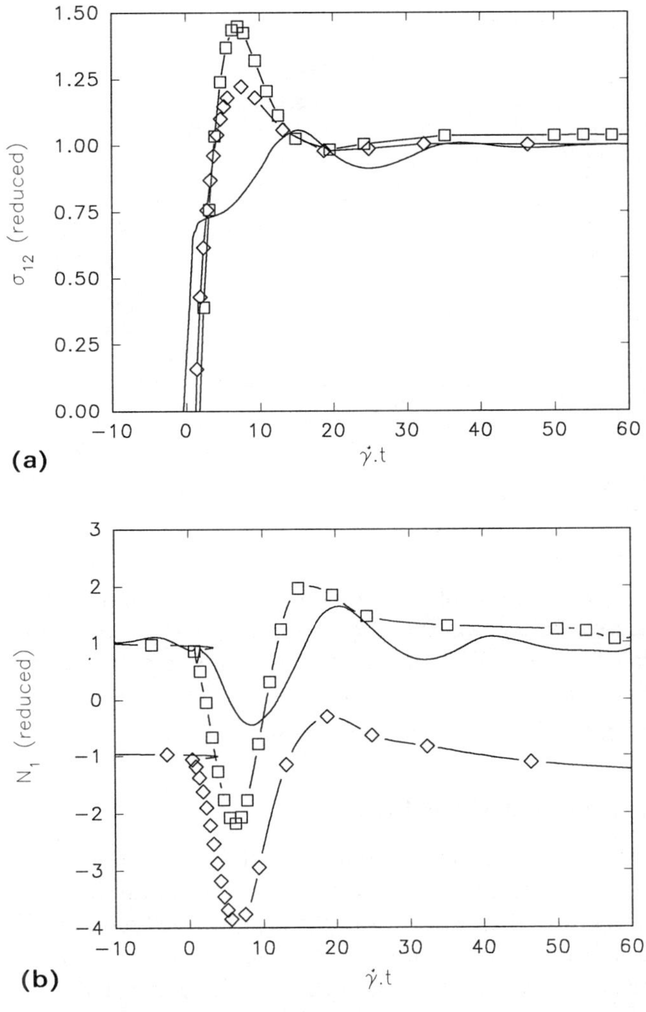

Figure 8.13 Reduced stresses: (a) shear stress, and (b) first normal stress difference, after flow reversal for 12% PBG at 283 K (flow aligning regime: ◇: 30 s^{-1}; □:50 s^{-1}); (tumbling regime:—).

Generally speaking, the stresses in lyotropic PLCs consist of an 'elastic' and a 'viscous' contribution. The elastic part refers to the stresses that arise due to the deviation of the orientational distribution from the equilibrium one. The viscous part is simply due to the frictional forces. The latter contribution has generally been ignored in PLCs as the elastic stress is assumed to dominate the viscous one. However, it turns out that it is necessary to include the viscous stresses, so as to obtain the correct counterclockwise spiral for the stress paths in flow reversal experiments

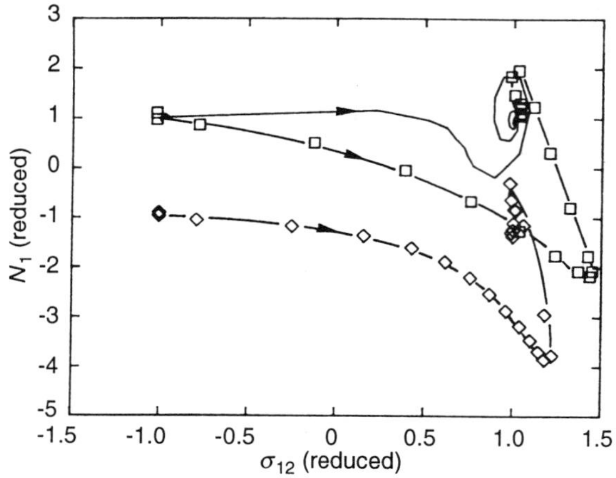

Figure 8.14 Stress paths for the data of Fig. 8.13, reprinted with permission [32].

[32]. Baek *et al.* [48] recently found that it was necessary to include the viscous stresses to explain the effect of concentration on the first normal stress difference of HPC solutions. At the same time, Smyth and Mackay [77] investigated the relaxation of the shear stress upon cessation of flow (see section 8.4) and came to the same conclusion.

8.3.4 Intermittent shear flow

Anisotropy is a key feature of liquid crystals. There are, however, not many data available in the literature on this crucial aspect. Taratuta *et al.* [78] determined the Miesowicz viscosities of a well-aligned PBG sample by means of quasi-elastic light scattering whereas Malkin *et al.* [79] demonstrated the anisotropy of a PLC under normal flow conditions. Transient flows offer an alternative way to demonstrate the flow-induced anisotropy of liquid crystals and its subsequent decay after the flow stops. A possible approach is to compare the rheological response from an intermittent forward flow experiment with that of an intermittent flow reversal experiment. The kinematics of such experiments are drawn schematically in Figs. 8.7(e) and 8.7(f). Any lack of symmetry with respect to the shear plane will show up in these tests.

In intermittent flow experiments, the transient stress response during start-up is used to track the evolution of the material during the previous rest period. Hence, this type of experiment has a somewhat intermediate position between transients during flow (this section) and those after stopping the flow (section 8.4). Because of the similarity with start-up flows, it will be discussed here. If the rest period in an intermittent flow

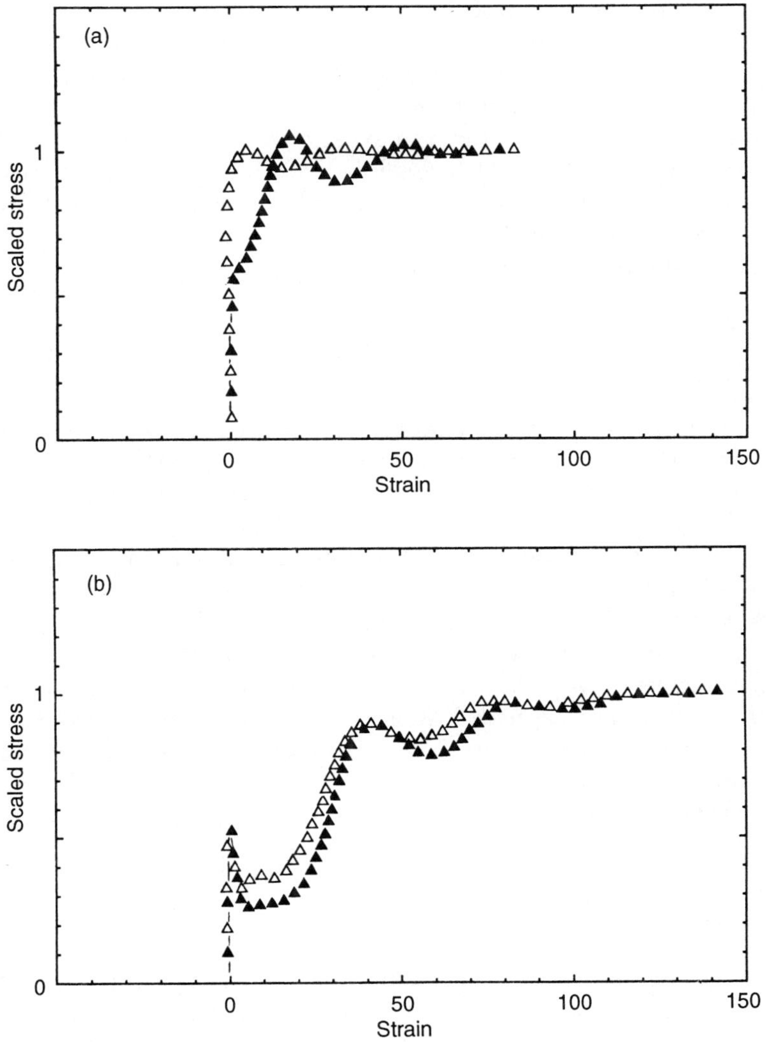

Figure 8.15 Comparison of the scaled shear stress in intermittent forward flow (IFF) and intermittent flow reversal (IFR) for a 25% PBDG ($\dot{\gamma} = 0.4\,s^{-1}$): (a) △: $t_{rest} = 16\,s$ (IFF); ▲: $t_{rest} = 18\,s$ (IFR); (b) △: $t_{rest} = 2000\,s$ (IFF); ▲: $t_{rest} = 2000\,s$ (IFR); reprinted with permission [24].

experiment is kept sufficiently short, it enables one to probe the anisotropy of the sample as induced by the flow. By systematically increasing the rest periods, the possible decay of the anisotropy can be followed. Figure 8.15(a) provides an example of the transient shear stress in an intermittent forward flow and an intermittent flow reversal experiment after a short rest period. The most remarkable feature in this figure is the phase shift of approximately 180° between the two stress signals, thus clearly demonstrating the anisotropy induced by the previous shearing. After a rest period of 2000 seconds (Fig. 8.15(b)), the transient shear stresses for the two experiments are not yet identical but they are already in phase. The sample actually requires several thousands of seconds before it reaches its equilibrium conditions and the memory for the earlier flow direction is decayed. Such slow structural transitions will be discussed further in the next sections.

8.4 TIME-DEPENDENT EFFECTS UPON CESSATION OF SHEAR FLOW

LCPs are known to possess a complex microstructure. Under steady-state shear flow, the average molecular orientation is affected by the shear rate, as is demonstrated by birefringence measurements on flowing systems [28, 49]. In addition, a texture develops which also depends on the shear rate. These textures have been studied using SALS and by microscopic observation between crossed polarizers [21, 36, 80]. Very often, a speckled texture is the dominating feature at low shear rates. Increasing the shear rate causes striations to appear in the flow direction; their number increases with shear rate. At the highest shear rates, the sample becomes uniformly aligned. When the flow stops, these textures will undergo prolonged variations before the material reaches its equilibrium structure in the quiescent state. This evolution after flow can be probed by recording various quantities of which the most commonly used ones will be discussed in the next sections.

8.4.1 Stress relaxation

When the flow is stopped, or alternatively when applying a step strain to the sample, the shear stress and the normal stress differences normally decay to zero. For isotropic polymeric materials numerous relaxation experiments have been recorded, both in the linear and the non-linear regime [68]. Surprisingly, only a very limited amount of systematic relaxation data on PLCs can be found in the literature. This may be, in part, due to the fact that gathering relaxation data on these relatively low viscosity samples is not straightforward. To measure reliable transients, inertia effects should be negligible and the transducer should be sufficiently stiff to avoid for example squeezing flow effects. In addition, the rheometer should be capable of fast data input and the drift of the base line should be very limited or should be corrected for. Finally, any unwanted filtering of the signals should also be avoided [81].

Moldenaers and Mewis studied the effect of previous shear rate on the stress relaxation of a PBLG solution [19, 20]. They concluded that the stress relaxation curves consisted of two portions: a fast initial decay of the shear stress, down to approximately half the initial value, followed by a slow tail. The two parts of the relaxation were of a different nature. It was established that the fast decay was shear rate independent in the Newtonian shear rate range, as would be expected for isotropic materials. The tail of the relaxation curves did, however, depend on the previous shear rate in a systematic way: it scaled with time multiplied with the preshear rate. Although this scaling is unexpected for stress relaxation curves, it turns out to be valid for many slow phenomena in PLCs; see sections 8.4.2 and 8.4.3. The different nature of the two parts of the stress relaxation curves was further confirmed by changing the temperature. It turns out that the fast stress decay scales with the steady-state viscosity, whereas the tail is insensitive to temperature.

Recently, Smyth and Mackay [77] investigated the stress relaxation upon cessation of flow for HPC/water solutions at various concentrations. Figure 8.16 shows the portion of the stress that does not vanish instantaneously when the flow stops. It is plotted as a function of shear rate for a series of HPC concentrations. Liang and Mackay used a specific extrapolation procedure to obtain the stresses at time zero [82]. The

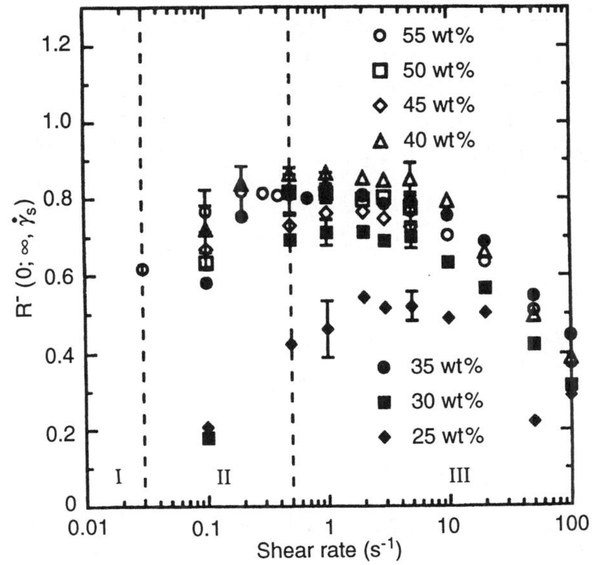

Figure 8.16 Portion of the shear stress retained at the instant of cessation of flow for HPC in water solutions as a function of shear rate; reprinted with permission [77].

highest concentrations in Fig. 8.16 represent fully liquid crystalline solutions. It is clear from this figure that the initial stress jump is considerable in all cases. In the intermediate shear rate range (region II), it could amount to approximately 20% of the total stress whereas, in the high shear range (region III), it becomes as large as 50%. The instantaneous decrease of the stress is associated with the viscous part of the stress during shear. The latter is attributed to the intermolecular friction and hence disappears immediately when there is no flow.

Based on different evidence, Baek *et al.* [48] and Maffettone *et al.* [32] also came to the conclusion that the viscous part of the stresses cannot be neglected, especially at high concentrations and high shear rates. These observations are of great importance for theoretical developments; most theories for PLCs so far only considered the elastic stress, neglecting the viscous contribution. Done and Baird [83] performed systematic step strain experiments on two thermotropic samples. Their relaxation curves also typically showed a rapid decay followed by a long relaxation tail. They argued that the presence of unmelted solid phase could cause the slow part of the relaxation curve. According to the data on the lyotropics, this is not necessarily the case.

8.4.2 Structural relaxation

For isotropic polymeric materials, the nonlinear stress relaxation does not always provide a sensitive means to probe flow-induced changes [84]. Hence, when it comes to probing structural effects, other techniques should be considered. Intermittent shear flow (see section 8.3.4) is a possibility but has the disadvantage that it is a destructive method. Non-destructive tests are of course preferred. Measuring the evolution of

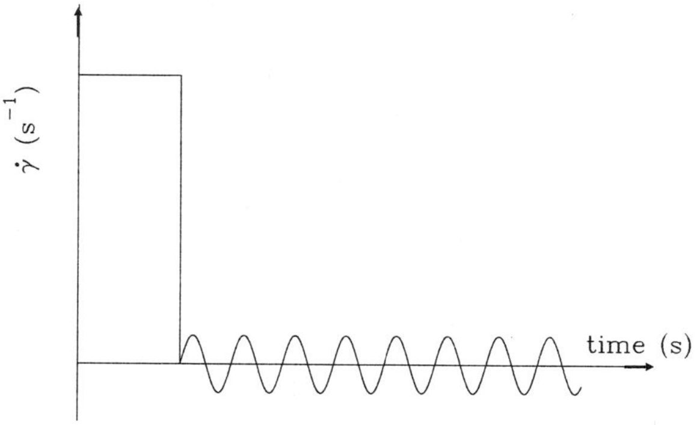

Figure 8.17 Kinematics of an experiment to measure the dynamic flow moduli upon cessation of flow.

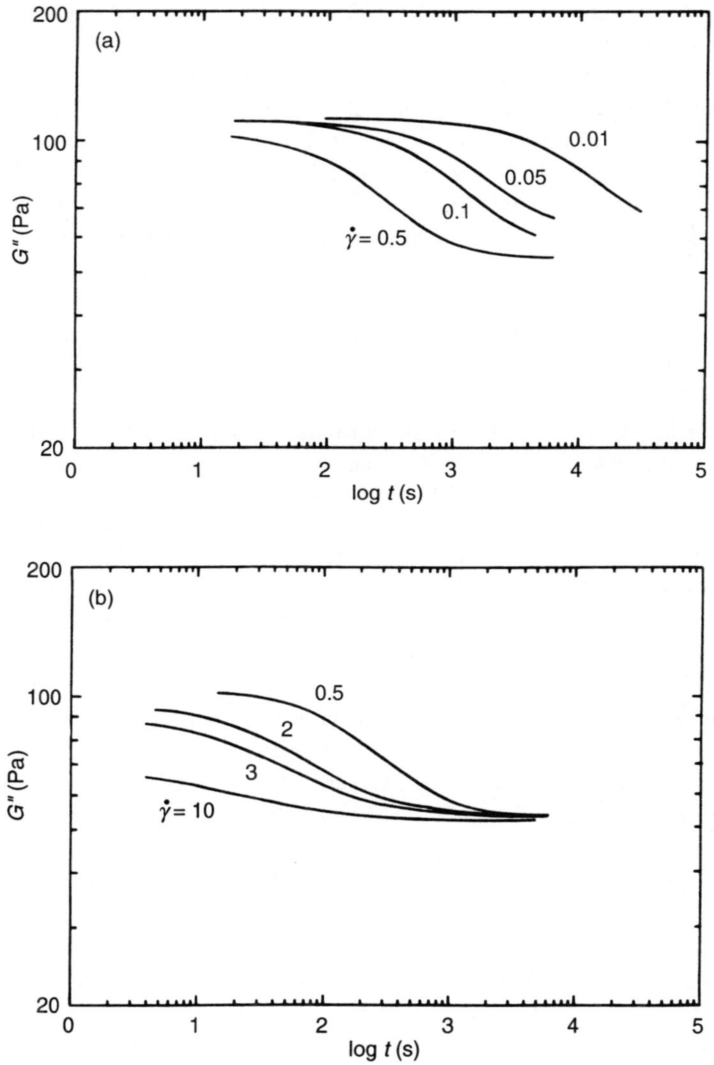

Figure 8.18 Effect of the initial shear rate on the transient moduli after cessation of flow for a 12% PBLG solution ($\omega = 10\,\mathrm{rad\,s^{-1}}$): (a) Newtonian region; (b) shear thinning region; reprinted with permission [14].

the dynamic moduli upon cessation of flow has proven to be very successful for PLCs. The kinematics of such an experiment are shown in Fig. 8.17. The material under investigation is sheared until steady-state conditions are reached. Upon stopping the flow, an oscillatory flow is applied with a fixed amplitude and frequency. The resulting moduli, G'

and G'', are measured as a function of rest time. Provided the amplitude of the oscillation is within the linear response regime of the material, the structure development can be probed non-destructively.

Figure 8.18 shows an example of the evolution of the linear loss modulus upon cessation of flow for a 12% PBG solution [14, 17]. Each curve corresponds to a different previous shear rate. Several characteristics can be deduced from Fig. 8.18. First, the loss modulus is seen to decrease in time. This is an unusual result as shear normally breaks down the structure. The subsequent structural recovery after the flow stops then causes the moduli to increase with time. A decrease has however been reported for liquid crystalline PBG solutions of different concentrations and molecular weights [14, 17, 20, 29]. Second, the moduli in Fig. 8.18 evolve over a time-scale which is much longer than the one observed in stress relaxation experiments. The time-scale for structural recovery as measured with the dynamic moduli corresponds, however, to the one measured with intermittent flow experiments (see section 8.3.4) [24]. In addition, the time-scales recorded by means of either the dynamic moduli or intermittent flow are both insensitive to temperature effects [14, 24], indicating that they might be related to the same structural features. Finally, in the shear rate region of constant viscosity, each curve has the same initial and final values but evolves over quite different time-scales (Fig. 8.18(a)): the lower the previous shear rate, the longer it takes to reach the values of the quiescent state. For solutions of PBG, the rate of structural recovery is generally found to be proportional to the previous shear rate. The recorded changes in the dynamic moduli in PLCs indicate that the region of constant viscosity does not correspond to a limiting Newtonian low shear region as found in isotropic polymer fluids, but that different microstructures must correspond to each shear rate.

Although a decrease of the moduli after stopping the flow seems to be a general characteristic of liquid crystalline PBG solutions, it has not been reported for other PLCs. On the contrary, the moduli have been observed to increase for several other lyotropics such as HPC and aromatic polyamides as well as for thermotropes [44, 46, 49, 85]. The evolution of the moduli for an HPC solution are shown as an example in Fig. 8.19. In this case, the moduli clearly increase upon cessation of flow, although not totally monotonically. As for PBG solutions, the time necessary to reach a constant value for G' and G'' is much longer than the time it takes for the shear stress to relax. The previous shear rate again has an effect on the kinetics of the evolution but the scaling with preshear rate is only valid over a limited shear rate range [46]. More importantly, it can be seen in Fig. 8.19 that the final value of the modulus depends strongly on the preshear rate, unlike the behaviour reported in Fig. 8.18 for PBLG where the final value of the moduli under quiescent conditions

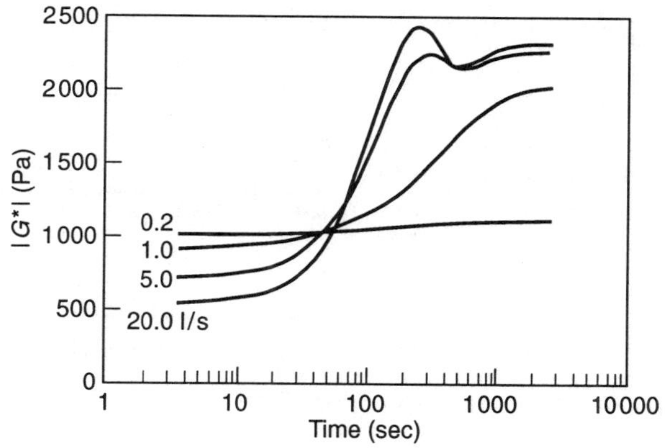

Figure 8.19 Evolution of the complex modulus upon cessation of flow for a 50% HPC solution in water ($\omega = 12.5\,\mathrm{rad\,s^{-1}}$); reprinted with permission [49].

was independent of the kinematic history. It should be pointed out that Larson and Mead [17] also observed some effect of the previous shear rate on the final moduli for a PBLG sample.

To understand the mechanisms that govern the slow structural changes, rheo-optical investigations have been used to supplement the rheological studies. Hongladarom *et al.* [28, 29] measured the average molecular orientation by means of birefringence under steady-state conditions as well as upon cessation of flow. For PBG solutions under steady-state shear flow, they recorded a constant value for the birefringence in the region of constant viscosity (region II). The birefringence was about half that of fully oriented monodomain solutions. Upon increasing the shear rate in the non-Newtonian region, Hongladarom *et al.* [28] observed the average orientation to increase up to about 90% of the monodomain value.

The fact that the average molecular orientation is independent of shear rate in region II for PBGs fits in with some observations of Moldenaers and Mewis [31]. They measured the linear parallel superposition moduli for a series of shear rates and found the moduli to be independent of shear rate in the Newtonian region. It is known that the texture length scale is refined by increasing the shear rate in this region, as was observed experimentally by scattering dichroism [18, 22]. Hence, the constant moduli seem to indicate that they are determined by molecular mechanisms rather than by the elastic forces (Frank elasticity) at the domain boundaries.

If moduli and average molecular orientation are linked, the evolution of the birefringence upon cessation of flow should directly reflect

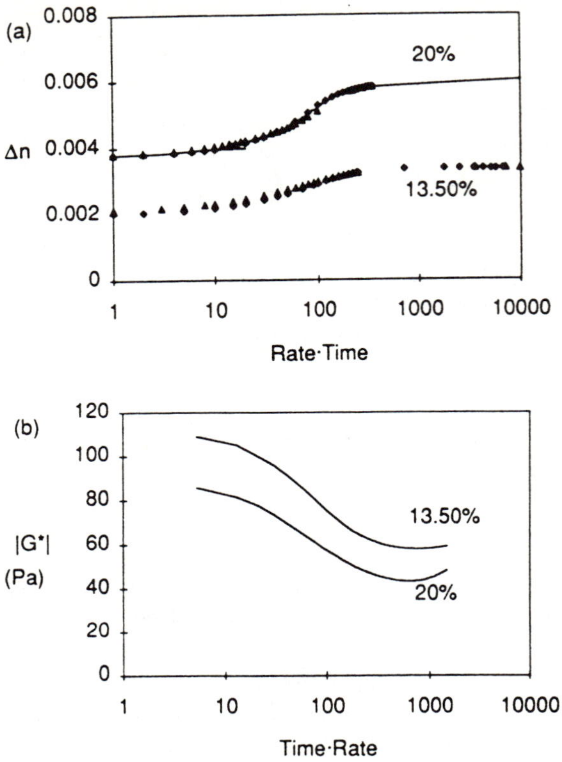

Figure 8.20 Relaxation upon cessation of flow for PBG solutions: (a) birefringence versus scaled time (initial shear rate: ◆: $0.2\ s^{-1}$; —: $0.4\ s^{-1}$; ▲: $1\ s^{-1}$; (b) complex moduli versus scaled time ($\omega = 12.5\ rad\ s^{-1}$); reprinted with permission [29] © 1993 by American Chemical Society.

the anomalous decrease of the moduli upon cessation of flow in PBG as illustrated in Fig. 8.18. Hongladarom and Burghardt [29] performed a series of birefringence 'relaxation' experiments, of which an example is given in Fig. 8.20. Figure 8.20(a) shows the birefringence data recorded upon cessation of flow for three shear rates in the region of constant viscosity for two concentrations of PBG. The birefringence is indeed seen to increase substantially from the initial steady-state value during shear to a value close to that of the monodomain structure, meaning that the molecular orientation in the flow direction increases after stopping the flow. Moreover, the time-scale over which the birefringence evolves depends strongly on the previous shear rate. The curves could again be superimposed by plotting them versus time multiplied by the previous shear rate. These

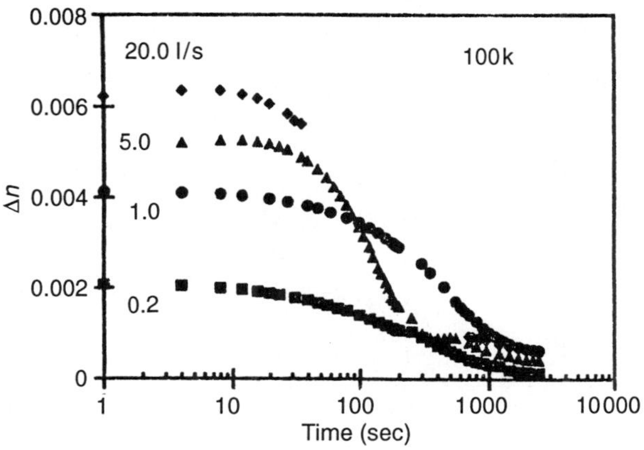

Figure 8.21 Evolution of the birefringence upon cessation of flow for a 50% HPC solution in water at four different preshear rates; reprinted with permission [49].

observations fit nicely with the mechanical data represented in Fig. 8.20(b).

Recently, it was verified that the relation between molecular orientation and dynamic moduli is more generally valid for PLCs. Birefringence data upon cessation of flow for a 50% HPC solution [49] are presented in Fig. 8.21 and should be compared with the rheological data of Fig. 8.19. It is clear from this comparison that the increase in moduli in this material indeed reflects a decrease of the average molecular orientation in the flow direction. The moduli and the birefringence also change over comparable time-scales. However, the average orientation of the HPC solutions decreases after all shear rates to an almost globally isotropic condition. If the moduli would only be determined by the average molecular orientation, the final values of the moduli would always be the same. This is not observed experimentally, as the moduli under quiescent conditions in Fig. 8.19 clearly depend on the previous shear rate. Hongladarom *et al.* [49] hypothesize that the transition from a flow-induced nematic phase to a cholesteric one under quiescent conditions interferes with the evolution of the birefringence.

Finally Picken *et al.* [56, 57] quantified the evolution of the orientation in the flow direction after cessation of flow in a nematic polyaramid solution by using X-ray scattering. Figure 8.22 shows their results for the evolution of the average orientation parameter for two widely different shear rates. On the basis of these results, it can be concluded that the average orientation was decreasing to zero after stopping the flow. Picken *et al.* could not attribute accurate time constants (τ) to these

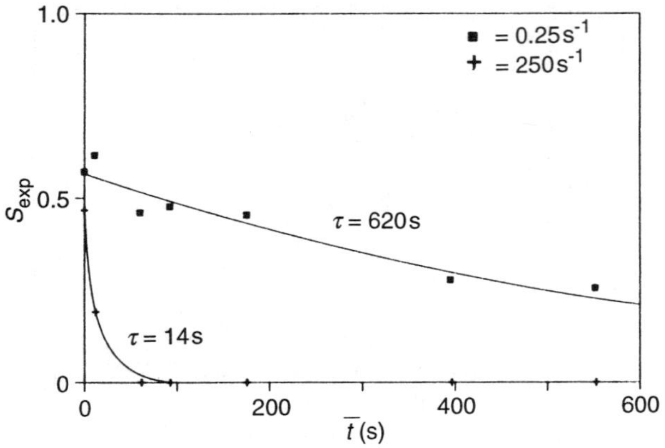

Figure 8.22 Orientation relaxation for an aromatic polyamide solution as a function of the average time ($\bar{t}$) after cessation of flow for two preshear rates; reprinted with permission [56] © 1990 by American Chemical Society.

evolutions due to the rather long time interval required for the collection of a single data point but, nevertheless, it is clear that the variation is faster with increasing previous shear rate. The data in Fig. 8.22 are again consistent with the evolution of the dynamic moduli as G' and G'' were observed to increase upon stopping the flow [85]. As in the case of PBG solutions, the final values of the moduli under quiescent conditions were independent of the previous shear rate for PPTA.

8.4.3 Recoil

Another way of determining time effects in viscoelastic media consists of measuring the recoil. In such an experiment, a constant stress is applied to the sample until steady-state conditions are reached. The stress is then suddenly removed and the backwards displacement of the material, due to elastic forces, is recorded as a function of time. For PLCs, only a limited number of recoil data have been published but the conclusions seem to be sound. As an illustration, the recoil behaviour of a PBLG solution is represented in a semi-logarithmic plot in Fig. 8.23 for a series of stresses [17].

The data are unusual in three respects:

1. The magnitude of the recoil in Fig. 8.23 is substantial. A final recoil of 3 indicates that there is a significant storage of elastic energy. This observation does not seem to be totally in agreement with the fact that LCPs show little or no extrudate swell [51, 86]. It should,

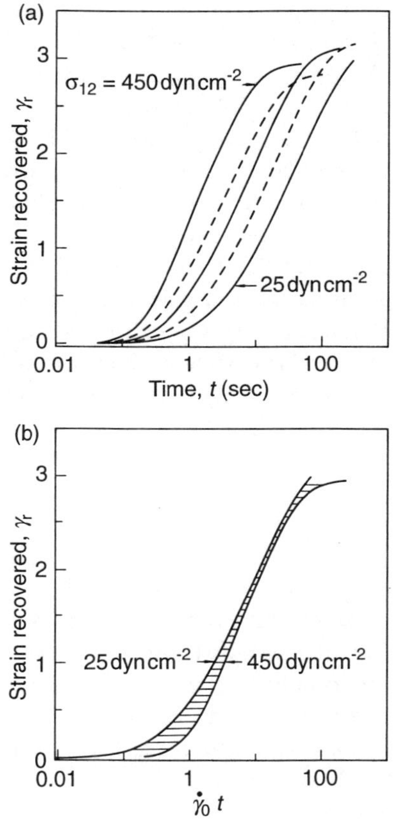

Figure 8.23 (a) Recoil versus time for a 30% PBLG solution (applied shear stresses before recovery from right to left: 2.5, 5, 10, 20, and 45 Pa); (b) same data plotted versus time multiplied with the previous shear rate; reprinted with permission [17].

however, be noted that recoil and extrudate swell are recorded in different time and/or shear rate regions.

2. The magnitude of the recoil in Fig 8.23 is independent of the applied shear stress over a wide range of shear stresses.
3. The recoil curves can be superimposed by plotting them versus time multiplied by the previous shear rate, as is illustrated in Fig8.23b. This scaling of the recoil curves is the same as the one observed for the evolution of the dynamic moduli as well as for the tail of the stress relaxation curves; see sections 8.4.1 and 8.4.2.

The general characteristics reported here for a PBLG solution [17, 22] have also been recorded for other lyotropic solutions such as HPC [17] and aromatic polyamides [85]. The only difference is the absolute value of the recoil which is slightly material dependent, although always large (of order 3–4).

Burghardt and Fuller [22] supplemented their recoil experiments with the determination of the transient dichroism in intermittent shear flow experiments. They reported a damped oscillatory pattern for the dichroism upon resuming the flow. At various shear rates, these dichroism transients obeyed the same scaling law as did the recoil. In addition, they observed that the time-scales for the recoil and the intermittent dichroism are the same, thus suggesting that the phenomena have a common origin. Moldenaers *et al.* performed similar experiments [24] and although they observed the same scaling for various shear rates, their intermittent dichroism changed with the rest period over substantially longer times than the recoil did.

8.4.4 Banded textures

Upon cessation of flow, the flow-induced molecular orientation and the texture in PLCs changes slowly. This was illustrated in the previous sections by studying the evolution of a number of rheological and rheo-optical quantities. When the sample is viewed between crossed polarizers, another particular phenomenon of PLCs shows up: the formation of a banded texture perpendicular to the previous shear direction when the flow is halted. This banded texture has been observed in lyotropics [25, 26, 42, 43, 87, 88, 89] as well as in thermotropics [90, 91]. Bands are associated with a periodic variation of the optical axis around the shear axis [91]. As an illustration, a well-developed banded texture in a BPG solution is shown in Fig. 8.24. Although the majority of the reports on bands deal with shear flow, they have been observed in elongational flow as well [92].

Bands do not always emerge when the shear flow is stopped. They actually only appear over an intermediate range of previous shear rates. It has been observed by various groups [25, 26, 43, 88] that there is a critical lower shear rate below which no bands are formed upon cessation of flow. Only a coarsening of the texture with time is observed under these conditions. Above this critical shear rate, a well-defined banded texture develops, the only prerequisite being that the flow has been applied for a sufficiently long time. Upon further increasing the shear rate, a flow-aligned monodomain structure is induced by the flow. This alignment persists for a long time after the flow has been stopped. As a consequence, no bands are formed after shearing at high shear rates [80, 89]. Banded structures are not uniquely associated with stress relaxation. They also can emerge as a transient phenomenon in experiments such as recoil or upon inception of flow.

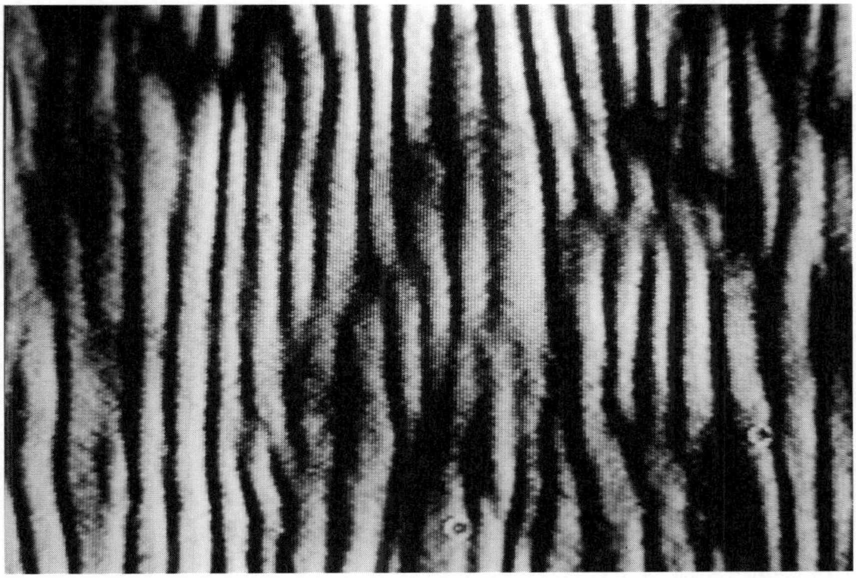

Figure 8.24 Typical banded texture in a PBG solution (25%) after the flow has been stopped at an intermediate shear rate (area shown 400×300 μm).

When bands are formed upon cessation of flow, their characteristics have been reported to change with time. This evolution has been studied in detail for PBG solutions [25, 26, 89]. It turns out that both the band spacing (width) and the band length increase with time. The two growth processes depend, however, in a different manner on the previous shear rate. Figure 8.25 shows the characteristic time constants for both the width and the length growth as a function of the previous shear rate. For the growth of the band spacing, a slope of −0.45 is obtained on a double-logarithmic plot whereas, for the length growth, the slope is −1. The latter scaling is the one that has been observed in the previous sections for several rheological and optical properties. Extrapolating the curves in Fig. 8.25 to low shear rates, it is expected that the two time constants approach each other at a finite value of the shear rate. Under these conditions, no band formation but rather a coarsening of the texture can be expected, in agreement with the existence of a lower critical shear rate as indicated above.

It has been reported that changing either the concentration of the PBG or the temperature does not substantially affect the time constants for the two growth processes in band formation [89]. Also the equilibrium structure, nematic or cholesteric, had no measurable effect on the band kinetics for PBGs. Increasing the gap size of the instrument, however,

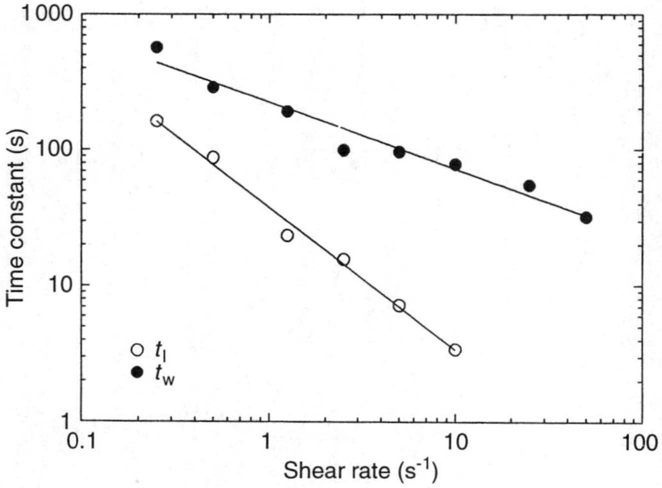

Figure 8.25 Time constant for both the width growth (t_w) and the length growth (t_l) as a function of shear rate for a 15% PBDG solution in m-cresol; reprinted with permission [89].

caused a faster growth of the bands in both directions. In all cases, the band width grows until it approximately equals the gap size of the instrument. At relatively high shear rates, probably in the wagging regime, the two-dimensional growth process described above becomes one-dimensional: long bands develop instantaneously when the flow stops. They subsequently grow wider with time [89].

For HPC solutions, the bands behave rather differently. First, the characteristic size of the texture is much smaller than the one observed with PBGs [43]. Second, the bands do not seem to widen with time. Third, they do not remain parallel but adopt a wavy pattern before transforming into elongated domains. The difference between PBG and HPC with respect to the evolution of the bands might explain the different evolution of the moduli in these two systems.

REFERENCES

1. de Gennes, P.G. (1974) *The Physics of Liquid Crystals*, Clarendon Press, Oxford.
2. Bowden, F.C. and Pirie, N.W. (1937) *Proc. Roy. Soc.*, **B123**, 274.
3. Elliott, A. and Ambrose, E.J. (1950) *Discuss. Faraday Soc.*, **9**, 246.
4. Flory, P.J. (1956) *Proc. Roy. Soc.*, **A234**, 73.
5. Doi, M. (1981) *J. Polym. Sci.: Polym. Phys. Ed.*, **19**, 229.
6. Lin, Y.G. and Winter, H.H. (1988) *Macromolecules*, **21**, 2439.
7. Kalika, D.S., Giles, D.W. and Denn, M.M. (1990) *J. Rheol.*, **34**, 139.

8. Cocchini, F., Nobile, M.R. and Acierno, D. (1991) *J. Rheol.*, **35**, 1171.
9. Guskey, S.M. and Winter, H.H. (1991) *J. Rheol.*, **35**, 1191.
10. Kiss, G. and Porter, R.S. (1978) *J. Polym. Sci.: Polym. Symp.*, **65**, 193.
11. Kiss, G. and Porter, R.S. (1980) *J. Polym. Sci.: Polym. Phys. Ed.*, **18**, 361.
12. Asada, T., Onogi, S. and Yanase, H. (1984) *Polym. Eng. and Sci.*, **24**, 355.
13. Asada, T., Tanaka, T. and Onogi, S. (1985) *J. Appl. Polym. Sci.: Appl. Polym. Symp.*, **41**, 229.
14. Moldenaers, P. and Mewis, J. (1986) *J. Rheol.*, **30**, 567.
15. Mewis, J. and Moldenaers, P. (1987) *Chem. Eng. Comm.*, **53**, 33.
16. Mewis, J. and Moldenaers, P. (1987) *Mol. Cryst. Liq. Cryst.*, **153**, 291.
17. Larson, R.G. and Mead, D.W. (1989) *J. Rheol.*, **33**, 1251.
18. Moldenaers, P., Fuller, G. and Mewis, J. (1989) *Macromolecules*, **22**, 960.
19. Moldenaers, P. and Mewis, J. (1990) *J. Non-Newtonian Fluid Mech.*, **34**, 359.
20. Moldenaers, P., Yanase, H. and Mewis, J. (1990) in *Liquid-Crystalline Polymers*, (eds R.A. Weiss and C.K. Ober) ACS Symposium Series, Washington DC, Ch. 26.
21. Takebe, T., Hashimoto, T., Ernst, B. *et al.* (1990) *J. Chem. Phys.*, **92**, 1386.
22. Burghardt, W.R. and Fuller, G.G. (1991) *Macromolecules*, **24**, 2546.
23. Magda, J.J., Baek, S.-G., DeVries, K.L. *et al.* (1991) *Macromolecules*, **24**, 4460.
24. Moldenaers, P., Yanase, H. and Mewis, J. (1991) *J. Rheol.*, **35**, 1681.
25. Gleeson, J.T., Larson, R.G., Mead, D.W. *et al.* (1992) *Liquid Crystals*, **11**, 341.
26. Picken, S.J., Moldenaers, P., Berghmans, S. *et al.* (1992) *Macromolecules*, **25**, 4759.
27. Baek, S.-G., Magda, J.J. and Larson, R.G. (1993) *J. Rheol.*, **37**, 1201.
28. Hongladarom, K., Burghardt, W.R., Baek, S.-G. *et al.* (1993) *Macromolecules*, **26**, 772.
29. Hongladarom, K. and Burghardt, W.R. (1993) *Macromolecules*, **26** (4), 785.
30. Larson, R.G., Promislow, J., Baek, S.-G. *et al.* (1993) in *Ordering in Macromolecular Systems*, (A. Teramoto ed.) Springer, New York.
31. Moldenaers, P. and Mewis, J. (1993) *J. Rheol.*, **37**, 367.
32. Maffettone, P.L., Marrucci, G., Mortier, M. *et al.* (1994) *J. Chem. Phys.*, **100**, 7736.
33. Hongladarom, K. and Burghardt, W.R. (1994) *Macromolecules*, **27**, 483.
34. Moldenaers, P., Mortier, M. and Mewis, J. (1994) *Chem. Eng. Sci.*, **49**, 699.
35. Yan, N.X., Labes, M.M., Baek, S.-G. *et al.* (1994) *Macromolecules*, **27**, 2784.
36. Vermant, J., Moldenaers, P., Picken, S.J. *et al.* (1994) *J. Non-Newtonian Fluid Mech.*, **53**, 1.
37. Walker, L.M., Wagner, N.J., Larson, R.G. *et al.* (1995) *J. Rheol.*, **39**, 925.
38. Sakamoto, R. (1984) *Coll. & Polym. Sci.*, **262**, 788.
39. Marrucci, G. and Maffettone, P.L. (1989) *Macromolecules*, **22**, 4076.
40. Larson, R.G. (1990) *Macromolecules*, **23**, 3983.
41. Larson, R.G. and Mead, D.W. (1991) *J. Polym. Sci.: Part B: Polym. Phys.*, **29**, 1271.

42. Navard, P. (1986) *J. Polym. Sci.: Polym. Phys. Ed.*, **24**, 435.
43. Ernst, B. and Navard, P. (1989) *Macromolecules*, **22**, 1419.
44. Grizzuti, N., Cavella, S. and Cicarelli, P. (1990) *J. Rheol.*, **34**, 1293.
45. Baek, S.-G., Magda, J.J. and Cementwala, S. (1993) *J. Rheol.*, **37**, 935.
46. Grizzuti, N., Moldenaers, P., Mortier, M. *et al.* (1993) *Rheol. Acta.*, **32**, 218.
47. Sigillo, I. and Grizzuti, N. (1994) *J. Rheol.*, **38**, 5389.
48. Baek, S.-G., Magda, J.J., Larson, R.G. *et al.* (1994) *J. Rheol.*, **38**, 1473.
49. Hongladarom, K., Secakusuma, V. and Burghardt, W.R. (1994) *J. Rheol.*, **38**, 1505.
50. Walker, L. and Wagner, N. (1994) *J. Rheol.*, **38**, 1525.
51. Metzner, A.B. and Prilutski, G.M. (1986) *J. Rheol.*, **30**, 661.
52. Grizzuti, N., Guido, S., Nastri, V. *et al.* (1991) *Rheol. Acta.*, **30**, 71.
53. Papkov, S.P., Kulichikhin, V.G., Kalmykova, V.D. *et al.* (1974) *J. Polym. Sci.: Polym. Phys. Ed.*, **12**, 1753.
54. Kulichikhin, V.G. (1989) *Mol. Cryst. Liq. Cryst.*, **169**, 51.
55. Doppert, H.L. and Picken, S.J. (1987) *Mol. Cryst. Liq. Cryst.*, **153**, 109.
56. Picken, S.J., Aerts, J., Visser, R. *et al.* (1990) *Macromolecules*, **23**, 3849.
57. Picken, S.J., Aerts, J., Doppert, H.L. *et al.* (1991) *Macromolecules*, **24**, 1366.
58. Yang, H.H. (1993) *Kevlar Aramid Fiber*, Wiley, Chichester.
59. Onogi, S. and Asada, T. (1980) in *Rheology*, Vol. 1, (G. Astarita, G. Marrucci and L. Nicolais eds) Plenum Press, New York, p. 127.
60. Wissbrun, K.F. (1981) *J. Rheol.*, **25**, 619.
61. Guido, S., Di Maio, M. and Grizzuti, N. (1994) in *Proceedings of the Fourth European Rheology Conference*, (C. Gallegos ed.) Steinkopff, Darmstadt, p. 308.
62. Picken, S.J. (1990) Orientational Order in Aramid Solutions, Ph.D thesis, Utrecht, The Netherlands.
63. Kim, S.S. and Han, C.D. (1993) *J. Rheol.*, **37**, 847.
64. Marrucci, G. and Greco, F. (1993) in *Advances in Chemical Physics*, (I. Prigogine and S. Rice eds) Wiley, New York, pp. 331–404.
65. Marrucci, G. (1984) in *Advances in Rheology*, Vol. 1, (B. Mena, A. Garcia-Rejon and C. Rangel-Nafaile eds) Universidad Nacional Autonoma De Mexico, Mexico, p. 441.
66. Wissbrun, K.F. (1985) *Faraday Discuss. Chem. Soc.*, **79**, 161.
67. Larson, R.G. and Doi, M. (1991) *J. Rheol.*, **35**, 539.
68. Bird, R.B., Armstrong, R.C. and Hassager, O. (1987) *Dynamics of Polymeric Liquids*, Wiley, New York.
69. Viola, G.G. and Baird, D.G. (1986) *J. Rheol.*, **30**, 601.
70. Yang, I.-K. and Shine, A.D. (1993) *Macromolecules*, **62**, 1529.
71. Leslie, F.M. (1979) in *Advances in Liquid Crystals*, Vol. 4, (G.H. Brown ed.) Academic Press, p. 1.
72. Burghardt, W.R. and Fuller, G.G. (1990) *J. Rheol.*, **34**, 959.
73. Marrucci, G. and Maffettone, P. L. (1990) *J. Rheol.*, **34**, 1231.
74. Moldenaers, P. (1987) Ph.D thesis, Katholieke Universiteit Leuven, Belgium.
75. Doi, M. and Ohta, D. (1991) *J. Chem. Phys.*, **95**, 1242.
76. Chow, A.W., Hamlin, R.D. and Ylitalo, C.M. (1992) *Macromolecules*, **25**, 7135.
77. Smyth, S.F. and Mackay, M.E. (1994) *J. Rheol.*, **38**, 1549.

78. Taratuta, V.G., Lonberg, F. and Meyer, R.B. (1988) *Phys. Rev. A*, **37**, 1831.
79. Malkin, A. Ya., Vasil'eva, N.V., Belousova, T.A. *et al.* (1979) *Koll. Zh.*, **41**, 200.
80. Larson, R.G. and Mead, D.W. (1993) *Liq. Cryst.*, **15**, 151.
81. Mackay, M.E., Liang, C.-H. and Halley, P.J. (1992) *Rheol. Acta.*, **31**, 481.
82. Liang, C.-H. and Mackay, M.E. (1993) *J. Rheol.*, **37**, 149.
83. Done, D. and Baird, D.G. (1990) *J. Rheol.*, **34**, 749.
84. de Cleyn, G. and Mewis, J. (1981) *J. Non-Newtonian Fluid Mech.*, **9**, 91.
85. Moldenaers, P. *et al.* (to be published).
86. Ernst. B., Navard, P. and Haudin, J.M. (1987) *J. Polym. Sci.: Part C: Polym. Lett.*, **25**, 79.
87. Kiss, G. and Porter, R.S. (1980) *Mol. Cryst. Liq. Cryst.*, **60**, 267.
88. Marsano, E., Carpaneto, L., Ciferri, A. (1988) *Mol. Cryst. Liq. Cryst.*, **158B**, 267.
89. Vermant, J., Moldenaers, P., Mewis, J. *et al.* (1994) *J. Rheol.*, **38**, 1571.
90. Donald, A.M., Viney, C. and Windle, A.H. (1983) *Polymer*, **24**, 155.
91. Viney, C., Donald, A.M. and Windle, A.H. (1983) *J. Mater. Sci.*, **18**, 1136.
92. Peuvrel, E. and Navard, P. (1991) *Macromolecules*, 24, 5683.

9

Processing and properties of rigid rod polymers and their molecular composites

W.-F. Hwang

9.1 INTRODUCTION

Innovative research on new nonmetallic materials for structural applications has been highly successful. The concept of a new approach to providing improved structural materials in terms of lightweight, thermal and thermoxidative stability, toughness and dimensional stability has been explored – the newly discovered compositions of rigid-rod molecular composites [1].

It is well known, in the art of fibre composites, that the most important factor in controlling the strength and rigidity of a fibre composite is the length-to-diameter ratio of its reinforcing fibres. Increasing this aspect ratio can have the effect of increasing the modulus by as much as one to two orders of magnitude. There are two ways of increasing this ratio: (i) by increasing the length of the fibre or (ii) by decreasing its diameter. Following the second line of reasoning, as pointed out by Lindenmeyer [2], one can at least conceptually envisage the ultimate reinforcing fibre to be a single polymer molecule in the extended chain conformation. This is probably the origin of the concept of 'molecular composites'.

A 'molecular composite', according to Lindenmeyer, 'is a polymer material in which a fraction of the molecules are extended to form reinforcing fibers and the remainder are coiled or folded to form the matrix'.

Rheology and Processing of Liquid Crystal Polymers
Edited by D. Acierno and A.A. Collyer
Published in 1996 by Chapman & Hall
ISBN 0 412 59640 7

This molecular composite is basically a semi-crystalline polymer in which the crystalline region serves as the reinforcing element.

Takayanagi [3] and Krigbaum [4] pointed out that the properties of semi-crystalline polymers can be explained using an empirical two-phase model. It was later suggested by Kardos [5] that molecular composites should be amenable to an engineering composite approach to the prediction of their mechanical properties such as stiffness, strength, and thermal expansion coefficient based on their constituent's properties. Thus, one can envisage that a molecular composite is a molecular analog of a chopped-fibre composite.

In the past decade, polymer synthesis efforts [6] in the US Air Force Ordered Polymer Program have provided a new class of polymers which are para-configured and are intrinsically rigid-rod molecules. An intrinsically rigid-rod polymer is a polymer whose persistence length is as long as its contour length. PBZ polymers have been proven to be such molecules [7]. By replacing extended polymer chains with these rigid-rod polymer molecules, such as PBZ polymers synthesized by Wolfe [8], in a matrix leads one to the concept of 'rigid-rod molecular composites [1]'. Hwang *et al.* [9] were the first to demonstrate this concept successfully. Elsewhere, Hwang *et al.* [10] and Takayanagi [11] were able to illustrate that the reinforcement efficiency of molecular composites can be modelled by the Halpin-Tsai relationships [12].

Rigid-rod molecular composites can be classified into two types: (i) high temperature molecular composites (HTMC) [9], (ii) thermoplastic molecular composites (TPMC) [13, 14]. HTMCs are systems composed of rigid-rod PBZ molecules dispersed in their analogous flexible ABPBZ polymers. This type of molecular composites has been extensively studied, and proven to possess excellent mechanical properties, chemical resistance, thermal and thermoxidative stability, and dimensional stability. TPMCs are systems composed of PBZs dispersed in a thermoplastic host.

TPMCs are, in principle, mouldable molecular composites from which large structural components may be built. Systems using a thermoplastic polyamide as a matrix have been prepared [15]. The reported properties of the composite, however, were much lower than expected from the rule of mixtures due to possible phase separation. In another report [13], the idea of processing molecular composites from highly concentrated solutions above the nematic-to-isotropic transition temperature, T_{N-I}, to accomplish the same purpose was suggested. Subsequently, Nishihara *et al.* [16] adopted these suggestions and were able to demonstrate these concepts and process thermoplastic molecular composites with excellent properties.

This chapter examines the various types of MCs with regard to their processing, properties, and limitations, starting with, however, the class of new lyotropic LCP mentioned in Chapter 1 and known as polybenzazoles (PBZ) with regards to their processing, properties and limitations.

9.2 LYOTROPIC LCPs

9.2.1 Background

Because of the stiff rod-like nature of the lyotropic LCPs, they usually are not thermoplastic and cannot be processed by conventional thermal processes. Mostly, they are processed from nematic liquid crystalline solution in strong acids. In this type of liquid crystalline solution, the rod-like PPTA molecules align side by side into highly ordered nematic domains which can be oriented easily by shear or elongational stress during the spinning process. Typically, they are spun into uniaxially oriented, high modulus, high strength, fibres by the conventional dry-jet wet-spinning method. A classical example of this type of LCP is poly(p-phenylene terephthalamide) (PPTA); see Fig. 9.1.

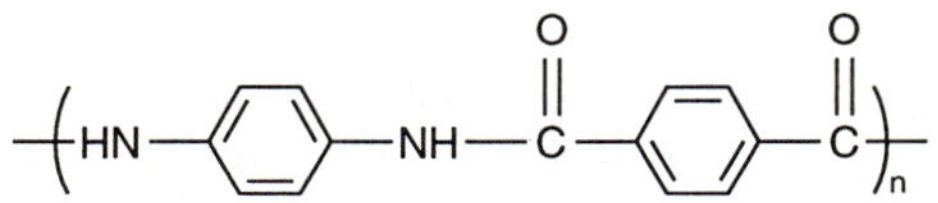

Figure 9.1 Poly(p-phenylene terephthalamide) (PPTA).

Du Pont's Kevlar™ fibres are spun from nematic liquid crystalline solution of PPTA in 100% sulphuric acids. Since its introduction some ten years ago, Kevlar™ has emerged as the leading high performance organic fibres for various high strength applications.

Polyaramide fibres with modulus as high as 180 $GN\,m^{-2}$(greater than 90% of the theoretical value) and tenacity as high as 3.6 $GN\,m^{-2}$ have been produced from liquid crystalline solution of poly(p-benzamide). However, closer examination of its molecular geometry shows that the polyaramide molecule still possesses a finite degree of flexibility which introduces deviation from perfect rigidity. This consideration led to the increasing activity in the research and development of new lyotropic LCPs of wholly aromatic, perfectly rectilinear rigid rods. This class of polymers, called polybenzazoles (PBZ), holds promise of extending not only the mechanical properties but also the thermal capability closer to the limit of organic materials. Specifically, the US Air Force Ordered Polymer Program has devoted a considerable effort since the mid-1970s to the development of trans-poly(p-phenylenebenzobisthiazole) (PBT)

(Fig. 9.2) and cis-poly(p-phenylenebenzobisoxazole) (PBO) (Fig. 9.3) for high performance applications.

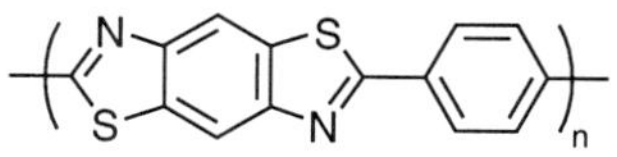

Figure 9.2 Trans-poly(p-phenylenebenzobisthiazole) (PBT).

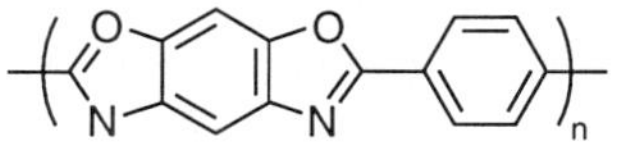

Figure 9.3 Cis-poly(p-phenylenebenzobisoxazole) (PBO).

The following section reviews the recent development of rigid-rod PBZ polymers with regards to their rigidity, and processing and properties of PBZ fibres.

9.2.2 Rigidity of PBZ

It is well known that the aspect ratio of the reinforcing entities is the most important factor in determining the stiffness properties of a composite. In the chopped-fibre composites, the aspect ratio of the fibre is simply defined as two times its length-to-diameter ratio. In a molecular composite, it depends greatly on the rigidity of the reinforcing rigid-rod polymers.

There are several parameters that have been used as measures of the rigidity of a polymer chain. One such measure is the **persistence length parameter**, ρ, which can be calculated theoretically using Flory's virtual bond model [17]. This type of calculation has been successfully demonstrated for several polymer chains [9, 18] including semi-flexible poly-2,5(6)benzimidazole (AB-PBI), rod-like PPTA, and rigid rod poly-p-(phenylenebenzobisthiazole) (PBT). The calculated results are listed in Table 9.1 and they are consistent with the observation that PBT is more rigid than PPTA, which in turn is more rigid than AB-PBI as evidenced by the relative ease in the formation of liquid crystalline solutions. The most flexible AB-PBI never (up to its solubility limit) forms a liquid crystalline solution in its good solvent methanesulfonic acids (MSA), while both PBT and PPTA form a nematic liquid crystalline solution readily in strong acids. Experimentally, the persistence length of a polymer chain can be determined by the light scattering technique [7, 19]. For example, light scattering characterization of PBT in

Table 9.1 Calculated persistence length of rod-like polymer chains

Polymers	*Persistence length,* ρ *(nm)*
AB-PBI	8.8[9]
PPTA	80.0[9]
PBT	120.0[7]

dilute MSA solution indicates that it has a persistence length, ρ, comparable or longer than its contour length, L, i.e. $L/\rho < 0$. Similar study of PPTA shows that it has a finite value of $L/\rho > 0$.

Another relative measure of chain rigidity is **Huggins' coefficient** [20]. The Huggins' coefficient for PBT polymers in MSA was determined [21] to vary from 0.41 to 0.45 with increasing molecular weight, while the Huggins' coefficient for PPTA was found to be 0.35 in the same solvent. This indicates that the PBT molecule is more rigid than the PPTA molecule. (The results are summarized in Table 9.2.) PBT polymer and its structural analogue PBO can be classified as intrinsic rigid-rod polymers, i.e. the persistence length of these molecules is as long as their contour length; these have been used extensively as the reinforcing entity in the molecular composite systems described in this Chapter.

For lyotropic liquid crystalline, rigid-rod polymers such as PBT and PBO, and the rod-like PPTA, there is a simple and useful method for determining their effective axis ratio, χ. As stated previously, the effective axis ratio of the reinforcing rigid-rod molecules in a molecular composite is the primary factor affecting its mechanical properties. Experimentally, it involves the determination of the critical concentration point, C_{cr}, of the rigid-rod polymers in their solvent such as MSA or sulphuric acids. C_{cr} is defined as the minimum polymer concentration (at any temperature) necessary for the separation of a stable optically anisotropic liquid crystalline phase in coexistence with an optically isotropic phase in the polymer solution. The rigid-rod molecules in the liquid crystalline state usually have the short range order of nematic mesophases in which

Table 9.2 Relative rigidity of PBT versus PPTA

Parameters	*PPTA*	*PBT*
M_w	40 000	20 000
C_{cr} (wt%) @ RT	9–9.5 in H_2SO_4	3–3.2 in MSA
Axis ratio, χ	85	260
Huggins' coefficient	0.35	0.45

Table 9.3 C_{cr} and χ of PBZ of various molecular weight

Rigid-rod	*Inherent viscosity** ($dL\,g^{-1}$)	C_{cr} (wt%)	χ
PBO	6.4	7.0	112
PBO	11.8	5.9	135
PBT	18.0	2.5	300

*Single-point IV, concentration = 0.05 g dL^{-1} at 25°C in MSA.

the rigid-rod molecules are aligned side-by-side. Visually, opalescence is observed when such a solution is stirred, so called 'stirred-opalescence' phenomenon, because of the alignment of nematic domains under shearing condition. A PBZ solution in MSA at above its critical concentration point shows typical nematic texture when examined under a polarizing optical microscope. Conversely, when the polymer concentration is below its C_{cr}, the solution is optically isotropic. It appears to be totally dark when examined under the cross-polar condition of the optical microscope.

The experimentally determined critical concentration of a rigid rod or a rod-like polymer is a direct measure of its effective axis ratio, χ. For a high molecular weight ($M_w \to \infty$) rigid rod or rod-like molecule, the relationship between C_{cr} and χ can be approximated by Flory's lattice model [22].

$$C_{cr} \approx \frac{8}{\chi}\left(1 - \frac{2}{\chi}\right)$$

Using this formula, the effective axis ratio, denoted as χ, of a rigid rod or rod-like polymer molecule can be calculated from the experimentally determined C_{cr}. Experimentally determined C_{cr} and calculated χ for PBZ of various molecular weight are summarized in Table 9.3.

As shown in Tables 9.1, 9.2 and 9.3, it is clear that PBZ is indeed more rigid than PPTA as reflected in its higher persistence length, higher axis ratio, even when its molecular weight is lower than that of PPTA. This has significant implications in the properties of their corresponding fibres as will now be discussed.

9.2.3 Processing and properties of lyotropic PBZ

Lyotropic rigid-rod PBZ polymers have been processed into high performance fibres [21, 23, 24] and films [23, 25]. PBZ fibres have been studied [26] extensively and are reviewed here, the emphasis being placed on the potential and limits of these fibres for advanced applications.

There are two fundamental methods of solution spinning of polymers when they do not form thermally stable, viscous melt but can be dissolved in high enough concentration in suitable solvents. When a polymer is soluble in a volatile solvent, a **dry-spinning method** is used. Dry-spinning of fibres from polymer solutions consists of extruding the solution into a vertical cell where jets of spinning solution leave the spinneret and come into contact with a stream of hot gas (such as air). In the gas stream, the solvent evaporates, the polymer concentration in the filament increases, and the filament solidifies into fibre. The other fundamental method, **wet-spinning**, is applied to polymers such as the lyotropic PBZ and PPTA which do not melt, and will dissolve only in non-volatile or thermally unstable solvents. Wet-spinning of lyotropic PBZ and PPTA consists of extrusion of the polymer solutions, usually in acids such as MSA, sulphuric acids, or polyphosphoric acids (PPA), through a spinneret into a liquid bath containing low molecular weight liquid which is miscible with the solvent but immiscible with the polymer. In this 'coagulation' bath, the polymer is precipitated from the solution and solid gel filaments are formed. Usually, there is an air gap between the spinneret and the coagulation bath so that the spinning jets of the solution are dry; thus, the term '**dry-jet wet-spinning**' is used for this process. At the air gap, the spinning solution jets are being drawn down by an elongational stress. The solid gel filaments are then washed with water to rid them of the residual acids. Sometimes, they are also neutralized in aqueous base such as NH_4OH before or after the water washing. The wet fibres are then air dried at room temperature or in a vacuum before the final heat treatment at elevated temperatures to prevent blistering of fibres. A final heat treatment at 500–650°C is usually carried out to further the mechanical properties of the fibres.

Processing of PBZ fibres

This has been reported previously [23, 24]. However, these reports focused on the process itself without any discussion on the spinnability of PBZ dopes. In this section, the processing of PBZ dopes, i.e. liquid crystalline solutions of PBZ in polyphosphoric acids (PPA) containing approximately 83 wt% P_2O_5, is illustrated using PBO and ABPBO as examples, with the emphasis being on the spinnability, properties and their relationships. In this respect, a study [27] was carried out at US Air Force Materials Laboratory between 1982 and 1983. Included in that study was the processing of poly(2,5-benzoxazole), ABPBO, dopes. ABPBO, as shown in Fig. 9.4, is the semi-rigid analogue of PBO.

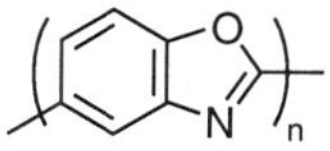

Figure 9.4 Semi-rigid analogue of PBO.

The spinnability of a polymer dope can be defined as follows: 'A fluid is deemed spinnable, when under a given deformation condition, a steady state, continuous elongation of the fluid jet can proceed without breakage of any kind' [28]. Spinnability of a polymer dope is dependent on its rheological characteristics as a function of time, temperature and pressure. As described by the deformation characteristics of a polymer dope, spinnability is limited by two breakage mechanisms: capillary breakage (due to surface tension instability) and cohesive brittle fracture. For wet-spinning of high concentration polymer solutions, the latter mechanism is the dominant one. In a simplified isothermal theory of cohesive fracture, four dominant factors controlling the spinnability are: deformation gradient (ξ), extrusion velocity (V_o), relaxation time (t) and the tensile strength of the polymer dope (s). In an actual spinning process, the deformation gradient is not usually a constant, and an 'effective' deformation gradient can be approximated by [28]

$$\xi_{\text{eff}} = \frac{\ln \frac{V_L}{V_o}}{L_s} = \frac{\ln (\text{SDR})_{\text{app}}}{L_s}$$

where V_L is the take-up velocity at the end of the jet, V_o is the average extrusion velocity (as determined by simple mass transfer), L_s is the length of the solidification path, i.e. the part of the spinning path where solidification of the filament is complete, and $(\text{SDR})_{\text{app}}$ is the apparent spin-draw ratio. V_L and V_o can be determined experimentally, and L_s is known. Thus, the apparent spin-draw ratio, as determined by the above relationship, is a measure of spinnability of PBZ dopes. Since most polymer dopes, especially LCP PBZ dopes, are non-newtonian fluids, the radius of the fluid jets leaving the spinneret channel of radius R_o expand to some value R_{max} due to die swell. The die swell ratio is defined as R_{max}/R_o, which can also be determined experimentally.

Apparatus for the dry-jet wet-spinning of PBZ fibres has been described previously [23, 24, 27]. They all consist of five basic elements: (i) extrusion device, (ii) dope reservoir, (iii) spinneret, (iv) coagulation bath and (v) take-up device. Spinning of PBO and ABPBO fibres was usually carried out with a temperature range of 50–170°C. The spin-draw ratio was in the range 10–300. The coagulation of PBO and ABPBO fibres was usually carried out in water at about room temperature. The coagulated fibres were washed for several days in running water until neutralized at room temperature. Further neutralization of coagulated fibres in aqueous base, e.g. 50/50 H_2O/NH_4OH was recommended to rid them of residual acids. Fibres were typically air dried in room conditions for at least eight hours, then vacuum dried for at least 24 hours before heat treatment at elevated temperatures. This drying step is critical in obtaining fibres with the highest possible tensile strength, because drying of wet fibres at elevated temperature produces damaged fibres with blisters on them. Heat treatment of PBO and ABPBO fibres was usually carried out by passing them from a rotating feed drum through a cylindrical oven and a set of godets on to a take-up winder. This was done under constant tension controlled by a set of tensioning rolls. The residence time in the oven was typically 10–30 seconds, in an environment of either air or nitrogen at 550–650°C. Heat treatment of these fibres only increases their tensile strength slightly, but doubles their tensile modulus.

ABPBO fibres discussed in this section were spun from a PPA dope containing 16.9 wt% of ABPBO with P_2O_5 content of 83.8%. The measured inherent viscosity of ABPBO was 11.8 dL g^{-1}. For this dope, the proper ranges of spinning (extrusion) temperatures were determined to be 70–100°C. Outside this range, the dope was so weak that a high spin-draw-ratio (greater than 20) could not be achieved. The existence of an optimum processing temperature range for ABPBO dopes indicated that the capillary breakage mechanism cannot be ignored. At higher temperatures, the bulk viscosity of the dope is low, and the capillary breakage mechanism dominates. At lower temperatures, the spinnability is solely determined by cohesive fracture. Furthermore, in the spinnable temperature range, the spinnability (experimentally, SDR_{max}) showed a decreasing trend as a function of free-jet velocity, as illustrated in Table 9.4.

These results indicated that, at constant viscosity and temperature, the spinnability of ABPBO dopes may have been controlled by a composite breakage mechanism.

Table 9.4 Spinnability (maximum spin-draw ratio) of ABPBO dope as a function of V_f

Spinning temperature (°C)	V_f *(cm min^{-1})*	SDR_{max}
90	4.45	103
90	5.36	85
90	17.78	50
90	44.45	20
110	2.79	97
110	19.56	46
110	46.99	19

Table 9.5 Spinning data and tensile properties of PBO fibres

Spinning temperature (°C)	*Spinning pressure ($Kg\ cm^{-2}$)*	*V_f ($cm\ min^{-1}$)*	*SDR*	*modulus (GPa)*	*Strength (GPa)*	*E_b (%)*
80	42.2	3.3	115	145	3.66	3.3
Heat-treated		–	–	256	4.09	1.8
90	21.8	2.95	230	235	3.56	2.4
Heat-treated		–	–	366	4.10	1.6

The spinning of PBO dopes was carried out in the same way as that used for ABPBO dopes, but was more straightforward. Extremely high spin-draw ratios (up to 300) were obtained. Unlike ABPBO dopes, no optimum spinning temperature range was found. In general, the spinnability (SDR_{max}) increases with spinning temperature. Also, increases in extrusion rate, V_f, generally caused a decrease in spinnability. Furthermore, a decrease in the solidification path, achieved by moving the take-up roll closer to the idle roll, resulted in a decrease in spinnability. Qualitatively, these observations were in good agreement with the spinning theory discussed above.

Properties of PBO and ABPBO fibres

This has been extensively reported [26]. This section briefly examines the processing-property relationships for PBO and ABPBO fibres. Tensile properties of these fibres were determined by the ASTM D3379 method for single filament measurements. Typical spinning data and tensile properties of as-spun PBO fibres are listed in Table 9.5.

As expected, the modulus of as-spun PBO fibres increases with spin-draw ratio and heat treatment. The typical spinning data and tensile properties of ABPBO fibres are summarized in Table 9.6. Again, heat

Table 9.6 Spinning data and tensile properties of ABPBO fibres

Spinning temperature (°C)	*Spinning pressure ($Kg\ cm^{-2}$)*	*V_f ($cm\ min^{-1}$)*	*SDR*	*modulus (GPa)*	*Strength (GPa)*	*E_b (%)*
90	42.2	4.32	110	60.8	2.83	4.8
Heat-treated		–	–	115.4	3.35	3.6
100	21.1	1.52	200	65.6	2.94	8.0
Heat-treated		–	–	134.0	3.45	4.0

treatment increases the tensile modulus and strength tremendously. The increases are due to better orientation and crystal growth along the fibre axis [26]. The tensile properties of ABPBO fibres are much lower than those of the PBO fibres. This is not surprising because PBO is more rigid than ABPBO.

Also, ABPBO and PBO fibres have the best thermal and thermoxidative stability among all known heterocyclic aromatic polymers. Based on TGA measurements, the degradation temperature in air is 635°C and 625°C for PBO and ABPBO, respectively. In inert atmosphere (He), the degradation temperature is 685°C and 640°C for PBO and ABPBO, respectively. One major drawback for the PBZ and ABPBZ fibres that prevents them from being widely used in the advanced composite arena is their low compressive strength as compared to pitch-based carbon fibres. This problem is inherent in any LCPs, either thermotropic or lyotropic.

9.3 MOLECULAR COMPOSITE SYSTEMS

As originally conceived, a molecular composite is a homogeneous, synergistic composite of molecularly dispersed rigid-rod molecules (single molecules or small bundle of molecules with lateral dimension of less than 5 nm) in a matrix material. The enhanced and desirable properties, such as superior chemical and environmental resistance, enhanced thermal and thermoxidative stability, toughness, and dimensional stability, resulting from the synergism between the reinforcing rigid rods and the matrix can only be realized in a true molecular composite. Without the development of a true rigid-rod molecular composite technology, the fundamentally detrimental interfacial problems (adhesion, different thermal expansion coefficients, etc.) encountered in conventional fibre composites cannot be averted. Such a development, however, has proved to proceed only with considerable difficulty.

The conception [1] of rigid-rod molecular composites arose out of the concerns over the property limitations of conventional fibre-composites, such as poor adhesion between the reinforcing fibres and the matrix, and the thermal, thermoxidative, and hydrolytic instability of the matrix material itself. In great contrast to the conventional fibre-composite practices, the idea has been to create a new and improved composite material (or an alloy) with distinct synergistic characteristics of the two constituent materials. Specifically, this material will have the mechanical reinforcement, chemical and hydrolytic stability, and thermal and thermoxidative stability attributable to the rigid-rod polymers, while retaining the malleability of the matrix material (in the case of thermoplastic matrix), albeit at higher temperature. The concept and the feasibility of forming rigid-rod molecular composites have been realized

[9, 10]. In addition to the tremendous molecular reinforcement of mechanical properties realized in a true molecular composite, synergistic effects such as enhancement in solvent resistance and elevation in glassy transition temperature (T_g) of a molecular composite over its corresponding thermoplastic matrix have also been demonstrated [13, 14]. These synergistic effects can only be accomplished through the intimate molecular level interaction between the rigid-rod polymer and the matrix in a true molecular composite. Thus, the key to the success of rigid-rod molecular composites technology lies in the molecular dispersity of rigid rods in a matrix material.

Extensive exploration of molecular composite systems comprised of rigid-rod polymers such as PBO and PBT and flexible thermoplastic host materials such as aliphatic polyamides and others, has been reported [29–31]. Based on the literature, the following conclusions can be derived:

1. Phase separation or segregation (with the lateral dimension of rigid-rod-rich phase in the range of a few tens nm to microns) is inevitable [29] in most of the physical blend systems during the solution processing of preforms such as films, fibres or powders, or during the subsequent consolidation and lamination process.
2. Phase separation kinetics is usually fast (minutes) in the physical blends [29]. For example [31], a transparent film sample prepared from an optically isotropic solution of 30/70 PBO/ABPBO-ether has an elevated single T_g of 280°C as compared to 230°C for that of the host matrix. After it was heated at 300°C for a few minutes, however, the sample became opaque and showed a T_g of 230°C indicating that phase separation had occurred.
3. Except in the uniaxially oriented sample prepared from the compression moulding [30] of uniaxially oriented fibres, a phase separated system usually has very low tensile properties as compared to those of conventional fibre composites. The difference is in the aspect ratio of the PBZ-rich phase. In the isotropic phase separated system, this aspect ratio is small because the PBZ-rich phase is in the form of more or less round aggregates. In the uni-oriented case, this aspect ratio is large because the PBZ-rich phase is in the form of fibrils. Furthermore, it was reported that uni-oriented parts made from PBT/polyamides had very low compressive strength [15].
4. In contrast to a true rigid-rod molecular composite, the chemical (solvent) resistance and temperature capability of the reported phase separated thermoplastic systems were not enhanced beyond that of its matrix material.

Extensive exploration of systems [31] comprised of 'physical blends' of rigid-rod polymers (such as PBO and PBT) and flexible host matrices such as ABPBO (non-thermoplastic) and various thermoplastics such as amorphous polyamides, ABPBO-ether, and PBO-sulfones resulted in the elucidation of some fundamental shortcomings of the blending approach. Such shortcomings include the inevitability of phase separation during either the coagulation or the post-forming processes, the reinforcement inefficiency in phase separated systems, the loss of toughness, strength, and environmental stability in phase separated systems, and the inability to consolidate systems in which the host matrix exhibits no T_g.

The following section briefly reviews some of the MC systems with respect to the phase separation phenomena and reinforcement efficiency and some of their potential and drawbacks.

9.3.1 Phase separation of rigid-rod/flexible coil blends

In his paper concerning the thermodynamics of a ternary blend of rigid-rod polymer/flexible coil polymer/solvent, Flory [32] predicts the inevitability of phase separation of the geometrically dissimilar polymer constituents in the blend when the total polymer concentration passes through a critical concentration point, C_{cr}. This prediction has been confirmed experimentally [13]. In the classical example of PBT/ABPBI blends [33], this phenomenon had been clearly illustrated. The rigid-rod PBT polymers have been shown to segregate into football-like domains of 2–4 microns long with aspect ratio of approximately three [9]. The film was processed by a vacuum casting technique [33]. During slow evaporation of solvent in the casting process, the polymer concentration, originally in an optically isotropic solution, passed through its critical concentration point and, as expected, phase separation occurred.

The mechanical properties of the above phase-separated PBT/ABPBI blend [33] is much lower than those of a true molecular composite of PBT/ABPBI [9]. The latter was prepared from an optically anisotropic solution with concentration lower than its C_{cr}; instead of slow sublimation, the solvent was removed quickly by coagulating in a water bath, thus maintaining the fine dispersion of PBT molecules in the ABPBI matrix.

In a rigid-rod/thermoplastic blend, the phase separation phenomenon is even more pronounced. It had been observed that PBT polymers phase-separated out during the coagulation process, even when the starting solution was optically isotropic. To circumvent this problem, one can partition the ternary solution into small droplets by spraying; they can then be coagulated faster in a stirred coagulation bath or in a blender. After neutralization, washing and drying, fine powders of PBO/thermoplastic blend were obtained. Consolidation of these powders, usually compression moulding, may induce phase separation dependent upon the consolidation temperature. As stated previously, the mechanical

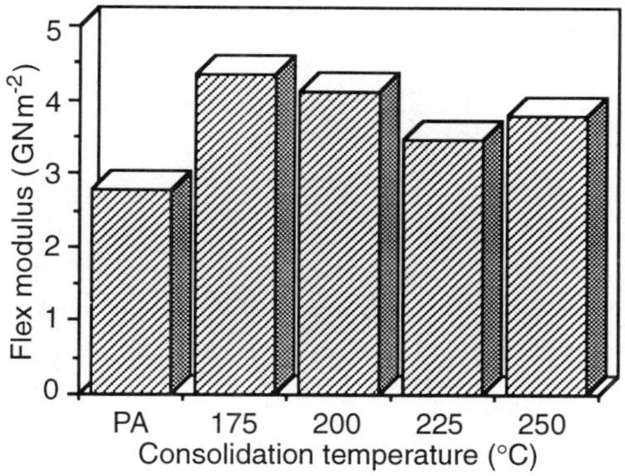

Figure 9.5 Flex modulus of 5/95 wt/wt% PBO/nylon as a function of consolidation temperature.

properties such as flexural modulus are affected by phase separation because of the reduction in aspect ratios of rigid rods when they are phase segregated. As shown in Fig. 9.5, the flexural modulus of 5/95 wt/wt% PBO/nylon generally decreases as the consolidation temperature becomes higher.

The decrease in the flexural modulus may be attributed to the increasing degree of phase separation of the rigid-rod PBO molecules out of the nylon matrix. However, with only 5 wt% PBO loading, the phase separation does not seem to be severe, based on the excellent properties of the MCs compared to that of the neat nylon. Also, it should be noted that the flexural modulus of these TPMCs, with only an equivalent of 3.7 vol.% rigid-rod reinforcements, is much higher (40–60% higher) than that of neat amorphous nylon. In fact, the reinforcing efficiency of the above TPMC is much higher than that of a corresponding chopped-fibre composite.

As shown in Fig. 9.6, the phase separation is more severe in a blend with higher rigid-rod PBO loading. This is reflected in the more pronounced reduction in flexural modulus as the consolidation temperature increases. Also, it is seen in the fact that the rise in PBO loading does not increase the flexural modulus of the TPMC blends. The inevitability of phase separation in the rigid-rod/thermoplastics blends has been experimentally confirmed by a recent study [29].

Clearly, the phase separation phenomenon in the TPMC blend is the single most detrimental factor in preventing the thermoplastic molecular composites from becoming viable structural composites. However, a TPMC system with low rigid-rod loading and with high enough flex modulus (1 Msi) may be useful as an advanced matrix for continuous fibre composites.

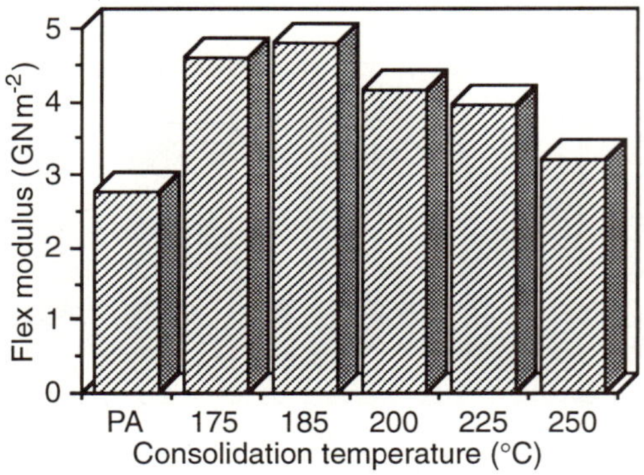

Figure 9.6 Flex modulus of 7.5/92.5 wt/wt% PBO/nylon as a function of consolidation temperature.

To minimize the phase separation, efforts [34, 35] have been focused in the preparation of block/segmented copolymers of rigid-rod PBO or PBT with flexible polymers. Tsai's system [34] is a PBT/ABPBI copolymer which has been shown to be a true molecular composite without gross phase separation. Also, the PBT/ABPBI copolymer system shows better mechanical properties than those of a corresponding physical blend. The other report [35] has focused on the preparation of block/segmented copolymers of rigid-rod PBO with flexible thermoplastics.

9.3.2 Block/segmented rigid-rod copolymer systems

The preparation of block/segmented copolymers containing rigid-rod cis-PBO with other flexible polymers has been extensively explored [35, 36]. Of particular interest is the preparation [36] of copolymers of PBO with thermoplastic polymers. The copolymers may be thermoplastic dependent upon their rigid-rod content. Articles such as oriented fibres or films, powders, or moulded parts can be fabricated [37, 38] from the thermoplastic block copolymers. In this section, we have a block/segmented copolymers of cis-PBO with a flexible thermoplastics, poly(aromatic ether ketone benzoxazole) (PEKBO) or poly(ether ether ketone) (PEEK); see Figs 9.7 and 9.8.

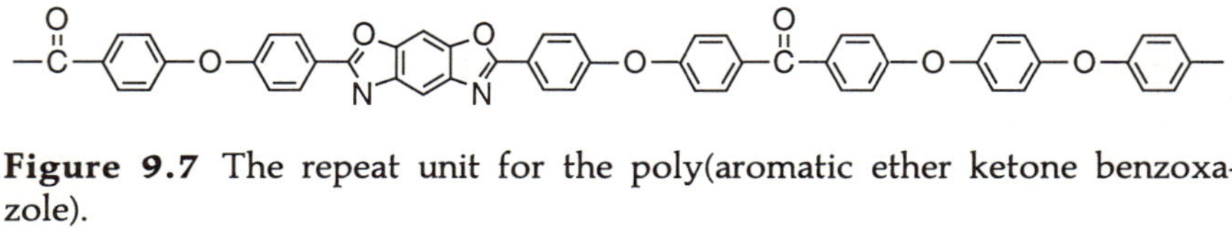

Figure 9.7 The repeat unit for the poly(aromatic ether ketone benzoxazole).

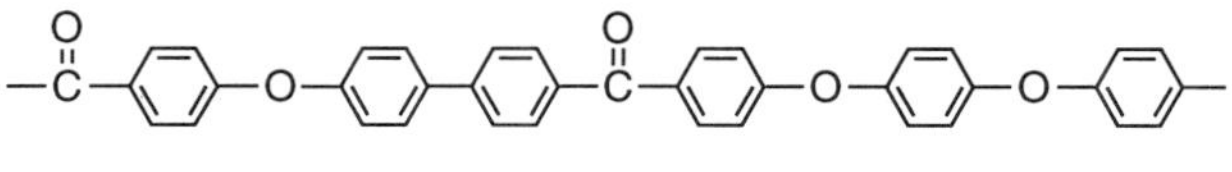

Figure 9.8 The repeat unit for the poly(ether ether ketone).

Oriented films or fibres made from these copolymer systems with rigid-rod PBO content of up to 70 wt% are optically transparent when observed under an optical microscope. Other studies [39] using dynamic SALS and small-angle x-ray scattering (SAXS) techniques have shown that, in contrast to the results [29] obtained for the rigid-rod/thermoplastic blends, no discernible phase separation occurred in these rigid-rod copolymer systems even when they were heated above their glassy transition temperature (T_g). The T_g of the copolymers is determined using a dynamic mechanical thermal analyser (DMTA). For the PBO/PEKBO copolymers system, the T_g as a function of rigid-rod content (vol.% PBO) is shown in Fig. 9.9. The copolymers with varying amounts of PBO all show a single, distinct, higher T_g than that of the neat PEKBO. This result also suggests that the rigid rods are reasonably well dispersed in the thermoplastic matrix without gross phase separation.

The mechanical properties of the copolymers are excellent. The tensile properties of heat-treated fibres of PBO/PEKBO copolymers are similar to those of the molecular composite fibres of PBT/ABPBI blends [9], or PBT/ABPBI copolymers [34]. They are shown in Figs 9.10, 9.11 and 9.12.

As expected for a typical molecular composite, the tensile modulus

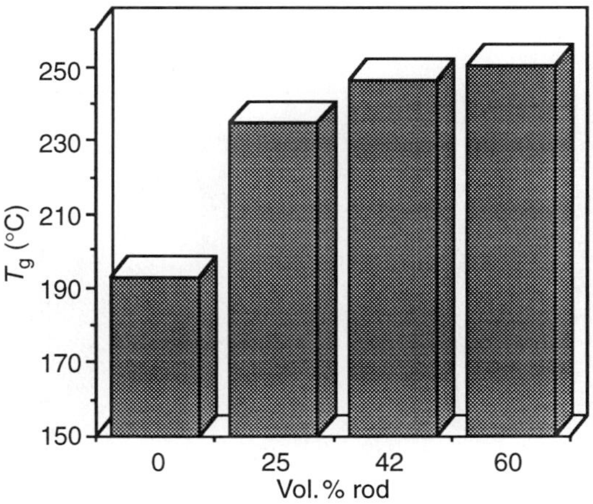

Figure 9.9 T_g of PBO/PEKBO copolymers as a function of PBO content.

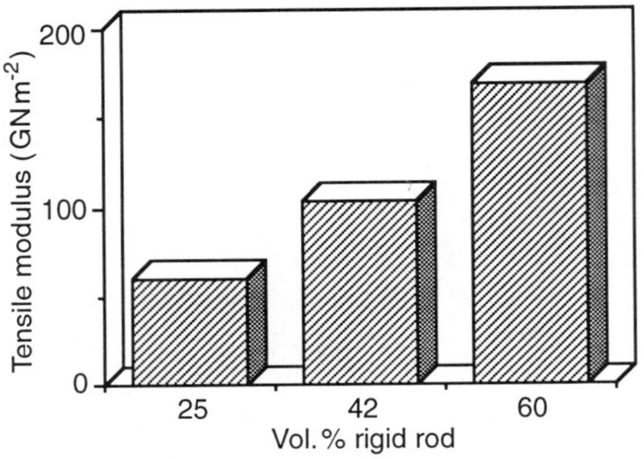

Figure 9.10 Tensile modulus of PBO/PEKBO fibres.

and strength are linearly proportional to the rigid-rod content, while the elongation-to-break is inversely proportional to the rigid-rod content. These results prove that tremendous mechanical properties can indeed be achieved for a molecular composite. In addition to the tensile properties, these TPMCs have excellent thermal and thermoxidative stability with degradation temperature higher than 500°C. With the exception of strong acids such as methanesulfonic acids, they also have excellent chemical/solvent resistance to most organics and caustics. This seems to suggest that the TPMCs of block/segmented rigid-rod copolymers discussed here may have the potential for advanced structural composite applications in the extreme environment. The key drawback, as one would encounter time and time again in the rigid-rod molecular composite arena, lies in the extremely high bulk viscosity ($\gg 10^4$ Ns m^{-2}, generally beyond the measuring capability of any rheometer) of these

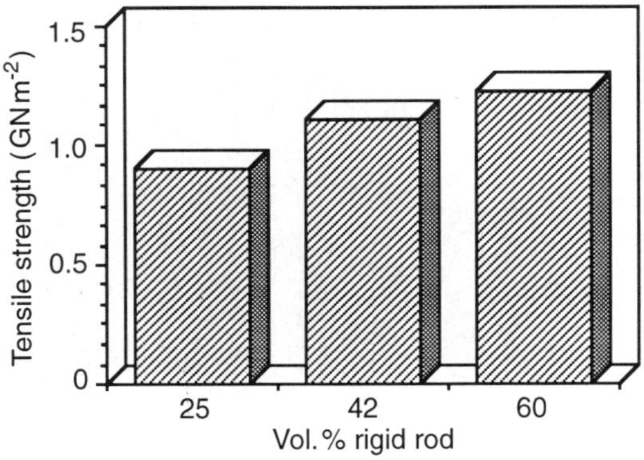

Figure 9.11 Tensile strength of PBO/PEKBO fibres.

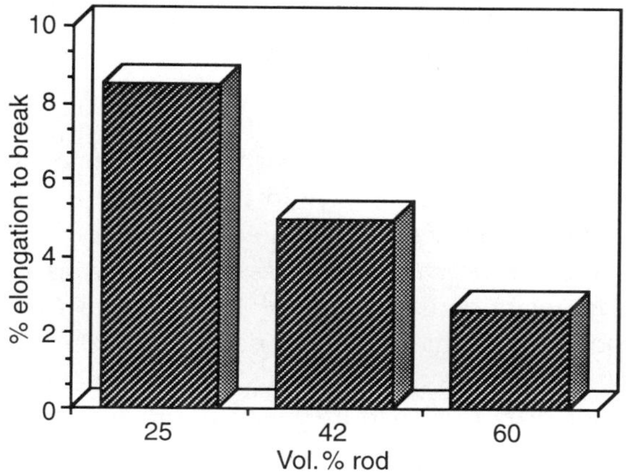

Figure 9.12 Elongation-to-break (%) of PBO/PEKBO fibres.

MC systems. The high dispersity of rigid-rod molecules in the thermoplastic matrix has greatly hindered the mobility of the flexible thermoplastics so that the composite becomes immobile even at temperatures higher than its T_g; whereas in a phase-separated system, the flexible thermoplastics can move freely and independently from the rigid rods. The T_g detected in the DMTA measurements probably originates from very small and localized motions characteristics of a TPMC. Because of the extreme high bulk viscosity associated with TPMCs with more than 10% rigid rods, these systems are not consolidatable by conventional means such as compression moulding or injection moulding. This, in effect, eliminates them from any structural applications where consolidated large parts are necessary. This impasse in the processability and, therefore, the practical utility of both the TPMCs and the HTMCs leads one to the consideration of thermoplastic molecular composite systems of copolymers containing less than 10 wt% rigid rods of finite length (lower molecular weight). Such a TPMC system, probably, would be more amenable to consolidation and would have much higher modulus than conventional thermoplastics; therefore, it may be useful as an advanced matrix for continuous-filament composites. The need for more rigid thermoplastics is based on the observation that a continuous-filament composite with soft thermoplastic matrix has lower compressive properties than those of a corresponding composite with a thermoset matrix. The hypothesis is that a rigid thermoplastic matrix would produce a thermoplastic fibre composite with improved compressive properties without the problem of brittleness associated with thermoset fibre composites, which are very susceptible to low velocity impact damage. The next section explores this idea.

9.3.3 Thermoplastic molecular composites as advanced matrices for continuous-filament composites

As discussed previously, one potential application is to use a TPMC of block copolymer with 10 wt% of rigid rods of finite length as a rigid matrix for continuous-filament composites. The system used to illustrate the hypothesized link between compressive properties of the thermoplastic fibre composite and modulus of the matrix is a 10/90 wt/wt% PBO/PEEK copolymer. The degree of polymerization for the rigid-rod PBO blocks or segments is 8. The details of the preparation of rigid-rod copolymers, their preforms such as powders or films, the consolidation and fabrication processes, have been described extensively [36, 37].

For this particular example, fine, isotropic powders were prepared. Specimens for flexural tests were prepared by compression moulding of the TPMC powders. Flexural measurements were carried out on the test specimens, cut from the moulding, in accordance with ASTM Standard D790. Flexural modulus was determined using an extensometer mounted on the test specimens to eliminate the effect of machine compliance. The flexural properties of 10/90 PBO/PEEK TPMC mouldings were determined to be: modulus of 6.77–7.60 GN m^{-2}, strength of 69–83 MN m^{-2}, and elongation of 1.0–1.6%. This represented one of the best moduli ever achieved by any engineering thermoplastic. This also confirmed the tremendous reinforcing efficiency only obtainable in a rigid-rod TPMC.

Uni-directional laminates were obtained by compression moulding from powder prepregs of the above TPMC copolymers with 60% of AS-4 carbon fibres. Compressive properties and flexural strength of the unilaminates of 60/40 AS-4/TPMC continuous-filament composites were then determined using ASTM test methods. The overall quality of the moulding was indicated by the 0° flexural strength of the above continuous fibre composites. As shown in Fig. 9.13, the 0° flexural strength of the AS-4/TPMC composite laminate compares very favourably with that of the uni-laminates of conventional thermoplastic fibre composites of identical AS-4 fibres loading. In this figure, PPS is polyphenylenesulfone, PEEK1 is PEEK 380G, PEEK2 is PEEK 150G, Avimid is Avimide K-III, and LARC is LARC-TPI. All are commercially available thermoplastic resins.

This result indicates that the AS-4/TPMC fibre composite is well consolidated; it also suggests that rigid TPMC matrix may be a better matrix material than the other advanced thermoplastic resins. Similarly, the compressive properties of uni-directional AS-4/TPMC fibre composites are impressive. As shown in Figs 9.14 and 9.15, the compressive strength of the TPMC fibre composite is much better than

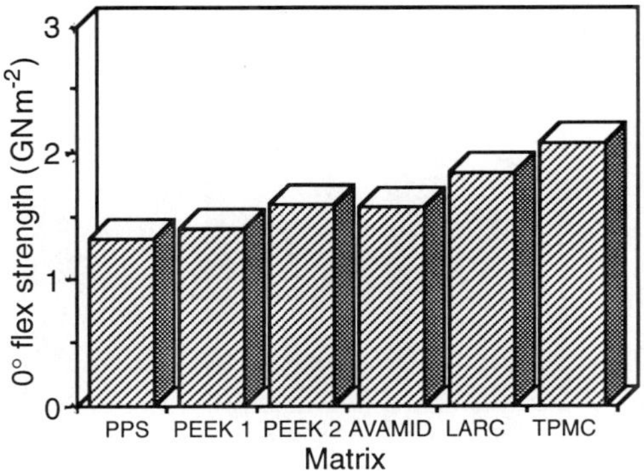

Figure 9.13 0° flexural strength of unilaminates of thermoplastic continuous fibre composites.

that of the other thermoplastic fibre composite, while its compressive modulus is also better than that of the PEEK fibre composites.

From these results, the rigid TPMC material seems to offer an excellent improvement over conventional engineering thermoplastics for use as matrix material in continuous-filament composites. One drawback is that the bulk viscosity of the 10/90 PBO/PEEK TPMC is rather high ($>10^4$ Ns m^{-2} at 400°C as measured by a Rheometrics RDS 2 dynamic mechanical spectrometer). The moulding of the TPMC and the AS-4/TPMC composites had to be carried out at the very high pressure of 351.5 Kg cm^{-2} and at 400°C. The extreme moulding conditions necessary for the TPMC materials, at present, are generally beyond the

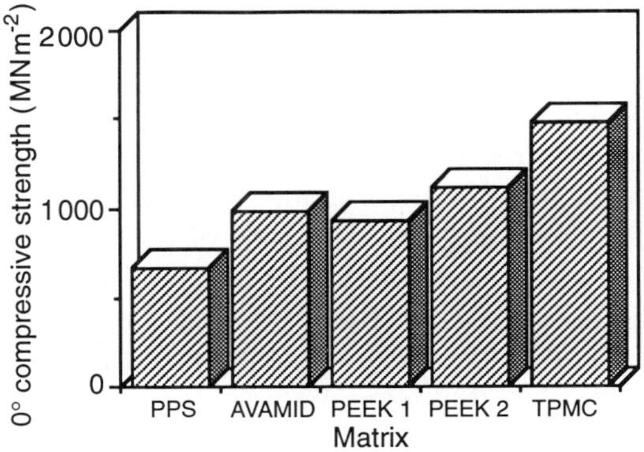

Figure 9.14 0° compressive strength of AS-4/thermoplastic fibre composites.

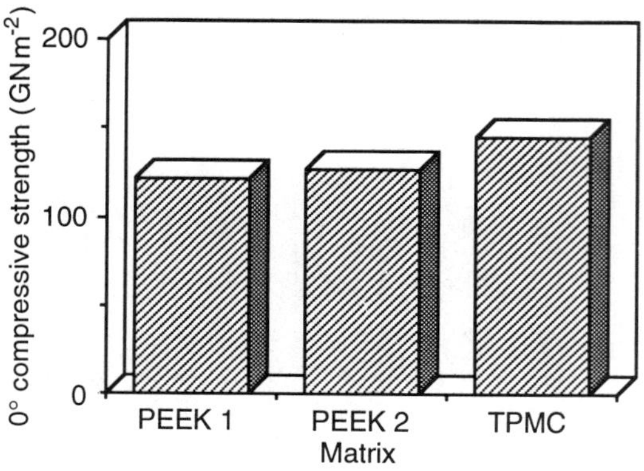

Figure 9.15 0° compressive modulus of AS-4/thermoplastic fibre composites.

capability of the composite industry. It has thus limited the TPMC's potential as a practical composite material.

9.4 RECENT DEVELOPMENTS

We have in this brief review illustrated the potential and the limitations of rigid-rod molecular composites as structural materials. However, the problem of processability and consolidability due to the high bulk viscosity, and the problem associated with processing MCs in strong acids remain illusive issues. These concerns have been the focus of more recent studies in the MC arena. This section briefly reviews some of these MC systems.

To circumvent the problem of using strong acids as solvent, one group of researchers has focused on developing MCs based on polyamides because their polyamic acid precursor is readily soluble in aprotic solvents such as N,N-dimethylacetamide (DMAC), N-methyl-pyrrolidinone (NMP), N,N-dimethylformamide (DMF). In one study, the preparation of molecular composite films of rigid aromatic polyimides, based on 3,3′4,4′-biphenyltetracarboxylic dianhydride (BPDA) and p-phenylene diamine (PPD), dispersed in a ductile matrix of flexible polyimides, based on BPDA and 4,4′-oxydianiline (ODA), was reported [40]. Since both the rigid and flexible components are polyimides of similar chemical composition, these MC films are transparent indicative of the lack of gross phase separation. Also, it is evidenced by the increase in the single T_g of the oriented MC films with the increasing content of rigid PI (BPDA/PPD). Tensile modulus as high as $50\,GN\,m^{-2}$ (7.2 Msi) was obtained for a 90/10 PI (BPDA/PPD)/PI (BPDA/ODA) MC film. Both the tensile modulus and tensile strength increase with rigid-rod content. This system probably suffers the same high bulk viscosity problem encountered in other high rod MCs. It does offer processability in a non-acidic solvent. In a similar study [41] of rigid/flexible PI blends, phase separation with domains of 1–3 microns of the soft component in the blends was observed. Phase separation occurred during the solvent evaporation process of the mixture of polyamic acids in NMP. In the later study [41], a more rigid PI prepared from pyromellitic dianhydride (PMDA) and PPD, and a much more flexible PI prepared from hexafluoroisopropylidene diphthalic anhydride (6FDA) and 2,2-bis(4-aminophenoxy-p-phenylene)-hexafluoropropane (BDAF) were used. The phase separation observed in this blend may be due to the bigger difference in the rigidity of the two PIs as compared to those in the former study [40]. Because of their excellent thermal stability and processability, PI MC systems are potential candidates for advanced dielectric and electronic packaging applications.

In another study [42], the above concept of using an aprotic solvent was adopted to the heterocyclic aromatic rigid-rod system. The idea is to prepare rigid-rod molecular composites from a homogenous solution of prepolymer of rigid-rod polymer, PBT (poly-p-phenylenebisbenzthiazole), and a flexible matrix polymer in an organic solvent. The prepolymer, as shown above, is alkyl group substituted, and is more flexible than the ring-closed rigid-rod PBT; therefore, it is readily soluble in an aprotic, organic solvent like NMP. The solution was then cast on a glass plate, and after solvent removal and ring-closure of the PBT prepolymer, as shown in Fig. 9.16, by heating according to an appropriate procedure [42], molecular composite films were obtained.

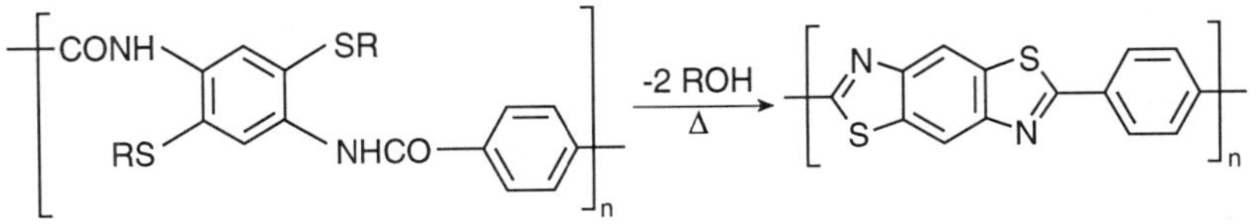

Figure 9.16 Ring closure of the PBT prepolymer.

In one example, the flexible matrix was a commercially available polyamide-imide (PAI) which is also readily soluble in NMP. The tensile properties of the MC film were improved tremendously over that of the neat PAI film. The tensile modulus increases linearly with PBT content. For a 40/60 wt% PBT/PAI, its tensile modulus of 7.6 GN m^{-2} is more than 2.6 times that of the neat PAI, while its tensile strength of 126 MN m^{-2} is more than 1.5 times that of PAI film. Although the tensile properties of this particular MC system are not as good as those of the block copolymer systems discussed previously, its reinforcement efficiency is still very impressive. More recently, a new class of rigid-rod polymers that are soluble in organic solvents has been developed [43, 44]. These organ-soluble rigid-rod polymers are basically substituted poly(p-phenylene). Poly(p-phenylene) is probably the most rigid polymer in existence, and its processable derivatives offer tremendous opportunity in the area of advanced materials. This type of polymer represents a new class of rigid-rod reinforcement; they have the potential to be a mouldable, high-performance resin, dependent on the kind of substituted groups on the rigid-rod backbone.

In this Chapter, I have reviewed various types of rigid-rod molecular composite systems. The unprecedented reinforcing efficiency of the rigid-rod molecules in a true molecular composite has been amply demonstrated over a wide range of rigid-rod contents. I have shown the importance of copolymerization of rigid rods with flexible coils in minimizing the detrimental effect of phase separation on the physical/mechanical properties of these materials. I have also illustrated some MC systems that are processable in organic solvents. From these studies, it is apparent that the trend is toward the development of processable, mouldable, consolidable TPMC systems utilizing a prepolymer of rigid rods of finite length. These materials also have to be amenable to conventional processes and with comparable properties to a metallic structural material such as aluminium. This intriguing issue remains the focus of many research efforts, and I hope that this brief review will serve as a source of references from which new ideas can be derived.

REFERENCES

1. Helminiak, T.E. (1979) *ACS Div. of Organic Coatings and Plastic Chem. Preprints*, **40**, 475.
2. Lindenmeyer, P. (1975) *Polym. Eng. Sci.*, **15**, 236.
3. Takayanagi, M. (1966) *J. Polym. Sci.*, **C15**, 263.
4. Krigbaum, W.R. (1966) *ibid.*, 251.
5. Kardos, J.L. and Raisoni, J. (1975) *Polym. & Eng. Sci.*, **15**, 183.
6. Wolfe, J.F., Loo, B.H. and Sevilla, E.R. (1981) *Polymer Preprints*, **22**, 60.
7. Berry, G.C., Casassa, E.F., Metzger, P. *et al.* (1979) *US Air Force Technical Report, AFML-TR-78-164*.
8. Wolfe, J.F., Sybert, P.D. and Sybert, J.R. (1985) US Patent 4,533,693.
9. Hwang, W.F., Wiff, D.R., Benner, C.L. *et al.* (1983) *J. Macromol. Science (Physics)*, **B22**, 231–57.
10. Hwang, W.F., Wiff, D.R., Price, G.E. *et al.* (1983) *Polym. Engin. Sci.*, **23**, 748.
11. Takayanagi, M. and Ogata, T. (1980) *J. Macro. Sci. (phys.)*, **B17**, 591.
12. Halpin, J.C. and Tsai, S.W. (1967) *US Air Force Technical Report, AFML-TR-67-423*.
13. Hwang, W.F., Wiff, D.R., Healminiak, T.E. *et al.* (1982) *Org. Coat. & Plas. Eng.*, **48**, 929.

14. Hwang, W.F., Wiff, D.R. and Verschoore, C. (1983) *Polym. Eng. Sci.*, **23**, 789.
15. Hwang, C.R., Malone, M.F. and Farris, R.J. (1988) *Polym. Mat. Sci. & Eng.*, **59**, 440.
16. Nishihara, T., Mera, H. and Matsuda, K. (1986) *Polym. Mat. Sci. & Eng.*, **55**, 821.
17. Flory, P.J. (1969) *Statistical Mechanics of Chain Molecules*, Wiley, New York.
18. Erman, B., Flory, P.J. and Hummel, J.P. (1980) *Macromol.*, **13**, 484.
19. Berry, G.C. (1989) *MRS Symposium Proceedings*, **134**, 181.
20. Brandrup, T. and Immergut, E.H. (eds) (1975) *Polymer Handbook*, Wiley-Interscience, 2nd edn.
21. Mammone, J.F. and Uy, W.C. (1982) *US Air Force Technical Report, AFWAL-TR-82-4145, Part II.*
22. Flory, P.J. (1956) *Proc. Roy. Soc., London*, **A234**, 73.
23. Chenevey, E.C. and Timmons, W.D. (1989) *MRS Symposium Proceedings*, **134**, 245.
24. Ledbetter, H.D., Rosenberg, S. and Hurtig, C.W. (1989) *MRS Symposium Proceedings*, **134**, 253.
25. Lusignea, R.W. (1989) *MRS Symposium Proceedings*, **134**, 265.
26. Krause, S.J. *et al.* (1985) *Polymer*, **29**, 1354–64, and the references cited therein.
27. Adams, W.W. *et al.* (1986) *US Air Force Technical Report, AFWAL-TR-86-4011.*
28. Ziabicki, A. (1976) *Fundamentals of Fiber Formation*, Wiley, Ch. 4.
29. Chuah, H., Kyu, T. and Helminiak, T.E. (1987) *Polym. Mat. Sci. & Eng.*, **56**, 58.
30. Uy, W.C. (1988) European Patent Pub. # 0-264-271-A2.
31. Hwang, W.F., Harris, W.J., Dixit, T. *et al.* (1987) *US Air Force Report, AFWAL-TR-87-4072.*
32. Flory, P.J. (1978) *Macromol.*, **11**, 1138.
33. Husman, G., Helminiak, T.E., Adams, W. *et al.* (1980) *ACS Symposium Ser.*, **132**, 303.
34. Tsai, T., Arnold, F. and Hwang, W.F. (1989) *J. Polym. Sci., Chem.*, **27**, 2839.
35. Harris, W.J. and Hwang, W.F. (1992) US Patent No. 5,098,985.
36. Harris, W.J. and Hwang, W.F. (1992) US Patent No. 5,151,489.
37. Hwang, W.F., Raspor, O.C., Harris, W.J. *et al.* (1993) US Patents No. 5,217,809 and 5,273,823.
38. Murphy, C.J. and Hwang, W.F. (1993) US Patent No. 5,244,617.
39. Kyu, T., University of Akron, private communications.
40. Yokota, R., Horiuchi, R., Kochi, M. *et al.* (1988) *J. Polym. Sci., Part C*, **26**, 215.
41. Rojstaczer, S.R., Yoon, D./Y., Volken, W. *et al.* (1990) *Mat. Res. Soc. Symp. Proc.*, **171**, 171.
42. Hattori, T. and Akita, H. (1992) European Patent 0474461 A1.
43. Morrocco, M.L., Gagne, R. and Trimmer, M. (1993) US Patent No. 5,227,457.
44. Wang, G.Y. and Quirk, R.P. (1944) *Polymer Preprints*, **35** (1), 503.

Index

Page numbers appearing in **bold** refer to figures and page numbers appearing in *italic* refer to tables.